Current Methods for Life Cycle Analyses of Low-Carbon Transportation Fuels in the United States

Committee on Current Methods for Life Cycle Analyses of Low-Carbon Transportation Fuels in the United States

Board on Environmental Studies and Toxicology

Board on Agriculture and Natural Resources

Division on Earth and Life Studies

Board on Energy and Environmental Systems

Division on Engineering and Physical Sciences

A Consensus Study Report of

The National Academies of
SCIENCES • ENGINEERING • MEDICINE

THE NATIONAL ACADEMIES PRESS
Washington, DC
www.nap.edu

NATIONAL ACADEMIES PRESS 500 Fifth Street, NW Washington, DC 20001

This activity was supported by contracts between the National Academy of Sciences and Breakthrough Energy, contract number 10005236. Any opinions, findings, conclusions, or recommendations expressed in this publication do not necessarily reflect the views of any organization or agency that provided support for the project.

International Standard Book Number-13: 978-0-309-27393-0
International Standard Book Number-10: 0-309-27393-5
Digital Object Identifier: https://doi.org/10.17226/26402

This publication is available from the National Academies Press, 500 Fifth Street, NW, Keck 360, Washington, DC 20001; (800) 624-6242 or (202) 334-3313; http://www.nap.edu.

Copyright 2022 by the National Academy of Sciences. All rights reserved.

Printed in the United States of America

Suggested citation: National Academies of Sciences, Engineering, and Medicine. 2022. *Current Methods for Life-Cycle Analyses of Low-Carbon Transportation Fuels in the United States*. Washington, DC: The National Academies Press. https://doi.org/10.17226/26402.

The National Academies of
SCIENCES · ENGINEERING · MEDICINE

The **National Academy of Sciences** was established in 1863 by an Act of Congress, signed by President Lincoln, as a private, nongovernmental institution to advise the nation on issues related to science and technology. Members are elected by their peers for outstanding contributions to research. Dr. Marcia McNutt is president.

The **National Academy of Engineering** was established in 1964 under the charter of the National Academy of Sciences to bring the practices of engineering to advising the nation. Members are elected by their peers for extraordinary contributions to engineering. Dr. John L. Anderson is president.

The **National Academy of Medicine** (formerly the Institute of Medicine) was established in 1970 under the charter of the National Academy of Sciences to advise the nation on medical and health issues. Members are elected by their peers for distinguished contributions to medicine and health. Dr. Victor J. Dzau is president.

The three Academies work together as the **National Academies of Sciences, Engineering, and Medicine** to provide independent, objective analysis and advice to the nation and conduct other activities to solve complex problems and inform public policy decisions. The National Academies also encourage education and research, recognize outstanding contributions to knowledge, and increase public understanding in matters of science, engineering, and medicine.

Learn more about the National Academies of Sciences, Engineering, and Medicine at **www.nationalacademies.org**.

The National Academies of
SCIENCES · ENGINEERING · MEDICINE

Consensus Study Reports published by the National Academies of Sciences, Engineering, and Medicine document the evidence-based consensus on the study's statement of task by an authoring committee of experts. Reports typically include findings, conclusions, and recommendations based on information gathered by the committee and the committee's deliberations. Each report has been subjected to a rigorous and independent peer-review process and it represents the position of the National Academies on the statement of task.

Proceedings published by the National Academies of Sciences, Engineering, and Medicine chronicle the presentations and discussions at a workshop, symposium, or other event convened by the National Academies. The statements and opinions contained in proceedings are those of the participants and are not endorsed by other participants, the planning committee, or the National Academies.

For information about other products and activities of the National Academies, please visit www.nationalacademies.org/about/whatwedo.

COMMITTEE ON CURRENT METHODS FOR LIFE CYCLE ANALYSES OF LOW-CARBON TRANSPORTATION FUELS IN THE UNITED STATES

Members

VALERIE M. THOMAS (*Chair*), Georgia Institute of Technology
AMOS A. AVIDAN (NAE), Bechtel Corporation (retired)
JENNIFER B. DUNN, Northwestern University
PATRICK L. GURIAN, Drexel University
JASON D. HILL, University of Minnesota, St. Paul
MADHU KHANNA, University of Illinois, Urbana-Champaign
ANNIE LEVASSEUR, École de technologie supérieure, Montreal, Quebec, Canada
JEREMY I. MARTIN, Union of Concerned Scientists, Washington, DC
JEREMY J. MICHALEK, Carnegie Mellon University
STEFFEN MUELLER, University of Illinois, Chicago
NIKITA PAVLENKO, International Council on Clean Transportation, Washington, DC
DONALD W. SCOTT, Scientific Certification Systems, St. Louis, Missouri
CORINNE D. SCOWN, Lawrence Berkeley National Lab
DEV S. SHRESTHA, University of Idaho, Moscow
FARZAD TAHERIPOUR, Purdue University
YUAN YAO, Yale University

Staff

CAMILLA YANDOC ABLES, Study Co-Director
BRENT HEARD, Study Co-Director
CLIFFORD S. DUKE, Board Director
TAMARA DAWSON, Program Coordinator
KYRA HOWE, Research Assistant

Report Editor

EUGENIA GROHMAN

Sponsor

BREAKTHROUGH ENERGY

BOARD ON ENVIRONMENTAL STUDIES AND TOXICOLOGY

Members

FRANK W. DAVIS (*Chair*), University of California, Santa Barbara
DANA BOYD BARR, Emory University, Atlanta, GA
ANN M. BARTUSKA, U.S. Department of Agriculture (retired), Washington, DC
GERMAINE M. BUCK LOUIS, George Mason University, Fairfax, VA
FRANCESCA DOMINICI, Harvard University, Boston, MA
GEORGE GRAY, The George Washington University, Washington, DC
R. JEFFREY LEWIS, ExxonMobil Biomedical Sciences, Inc., Annandale, NJ
LINSEY C. MARR, Virginia Polytechnic Institute and State University, Blacksburg, VA
MARIE LYNN MIRANDA, Children's Environmental Health Initiative, Notre Dame, IN
R. CRAIG POSTLEWAITE, U.S. Department of Defense, Burke, VA
REZA J. RASOULPOUR, Corteva Agriscience, Indianapolis, IN
JOSHUA TEWKSBURY, Smithsonian Tropical Research Institute, Panamá
SACOBY M. WILSON, University of Maryland, College Park, MD
TRACEY J. WOODRUFF, University of California, San Francisco, San Francisco, CA

Staff

CLIFFORD S. DUKE, Director
RAYMOND A. WASSEL, Scholar and Director of Environmental Studies
NATALIE ARMSTRONG, Associate Program Officer
KATHRYN GUYTON, Senior Program Officer
KALEY BEINS, Program Officer
LAURA LLANOS, Finance Business Partner
TAMARA DAWSON, Program Coordinator
LESLIE BEAUCHAMP, Senior Program Assistant
THOMASINA LYLES, Senior Program Assistant

Preface

Emissions of greenhouse gases, especially carbon dioxide from the combustion of fossil fuels, are changing the climate in ways that will have significant and long-term effects globally, including on the economy of the United States and the welfare of its people and environment. Transportation fuels are one of the largest sources of U.S. greenhouse gas emissions. The most widely used transportation fuels—gasoline, diesel, and jet fuel—are made from petroleum and emit carbon dioxide when combusted. The alternatives to petroleum transportation fuels, including electricity, biofuels, synthetic fuels, and hydrogen, also have emissions of carbon dioxide and other greenhouse gases, from the tailpipe, from the production processes, or from wider supply-chain contributions, depending on the fuel. Determining the total net emissions of these alternative fuels requires understanding how fuels are made and how they affect markets. Life-cycle assessment (LCA) is a method to account for the environmental impact of a product throughout its life cycle, including resource extraction, production, and all other supply-chain impacts. Existing LCAs of transportation fuels differ in their methods, such as data, system boundary, and assumptions, and have produced differing results.

Breakthrough Energy provided funding to the National Academies of Sciences, Engineering, and Medicine for a committee to conduct an assessment of the methods for life-cycle analyses of low-carbon transportation fuels in the United States, with the aim of developing a reliable and coherent approach for applying LCA to developing low-carbon fuel standards.

To gather information from experts on various fuel LCA questions, the committee invited presentations from Bo Weidema on consequential LCA, James Hileman on aviation fuels, Michael Wang on LCA of transportation fuels, Amgad Elgowainy on hydrogen fuels, Joule Bergerson on fossil fuels, Richard Plevin on uncertainties associated with LCA for transportation fuels, Tyler Lark and Seth Spawn on land use change, and John Field on soil carbon implications of biofuel production. I thank all of these individuals for sharing their time and expertise with the committee.

I thank the experienced scientists with deep expertise in LCA of fuels who served on this committee for their steadfast dedication to the work of preparing this report. Those efforts include hours of literature reviews, multiple committee meetings, working with and learning from numerous presenters with expertise related to LCA and transportation fuels, and writing working drafts with many edits to make the report readable and of high quality.

I also thank our wonderful team from the National Academies who worked diligently for many months to keep us on track and gave their total support throughout the entire process. On the committee's behalf, I especially thank our study co-directors, Camilla Yandoc Ables and Brent Heard, for their assistance through every aspect of the development of this report. On behalf of the committee, I also thank project staff members Tamara Dawson, Kyra Howe, Cliff Duke, Robin Schoen, John Holmes, and Ray Wassel.

Special thanks to the representative of Breakthrough Energy, Maria Martinez, and to the representatives of federal and state agencies for the information they provided to the committee. Lastly, I thank the members of the public who contributed to the committee's knowledge and understanding of issues important to the study and ultimately to the life cycle evaluation of greenhouse gas emissions from transportation fuels.

Valerie M. Thomas, *Chair*
Committee on Current Methods for Life Cycle Analyses of Low-Carbon
Transportation Fuels in the United States

Acknowledgments

This Consensus Study Report was reviewed in draft form by individuals chosen for their diverse perspectives and technical expertise. The purpose of this independent review is to provide candid and critical comments that will assist the National Academies of Sciences, Engineering, and Medicine in making each published report as sound as possible and to ensure that it meets the institutional standards for quality, objectivity, evidence, and responsiveness to the study charge. The review comments and draft manuscript remain confidential to protect the integrity of the deliberative process.

We thank the following individuals for their review of this report:

STEVEN BARRETT, Massachusetts Institute of Technology
MIGUEL BRANDÃO, KTH Sweden
JOHN FIELD, Oak Ridge National Laboratory
KEN GILLINGHAM, Yale University
DAVID HASSENZAHL, California State University, Chico
DON O'CONNOR, S&T Squared Consultants Inc.
MARK REID (NAS), Harvard & Smithsonian
NAGORE SABIO, ExxonMobil
NORBERT SCHMITZ, International Sustainability and Carbon Certification (ISCC)
TIMOTHY SEARCHINGER, Princeton University
MICHAEL WANG, Argonne National Laboratory

Although the reviewers listed above provided many constructive comments and suggestions, they were not asked to endorse the conclusions or recommendations of this report nor did they see the final draft before its release. The review of this report was overseen by **CHRIS HENDRICKSON** (NAE), Carnegie Mellon University and **JEFFERSON TESTER** (NAE), Cornell University. They were responsible for making certain that an independent examination of this report was carried out in accordance with the standards of the National Academies and that all review comments were carefully considered. Responsibility for the final content rests entirely with the authoring committee and the National Academies.

Contents

ACRONYMS .... **xv**

SUMMARY **1**

PART I: BACKGROUND AND POLICY CONTEXT FOR LIFE-CYCLE ANALYSIS

1   INTRODUCTION AND POLICY CONTEXT **13**
    Transportation Emission Reduction Policies in the United States, 13
    Life-Cycle Analysis to Assess Greenhouse Gas Emissions, 14
    The Committee's Charge and Its Approach, 14
    Organization of the Report, 16
    References, 16

2   FUNDAMENTALS OF LIFE-CYCLE ASSESSMENT **18**
    The Four Phases of Conducting a Life-Cycle Assessment, 18
    Two Broad Categories of Life-Cycle Assessment, 20
    Attributional Life-Cycle Assessment, 22
    Consequential Life-Cycle Assessment, 26
    Comparison of Attributional and Consequential Life-Cycle Analysis, 27
    References, 32

3   LIFE-CYCLE ASSESSMENT IN A LOW-CARBON FUEL STANDARD POLICY **35**
    Historical Context, 35
    Low-Carbon Fuel Policies in the United States, Europe, and Brazil, 36
    Considerations in Applying Attributional Life-Cycle Assessment and Consequential
        Life-Cycle Assessment in Low-Carbon Fuel Standards, 41
    Conclusions and Recommendation on the Use of Different Life-Cycle
        Assessment Approaches in Low-Carbon Fuel Standards, 44
    Recommendations for Research, 46
    References, 46

PART II: GENERAL CONSIDERATIONS FOR LIFE-CYCLE ANALYSIS

**4   KEY CONSIDERATIONS: DIRECT AND INDIRECT EFFECTS, UNCERTAINTY,
     VARIABILITY, AND SCALE OF PRODUCTION** **49**
    Direct and Indirect Effects, 49
    Uncertainty and Variability, 54
    Scale of Production, 67
    References, 69

5   VERIFICATION **74**
    The Importance of Verification, 74
    Current Use of Verification, 75
    Challenges in Implementing Verification Approaches, 81
    Future Technology for Verification, 85
    References, 88

6 SPECIFIC METHODOLOGICAL ISSUES RELEVANT TO A LOW-CARBON FUEL STANDARD 93
Allocation to and from Other Products, 93
Negative Emissions, 97
Biogenic Emissions, 98
Indicators, Other Climate Forcers, and Timing of Emissions, 106
Vehicle-Fuel Combinations and Efficiencies, 109
References, 116

PART III: SPECIFIC FUEL ISSUES FOR LIFE-CYCLE ANALYSIS

7 FOSSIL AND GASEOUS FUELS FOR ROAD TRANSPORTATION 126
Liquid Fossil Fuels, 126
Gaseous Fossil Fuels and Hydrogen, 130
References, 139

8 AVIATION AND MARITIME FUELS 142
Aviation Fuels, 142
Maritime Fuels, 152
References, 154

9 BIOFUELS 157
Biofuel Feedstocks and Finished Fuels, 158
Feedstocks for Biofuel Production, 160
Key Considerations in Biofuel Life-Cycle Analysis, 161
Carbon Emissions and Sequestration from Biorefineries, 164
Quantifying Market-Mediated Emissions from Biofuels, 170
References, 177

10 ELECTRICITY AS A VEHICLE FUEL 186
Comparing Attributional and Consequential Life-Cycle Assessment for Electricity, 186
Approaches to Consequential Life-Cycle Assessment for Plug-In Vehicle Charging Emissions, 190
Upstream Emissions, 195
Uncertainty and Dynamics, 195
Energy Efficiency, 196
Data Sources for Research, 196
Effects of Public Policy on Consequential Plug-In Electric Vehicle Emissions, 199
References, 199

APPENDIXES

A CONCLUSIONS AND RECOMMENDATIONS 203

B COMMITTEE MEMBERS' BIOGRAPHICAL SKETCHES 215

C OPEN SESSION AGENDAS 219

BOXES, FIGURES, TABLES

BOXES

S-1   Statement of Task, 2
1-1   Statement of Task, 15

2-1 Example of System Boundaries in ALCA and CLCA, 28
3-1 Attributional LCA Models Used for Transportation Fuel Policies, 43

FIGURES

2-1 Four phases of life-cycle assessment, 20
2-2 Illustration of how attributional and consequential LCA address different questions, 23
2-3 LCA approaches by research question and relationship to average and marginal emission, 29
2-4 Illustration of the relationships between attributional and consequential emissions and average and marginal emission factors for a single fuel, 31

3-1 Nested compliance categories within the EPA Renewable Fuel Standard (RFS2), 38
3-2 Illustration of potential consequential outcomes in regulatory impact assessment for a low-carbon fuel standard (LCFS), 42

4-1 Illustration of different delineations between direct emissions and indirect emissions for various corn ethanol LCA studies, 52

5-1 Hypothetical illustration showing reported average emission rates can be substantially lower than true average emission rates, 84

6-1 GHG emission flows across the life cycle of biofuel, 99
6-2 Soil carbon sequestration rate changes with time, 105
6-3 Range of vehicle energy consumption rates (inverse of efficiency) by fuel type, 111
6-4 Relationship between fuel consumption rate and ambient temperature, 112
6-5 Estimated average energy consumption (watt hour) per mile for a Nissan Leaf in different regions of the United States based on regional ambient temperature over the year, 113
6-6 A comparison between current and beyond Li-ion batteries for electrifying semi trucks, 116

7-1 Petroleum products made from a barrel of crude oil, worldwide, 2020, 126
7-2 Hydrogen produced from different technologies and feedstocks, 130
7-3 Global demand for pure hydrogen, 1975-2018, 131
7-4 U.S. Dry natural gas production, 2000-2050, 132
7-5 Location of shale gas plays in the continental United States, 132
7-6 Steps in the natural gas supply chain from recovery to delivery, 133
7-7 Bottom-up methane (CH_4) emissions estimation tool, 134
7-8 Evolution of the Intergovernmental Panel on Climate Change (IPCC) estimate of methane global warming potential from 1990-2021, 136

8-1 The principal emissions from aviation operations and the atmospheric processes that lead to changes in radiative forcing components, 148
8-2 Radiative forcing components in milliwatts per square meter from global aviation as evaluated from preindustrial times until 2018, 149
8-3 Aircraft power profiles and power to energy ratio by flight segment, 150

9-1 Land available for conversion in commonly-used models for assessing land use change in biofuel production, 172

10-1 Conceptual illustration of the difference in approaches for assessing power sector emissions from PEV charging, 187
10-2 Relationship of average and marginal power grid emission factors to attributional and consequential LCA, 188

10-3 Illustration of why the emissions implications of charging a PEV in a region can differ from the emissions of the average grid mix in that region, 190
10-4 Hypothetical dispatch curve, 191
10-5 Example of regression results identifying sources of generation changes in the MRO grid region, 192
10-6 Conceptual illustration of energy balance maintained at every time step of a simulated dispatch model, 193
10-7 Illustration of how the relative life-cycle GHG emissions of a particular PEV compared to with a gasoline vehicle can depend on many factors, 197

TABLES

2-1 Definitions of Attributional and Consequential LCA, 21
2-2 Relationship between System Boundary and LCA Type, 27
2-3 Summary of Major Approaches to LCA, 29

4-1 Definitions of Direct and Indirect Emissions in the Literature, 50
4-2 Relationship between Direct/Indirect Emissions and Attributional/Consequential LCA, 53
4-3 Examples of Factors Often Referred to as "Direct" Or "Indirect" in the Transportation Fuels Literature, 53

5-1 Studies Used in U.S. EPA's 2018 Triennial Report to Congress, 76

8-1 Summary of Approved and Pending Alternative Aviation Fuel Production Pathways, 143
8-2 Sample of Published LCAs of Aviation Fuels, 144
8-3 Comparison of Physical Properties across Different Energy Carriers Used for Aviation, 151
8-4 Marine Fuels That Could Be Included in an LCFS, 153

10-1 Sample of Published Studies Assessing PEV Emissions in the United States, 189
10-2 Comparison of Approaches to Estimate Grid Emissions from PEV Charging, 194
10-3 Comparison of Marginal AVERT Factors with eGrid, by Selected States and Regions, 198
10-4 Comparison of Estimated Emission Factors for Changing Electricity Load, 198

Acronyms

AAF	alternative aviation fuel
AEZ-EF	Agro-ecological Zone Emission Factor (model)
ALCA	attributional life-cycle assessment; attributional life-cycle analysis
ASTM	American Society for Testing and Materials
AVERT	AVoided Emissions and geneRation Tool
BCM	billion cubic meters
BEPAM	Biofuel and Environmental Policy Analysis Model
BEV	battery electric vehicle
CA-LCFS	California Low Carbon Fuel Standard
CAFE	corporate average fuel economy
CA-GREET	modified Greenhouse Gases, Regulated Emissions, and Energy Use in Technologies (GREET) model
CARB	California Air Resources Board
CARD	Center for Agricultural and Rural Development (Iowa State University)
CCLUB	Carbon Calculator for Land Use Change from Biofuels Production
CCUS	carbon capture utilization and storage
CEDM	Center for Climate and Energy Decision Making
CGE	computable general equilibrium (model)
CI	carbon intensity
CLCA	consequential life-cycle assessment; consequential life-cycle analysis
CO_2e	carbon dioxide equivalent
CORSIA	Carbon Offsetting and Reduction Scheme for International Aviation
EEIO	environmentally extended input output (LCA)
eGRID	Emissions & Generation Resource Integrated Database (of the US Environmental Agency)
EIO	economic input output (applied to LCA)
EISA	Energy Independence and Security Act
EPA	U.S. Environmental Protection Agency
EPA GHGI	EPA Greenhouse Gas Emissions Inventory
EPPA	Emissions Prediction and Policy Analysis (model)
EU	European Union
EV	electric vehicle
FAPRI-CARD	Food and Agricultural Policy Research Institute (model developed at Center for Agricultural and Rural Development of Iowa State University)
FASOM	Forest and Agricultural Sector Optimization Model
GHG	greenhouse gas
GLOBIOM	Global Biosphere Management model
GREET	Greenhouse gases, Regulated Emissions, and Energy use in Technologies model (developed by Argonne National Laboratory)
GTAP	Global Trade Analysis Project (model developed in 1994)
GTAP-BIO	Global Trade Analysis Project-BIO (model first modified in 2007 with subsequent changes over time)
GTP	Global temperature change potential

GWP	global warming potential
HEFA	hydroprocessed esters and fatty acids
ICAO	International Civil Aviation Organization
IEA	International Energy Agency
ILUC	indirect land use change
ISCC	International Sustainability and Carbon Certification
IPCC	Intergovernmental Panel on Climate Change
IRENA	International Renewable Energy Agency
ISO	International Organization for Standardization
ISOs	Independent System Operators
kWh	kilowatt-hour
LCA	life-cycle assessment; life-cycle analysis
LCFS	low-carbon fuel standard
LCI	life-cycle inventory
LNG	liquefied natural gas
LUC	land use change
MIRAGE	Modeling International Relationships in Applied General Equilibrium (model)
MISO	Midcontinent Independent System Operator
MJ	megajoule
MMb/d	million barrels per day
mW/m^2	milliWatts per square meter
OPGEE	Oil Production Greenhouse gas Emissions Estimator (Stanford)
PE	partial equilibrium (model)
PEV	plug-in electric vehicle
PHEV	plug-in hybrid electric vehicle
PRELIM	Petroleum Life Cycle Inventory Model
RED	Renewable Energy Directive (2009)
RED II	Renewable Energy Directive (updated 2018)
RFS	U.S. Renewable Fuel Standard
RFS1	Renewable Fuel Standard (established in 2005)
RFS2	Renewable Fuel Standard (established in 2007)
RINs	Renewable Identification Numbers
RTOs	Regional Transmission Operators
SETAC	Society of Environmental Toxicology and Chemistry
SOC	soil organic carbon
t*km	ton-kilometer
UNEP	United Nations Environment Programme
USDA	U.S. Department of Agriculture
USGS	U.S. Geological Survey

Summary

Greenhouse gas (GHG) emissions drive climate change. In the United States, transportation is the largest source of GHG emissions. Petroleum products account for about 90 percent of U.S. transportation fuels, with biofuels, natural gas, and electricity accounting for the rest.[1] To mitigate further effects of climate change, deployment of low-carbon energy technologies, such as fuels with low GHG emissions, is considered to be critical.

There are federal and state programs to reduce GHG emissions from transportation fuels. The Renewable Fuel Standard (RFS) program, which is administered by the U.S. Environmental Protection Agency (EPA), was enacted by Congress in 2005 and amended in 2007. The program aims to reduce life-cycle GHG emissions from transportation fuels, expand the U.S. renewable fuels sector, and reduce reliance on imported oil. At the state level, California and Oregon have adopted low-carbon fuel standards (LCFSs). Recent reports from staff of the House Select Committee on the Climate Crisis[2] and from a bipartisan network of former EPA career employees[3] indicate interest in a national LCFS.

STUDY PURPOSE AND COMMITTEE'S TASK AND APPROACH

If policies aim to promote low-GHG emissions fuels, the status and capabilities of the methods and assumptions to identify the GHG emissions of fuels need to be understood. A critical part of that understanding is how to use and interpret life-cycle assessment (LCA), that is, the total emissions from any proposed low-carbon fuel.

At the request of Breakthrough Energy, the National Academies of Sciences, Engineering, and Medicine appointed an ad hoc committee to assess current methods for estimating life-cycle GHG emissions associated with transportation fuels (both liquid and nonliquid) for potential use in a national low-carbon fuel program: see Box S-1 for the formal statement of task for the committee.

The committee organized its work by focusing on the methods of LCA and the capabilities needed for potential use in a national low-carbon fuels program. The committee examined general methodological approaches of LCA, key issues for evaluating GHG emissions, issues that arise for transportation fuels, and methodological issues that arise for characteristic types of transportation fuel. Some conclusions and recommendations are given in the next section and all are also provided in a table (sorted by topic) in Appendix A.[4]

---

[1] EIA 2020, see https://www.eia.gov/energyexplained/use-of-energy/transportation.php.
[2] Select Committee on the Climate Crisis, 2020.
[3] Environmental Protection Network, 2020.
[4] Committee member Jason Hill wishes to point out that many other equally important conclusions and recommendations, each of which is also supported by the entire committee, are not shown here. For example, the following three points are fundamentally important to the understanding, design, and application of LCA in LCFS policy:

> Conclusion 3-1: The carbon intensities of fuels used in an LCFS are not necessarily equivalent to the full climate consequences of their adoption. Increased use of a fuel with a low carbon intensity, as defined in an LCFS, could potentially decrease or increase carbon emissions relative to the baseline, depending on policy design and other factors. Regulatory impact assessments that use CLCA to project the consequences of policy can help assess the extent to which a given policy design with particular carbon intensity estimates will result in reduced GHG emissions.
>
> Recommendation 9-6: Beyond research on induced land use change and rebound effects, research should be done to identify and quantify the impacts of other indirect effects of biofuel production, including but not limited to market-mediated effects on livestock markets, land management practices, and dietary change of food type, quantity, and nutritional content.

> **BOX S-1** Statement of Task
>
> An ad hoc committee of the National Academies of Sciences, Engineering, and Medicine will assess current methods for estimating life-cycle greenhouse gas (GHG) emissions associated with transportation fuels (liquid and nonliquid) for potential use in a national low-carbon fuels program. In carrying out its assessment, the committee will identify the general characteristics and capabilities of GHG emissions estimation methods that would be commonly needed across various types of low-carbon fuels programs applied at a national level. The committee will include these considerations:
>
> - GHG emissions over the entire life cycle of a given transportation fuel, including feedstock generation or extraction, feedstock conversion to a finished fuel or blendstock, distribution, storage, delivery, and use of the fuel in vehicles.
> - Potentially significant indirect GHG emissions.
> - Key assumptions, input parameters, and data quality and quantity, including spatial variability, for application of life-cycle GHG emission models.
> - Needs for additional data, methods for data collection, standardized inputs for life-cycle analyses, and model improvements.

LIFE-CYCLE ASSESSMENT

LCA can address a range of questions regarding GHG emissions of low-carbon transportation fuels. There are two broad approaches to LCA: attributional life-cycle assessment (ALCA) and consequential life-cycle assessment (CLCA), and each require different analysis. ALCA evaluates the emissions that can be attributed to a given fuel while CLCA evaluates how emissions would change if a given policy or set of actions were followed.

Conclusion 2-1: The approach to LCA needs to be guided on the basis of the question the analysis is trying to answer. Different types of LCA are better suited for answering different questions or achieving different objectives, from fine tuning a well-defined supply chain to reduce emissions, to understanding the global, economy-level effect of a technology or policy change.

Recommendation 2-1: When emissions are to be assigned to products or processes based on modeling choices including functional unit, method of allocating emissions among co-products, and system boundary, ALCA is appropriate. Modelers should provide transparency, justification, and sensitivity or robustness analysis for modeling choices.

Recommendation 2-2: When a decision-maker wishes to understand the consequences of a proposed decision or action on net GHG emissions, CLCA is appropriate. Modelers should provide transparency, justification, and sensitivity or robustness analysis for modeling choices for the scenarios modeled with and without the proposed decision or action.

LIFE-CYCLE ASSESSMENT IN A LOW-CARBON FUEL STANDARD POLICY

Challenges in the application of LCA for regulatory purposes, such as an LCFS—including determinations of system boundaries, modeling choices, and uncertainty management—have long been recognized. Nevertheless, given the desire to design policies that achieve reductions in GHG emissions, LCA has been increasingly applied to policy development, and energy and biofuel policy in particular, in recent decades.

Recommendation 10-4: Analyses estimating the emissions implications of PEV adoption in future power grid scenarios should consider changes in power grid emissions caused by PEV charging in each power grid scenario.

Conclusion 3-2: More research is needed to evaluate effective methods to collectively leverage the strengths of CLCA, ALCA, and verification methods in achieving LCFS objectives.

Recommendation 3-1: When some emissions consequences of fuel use are excluded from carbon intensity values in an LCFS, the rationale, justification, and implications for these exclusions should be documented.

Recommendation 3-2: Public policy design based on LCA should ensure through regulatory impact assessment that, at a minimum, the consequential life-cycle impact of the proposed policy is likely to reduce net GHG emissions and increase net benefits to society. Regulatory impact assessments should consider changes in production and use of multiple fuel types (e.g., gasoline, electricity, biofuels, hydrogen).

Recommendation 3-3: LCA practitioners who choose to combine attributional and consequential LCA estimates should transparently document these choices and clearly identify the implications of combining these different types of estimates for the given application, scope and research question.

Recommendation 3-4: Research programs should be created to advance key theoretical, computational, and modeling needs in LCA, especially as it pertains to the evaluation of transportation fuels. Research needs include:

- Further development of robust methods to evaluate the GHG emissions from development and adoption of low-carbon transportation fuels, and development or integration of process-based, economic input-output, hybrid, and CLCA methodologies.
- Products could include the following:
  - development of national, open-source, transparent CLCA models for use in LCFS development and assessment
  - continued development of national, open-source attributional ALCA models from new or existing models
  - evaluation of different approaches to creating, using, or combining ALCA, CLCA, and verification for evaluation of policy outcomes
  - quantification of variation between marginal and average GHG emissions for various feedstock-to-fuel pathways; and
  - quantification and characterization of the implications of approximations and proxies in LCA, such as comparisons of marginal and average emissions.

DIRECT AND INDIRECT EFFECTS

GHG emissions associated with transportation fuels include emissions from producing the fuel, from combusting the fuel, and from the full supply chain for producing and distributing the fuel. GHG emissions also include market effects, including changes in land use, changes in electricity infrastructure and electricity system operations, and changes in the demand for fuel and other products. Researchers differ in what they consider as direct and indirect effects. There is, however, agreement that all emissions should be included in an LCA.

Recommendation 4-1: Because the terms "direct" and "indirect" are used differently in different contexts, these terms should be carefully defined and transparently presented when used in LCA studies or policy. Another option is to avoid using the terms "direct" and "indirect" altogether, as they are not considered necessary elements of LCA and may lead to greater confusion.

UNCERTAINTY AND VARIABILITY

LCAs are subject to considerable uncertainty and variability. LCA methods need to appropriately characterize uncertainty and variability to aid LCA stakeholders' interpretation of LCA results.

Recommendation 4-2: Current and future LCFS policies should strive to reduce model uncertainties and compare results across multiple economic modeling approaches and transparently communicate uncertainties.

Recommendation 4-4: When LCA results are used in policy design or policy analysis, the implications of parameter uncertainty, scenario uncertainty, and model uncertainty for policy outcomes should be explicitly considered, including an assessment of the degree of confidence that a proposed policy will result in reduced GHG emissions and increased social welfare.

Recommendation 4-5: Regulatory agencies should formulate a strategy to keep LCAs up to date, which may involve periodic reviews of key inputs to assess whether sufficient changes have taken place to warrant a re-analysis, and agencies should be aware that substantial changes to LCAs on timescales of less than a decade can occur.

Recommendation 4-6: LCA studies used to inform transportation fuel policy should be explicit about the feedstock and regions to which the study applies and to the extent possible should explicitly report sensitivity of results to variation in these assumptions.

Recommendation 4-9: Modelers should conduct sensitivity analysis to understand implications of variation.

Recommendation 4-10: To effectively inform policymaking, LCA studies should document results for a range of input values.

SCALE OF PRODUCTION

Scale of production can affect the life-cycle implications of a fuel or technology in nonlinear ways. When a fuel is produced at high volumes, it may be produced differently and have different effects on supply chains than when produced at low volume.

Recommendation 4-11: Researchers and regulatory agencies should identify additional information to assess impacts of large changes in fuel systems.

Recommendation 4-12: Because LCA-based carbon intensities in current LCFS policy are often not structured to capture nonlinear and non-lifecycle implications of large changes in fuel and fuel pathway production volume, policymakers should consider potential complementary policy mechanisms.

VERIFICATION

An LCA considers emissions across activities that occur in varied sectors and geographic locations. Because many GHG-emitting activities are not regularly monitored, LCAs rely on data from theoretical calculations, experimental measurements, or a small number of field measurements to approximate the magnitude of their emissions.

Confirming LCA results through direct measurement of all activities for an entire fuel pathway is impractical. However, effort can be focused on verification of emissions sources and effects that have the greatest impact for a given fuel.

Recommendation 5-2: The research and policy communities should develop frameworks and methodologies for use of satellite data to characterize national and international land use change that may be in part attributable to an LCFS. Examples of framing questions include:

- Should an LCFS include measures to mitigate undesirable international land use change, or is it sufficient to monitor international land use change that may be due to the LCFS and these GHG emissions to the associated fuel?
- What are the guardrails (e.g., amount and type of land converted to agriculture in a certain region) that a monitoring approach would put in place and, if approached or exceeded, what action would be undertaken as a result?
- How can satellite data and economic modeling be most effectively used synergistically to limit GHG emissions from international land use change?
- What public data sources will be used to track land use change?
- How should uncertainty in land use change estimates be reported?

Programs may use default baseline values for parameters that are to be certified, such as assuming a default amount of diesel consumed when harvesting corn. The certification process will then establish whether a farm consumes less or more than this default amount. The ability to verify lower or higher emissions can result in economic gain or loss for a supply chain actor, which may motivate them to pursue certification or to produce a fuel that complies with the policy.

Conclusion 5-9: Certification protocols that use verification strategies can complement initial fuel pathway modeling with LCA and associated models (e.g., economic models used to estimate land use changes) to lessen the impacts of uncertainty in LCA results and to inform policymakers of the effects of an LCFS as they unfold. This insight can aid in policy adjustments if undesirable effects arise over the course of the policy.

Recommendation 5-6: An LCFS should consider inclusion of a certification protocol with verification. The protocol and its implementation should be overseen by an agency or group of agencies with the complementary expertise sets needed for success. These expertise sets include insights into multiple energy systems and new technologies, economics, environmental effects of fuels and their production routes, agriculture, fossil fuel production, and electricity generation.

Recommendation 5-7: Certification protocols should be revisited periodically to adapt to the emergence of new verification technology, national and global trends in the energy, transportation, and agriculture sectors, and to update baselines as needed based on evolving common practice.

Recommendation 5-8: Economic modeling and verification processes are complementary to each other and should both be used. Verification processes to assess international- and national-level land use change should use state-of-the art remote sensing technologies, when appropriate, which are evolving toward increased frequency and spatial resolution.

NEGATIVE EMISSIONS

In life-cycle GHG calculations, it is not always just a matter of adding the emissions from different portions of the life cycle; there are some cases in which quantities are subtracted (i.e., negative emissions).

Fuels assigned net negative carbon intensity values raise important questions that warrant special scrutiny to distinguish between actual carbon dioxide removal and storage and fuels pathways that include credits for avoided emissions.

>**Conclusion 6-1:** The carbon intensity of fuels derived from methane that would otherwise be released (e.g., methane from manure or landfill) is strongly influenced by assumptions in the LCA of the alternative fate of methane pollution and is subject to dramatic change if relevant regulations or practices change.

Biomass removes atmospheric carbon (biogenic carbon) through the photosynthesis process and part or all of the biogenic carbon is released during biomass conversion, transportation, decay, and biofuel combustion. Fossil-based carbon may also be released in the same system, such as GHG emissions from burning fossil fuels to supply heat for biomass drying and conversion.

>**Recommendation 6-2:** All biogenic carbon emissions and carbon sequestration generated during the life cycle of a low-carbon fuel should be accounted for in LCA estimates.

>**Recommendation 6-3:** Research should be conducted to improve the methods for accounting and reporting biogenic carbon emissions.

Land use change, land management, and land management change (e.g., reducing tillage frequency, applying manure as a soil amendment) can alter soil carbon. Changes in soil organic carbon can be a significant contributor to the life-cycle GHG emissions of a biofuel.

>**Conclusion 6-3:** Given the importance of soil organic carbon changes in influencing life-cycle GHG emissions of biofuels, investments are needed to enhance data availability and modeling capability to estimate soil organic carbon. Capabilities to evaluate permanence of soil organic carbon changes should also be developed.

>**Recommendation 6-4:** Research should be conducted to collect existing soil organic carbon data from public and private partners in an open source database, standardize methods of data reporting, and identify highest priority areas for soil organic carbon monitoring. These efforts could align with the recommendations made in the 2019 National Academies report on negative emissions technologies to study soil carbon dynamics at depth, to develop a national on-farm monitoring system, to develop a model-data platform for soil organic carbon modeling, and to develop an agricultural systems field experiment network. These efforts should also be extended internationally.

>**Recommendation 6-5:** Research should be conducted to explore remote-sensing and in situ sensor-based methods of measuring soil carbon that can generate more data quickly.

Fuel production leads to the emission and uptake of carbon dioxide (CO_2) and other GHGs at every life-cycle stage. These emissions and uptakes are then aggregated into a common unit. To aggregate different GHG emissions into a common unit (i.e., carbon dioxide equivalent [CO_2e]), metrics expressing the relative contribution of GHGs to climate change are used.

>**Recommendation 6-6:** Use of more than one climate change metric should be considered in the analysis of low-carbon fuel policies.

>**Recommendation 6-7:** Further research should be conducted to better understand the suitability of different GHG metrics for LCA.

Recommendation 6-8: Further research should be conducted to develop a framework to include albedo effects from land cover change, and near-term climate forcers, in LCA of low-carbon fuels.

Recommendation 6-9: Further research should be conducted to better understand the climate implications of increased GHG emissions on the short-term (carbon debt) to support the selection of an appropriate approach to account for the timing of GHG emissions and uptakes in LCA.

VEHICLE-FUEL COMBINATIONS AND EFFICIENCIES

Life-cycle GHG emissions of transportation fuels can be compared on a per-unit-energy basis, but such a comparison can be incomplete or misleading without also considering how much energy is needed to propel a vehicle with each type of fuel as well as how much energy is required and emissions are created in the production and maintenance of each type of vehicle. Efficiency and production emissions can vary widely both within and across vehicle fuel type technologies, making fair comparisons with single point estimates challenging.

Conclusion 6-5: To make a meaningful comparison of the LCA of transportation fuels, the vehicles that use those fuels should be considered.

Recommendation 6-10: LCA of transportation fuels may include analysis using functional units based on the transportation service provided, such as passenger-mile or ton-mile, or otherwise be based on comparison of comparable transportation services. This may be reported in addition to an energy-based functional unit. LCAs should clearly describe their assumptions for the energy- and service-based functional units, such as through vehicle efficiency, market share, or other factors.

Recommendation 6-11: When comparing life-cycle emissions of different transportation fuels, LCA studies that assess or inform policy should consider the range of vehicle efficiencies within each fuel type to ensure that the comparisons are made on comparable transportation services, such as passenger capacity, payload capacity, or performance.

Recommendation 6-14: For regulatory impact assessment, LCA of transportation fuels and transportation fuel policy should consider a range of estimates for possible changes in the emissions of vehicle production required to convert transportation fuels into transportation services, and the resulting changes in vehicle fleet composition.

Recommendation 6-15: LCA comparing transportation fuels for weight-constrained applications should present a per-ton-mile functional unit and/or explicitly model the logistical implications of payload effects by fuel type.

FOSSIL AND GASEOUS FUELS FOR ROAD TRANSPORTATION

The life-cycle GHG emissions of petroleum fuels – gasoline, diesel, jet fuel – differ by source and by refinery, and could vary over time as petroleum sources change, and as refinery operations change as a result of lower consumption of some petroleum products. The committee recommends that these variations be explicitly included in a low-carbon fuels policy.

Recommendation 7-1: Policymakers may consider recognizing the variation in GHG emissions across different petroleum fuel pathways, and include mechanisms to reduce these emissions in fuel policies.

Conclusion 7-1: Additional data, reporting, and transparency are needed for petroleum sector operations, including improved information on venting and flaring of methane.

Conclusion 7-2: More emissions inventory data from natural gas systems are needed, particularly regarding emissions from storage tanks.

There are multiple steps involved in natural gas recovery and delivery to the point-of-use. In natural gas LCAs, it is important to consider direct methane emissions from each step of this supply chain, which may derive from venting, flaring, or leaking.

Recommendation 7-2: Further research should be done on the key parameters used to assess the climate impacts of natural gas production, such as methane leakage rates. These parameters will evolve as technology advances, data availability increases, and statistical methods may be used to translate the additional data into improved emissions estimates.

Hydrogen can be made from steam-methane reforming or autothermal reforming of methane with ("blue") or without ("grey") carbon capture and utilization or storage. In the case of blue and grey hydrogen, the LCA issues pertaining to natural gas can significantly influence hydrogen LCA results. "Green" hydrogen is to be primarily made using electrolysis powered by renewable electricity. This pathway takes into account any energy consumption and associated emissions for delivering and purifying the water.

Recommendation 7-5: In the context of an LCFS, LCAs of hydrogen should be well documented with choices of key parameters supported with facility-measured data or well-supported citations from the literature. These key parameters include the choice of energy source for steam-methane reforming or authothermal reforming, the carbon capture level from the waste gaseous stream, source of upstream electricity, and the rate of methane or CO_2 leakage. Where relevant, the approach to quantifying emissions of upstream natural gas production should align with those used elsewhere in an LCFS for other fuels produced from natural gas.

AVIATION FUELS

The life-cycle climate impacts of aviation fuels have been evaluated in the literature and as part of regulatory assessments for several fuels policies. There has been analysis of both conventional, petroleum-derived jet fuel (i.e., "jet fuel") and of a variety of alternative fuels produced through a wide array of conversion processes (i.e., alternative aviation fuels). Several key areas that may require special consideration beyond the approaches used for alternative fuels used in other sectors include: 1) the non-CO_2 effects of aviation fuels when combusted at high altitudes, 2) the impacts of alternative fuels on airplane efficiency, and 3) the impact of a flexible product slate on the life-cycle emissions calculations for aviation fuels.

Conclusion 8-1: Non-CO_2 effects from aviation fuels may be high but remain uncertain. The largest non-CO_2 impact from aviation fuel combustion may be aviation-induced cloudiness, with the remaining contributions being much smaller.

Recommendation 8-1: Because the non-CO_2 effects from aviation fuels remain uncertain, research should be done to clarify the magnitude and direction of these effects.

Recommendation 8-2: Alternative fuels and airframe combinations, particularly those with large density differences such as battery electric technology and hydrogen, may impact airplane efficiency and thus influence overall emissions. The comparative LCA of these technologies should use functional units based on the transportation service provided or otherwise be based on comparison of consistent transportation services.

Recommendation 8-3: Alternative aviation fuel LCA estimates developed for fuel policy should reflect existing practices at facilities or the expected behavior in response to future policies.

MARITIME FUELS

Marine fuels have similar supply chains to other transportation fuels (e.g., aviation, road transport). The unique aspects of their life cycle that is relevant to quantifying their emissions primarily come in the operations stage, such as methane slip from liquefied natural gas combustion in marine engines. Additionally, the non-CO_2 effects of the maritime sector warrant additional analysis. The contribution of aerosols to net radiative forcing from the sector via ship tracks—the clouds from ship exhaust—is highly uncertain and may require further research.[5] LCA methodological considerations for marine fuels are similar to those for other transportation fuels.

Recommendation 8-5: The baseline life-cycle GHG emissions for marine fuels should reflect current industry trends stemming from MARPOL Annex VI and potentially be updated after several years' time once the industry adjusts more fully to the new regulations through, for example, deployment of more liquid natural gas-fueled vessels.

BIOFUELS

Corn and soybeans are the most common feedstocks currently used to produce biofuels in the United States. LCA methods commonly used to estimate GHG emissions associated with crop production in conventional agricultural systems are largely similar regardless of the specific crop in question.

Conclusion 9-1: Improved data on biofuel feedstock production, including energy consumption, yield, and fertilizer application at fine spatial resolutions may be useful for some applications. Data quality improvements may support improved GHG accounting in biofuel feedstock production, especially should a performance-based LCFS be developed that accounts for spatially-explicit fertilizer and energy consumption, and land management practices like cover crop planting, land clearing, overfertilization, manure application, use of nitrification inhibitors, or noncompliance with long-term soil carbon storage incentives.

Woody biomass is one of the most abundant feedstock for bioenergy production in the United States. The GHG emissions associated with the production of woody biomass come from multiple sources, including the use of energy and materials (e.g., fertilizers and soil amendments) for forest management, harvesting, storage and transportation.

Recommendation 9-1: Additional research should be done to assess key parameters and assumptions in forest management practices induced by increased woody biomass demand, including: changes in residue removal rates, stand management and forest productivity, and changes in tree species selection during replanting.

Recommendation 9-2: Research and data collection efforts should be carried out for improved data and modeling related to forest feedstock production and storage, including energy use, yield, inputs, fugitive emissions, and changes in forest carbon stock should be supported.

GHG emissions associated with biomass conversion come from multiple sources, including on-site combustion of fuels (e.g., fossil fuels, biomass, or byproducts), direct emissions from conversion processes,

---

[5] Glassmeier F., F. Hoffmann, J. S. Johnson, T. Yamaguchi, K. S. Carslaw, and G. Feingold. 2021. Aerosol-cloud-climate cooling overestimated by ship-track data. *Science* 371(6528):485-489. https://doi.org/10.1126/science.abd3980.

and upstream emissions associated with the production of chemicals, enzymes, and electricity used by biorefineries.

Mass or energy balance are the most common methods used to estimate GHG emissions of biorefineries. Some of the attributional GHG emissions of upstream production of electricity and chemicals used in biomass conversion are available in many life-cycle inventory databases but have large variations depending on the production technologies and market mix. Research articles vary in their assumptions about the potential for carbon sequestration using biorefinery co-products as soil amendments, but many assume 80-85 percent of biochar is stable for at least 100 years whereas digestate and compost are not assumed to result in accumulation of stable carbon in soils.

> **Recommendation 9-3:** Policymakers should exercise caution in crediting biorefineries for GHG emissions sequestration as a result of exporting co-products such as biochar, digestate, and compost, as it risks over-crediting producers for downstream behavior that is not necessarily occurring. The committee recommends that any credits generated from these activities must be contingent on verification that these activities are being practiced.

> **Recommendation 9-4:** Applying credits for carbon sequestration to soil or reduced use of fertilizer should require robust measurement and verification to prove the co-products are applied in a manner that yields net climate benefits.

Large-scale production of biofuels has an effect on various markets at regional, national, or global scales and can affect prices in these markets. Changes in market prices can trigger other changes in production and consumption decisions that may have positive or negative effects on GHG emissions from those markets. These changes may be included in the modeling of indirect effects. These secondary effects on GHG emissions are of concern because they affect the savings in GHG emissions obtained by displacing fossil fuels by biofuels.

> **Recommendation 9-7:** Though the study of induced land use changes from biofuels has been the topic of intense study over the last decade, substantial uncertainties remain on many key components of economic models used to assess these impacts. Further work is warranted to update these estimates of market-mediated land use change and the models so as to inform the development and implementation of an LCFS.

> **Recommendation 9-8:** Assessment of the consequential effects from a future proposed policy, such as induced land use change, should be further developed in order to assess the risk of market-mediated effects and emissions attributable to the policy. Consequential assessment can inform the implementation of safeguards within policies such as limits on high-risk feedstocks, can inform the development of supplementary policies, identify hotspots, and reduce the likelihood of unintended consequences.

> **Recommendation 9-9:** To improve understanding of market-mediated effects of biofuels, research should be supported on different modeling approaches, including their treatment of baselines and opportunity costs, and to investigate key parameters used in national and international modeling based on measured data, including various elasticity parameters, soil carbon sequestration, land cover, and emission factors and others.

> **Recommendation 9-10:** Because other market-mediated effects of biofuel production, such as livestock market impacts, land management practices, and changes in diets and food availability may be linked to land use and biofuel demand assessed using induced land use change models, additional research should be done and model improvements undertaken to include these effects.

ELECTRICITY AS A VEHICLE FUEL

Plug-in electric vehicles (PEVs) use energy stored in an onboard battery for propulsion and charge the battery using electricity from the power grid. PEVs include battery electric vehicles, and plug-in hybrid electric vehicles. In ALCA approaches, a portion of total power grid GHG emissions is assigned to PEV charging. In contrast, in CLCA approaches power grid emissions are estimated with and without PEV charging, and the difference between emissions in the two scenarios is the consequential effect of PEV charging.

Conclusion 10-1: ALCA is sometimes applied to estimate emissions from electricity consumption because it is easy or because the modeler is interested in an attributional, rather than consequential, question. However, using average emission factors does not answer the question of how emissions will change if PEVs or PEV policy is adopted. CLCA aims to answer how PEV or PEV policy adoption would change emissions from the power sector.

Recommendation 10-1: Regulatory impact assessment or other analyses estimating the emissions implications of a change in PEV charging load should use a CLCA approach to estimate the implications of power grid emissions and clearly characterize uncertainty of estimates due to assumptions, especially for future scenarios.

Recommendation 10-2: Research should be conducted to estimate how upstream emissions in the power sector change in response to changes in generation.

Recommendation 10-3: Analyses estimating the emissions implications of changing PEV adoption or PEV policy should provide a transparent assessment of how sensitive or robust the results of the analyses are to reasonable variations in modeling assumptions and future scenarios.

Recommendation 10-5: LCA to estimate the change in GHG emissions induced by a policy or a change in technology adoption should consider how interaction with existing policies may affect outcomes. For cars and trucks, national fleet standards are key to understanding the net GHG outcomes of technology or policy actions.

Part I

Background and Policy Context for Life-Cycle Analysis

1
Introduction and Policy Context

Greenhouse gas (GHG) emissions drive climate change, and transportation is the largest source of GHG emissions in the United States and the fastest-growing emissions source globally.[1] In 2020, petroleum products accounted about 90 percent of the U.S. transportation sector energy use, biofuels accounted for about 5 percent, natural gas accounted for about 3 percent, and electricity accounted for less than 1 percent (U.S. Energy Information Administration, 2020). In order to mitigate further effects of climate change, the adoption of low-carbon energy technologies, such as fuels with low GHG emissions, will be of paramount importance (NASEM, 2021a).

TRANSPORTATION EMISSION REDUCTION POLICIES IN THE UNITED STATES

There are a number of federal and state policies that aim to reduce GHG emissions from transportation. This section briefly reviews those policies, with notes on their similarities, differences, or relationship to a low-carbon fuel policy.

Renewable Fuels Standard

The Renewable Fuel Standard (RFS) program,[2] which is implemented by the U.S. Environmental Protection Agency (EPA), aims to reduce life-cycle GHG emissions from transportation fuels, expand the U.S. renewable fuels sector, and reduce reliance on imported oil. This standard specifies volumes of renewable fuels to be blended into domestic transportation fuels. Although the RFS addresses only biofuels and does not include other low-carbon fuels, such as electricity, it has provided regulatory experience with life-cycle assessment (LCA)[3] of GHG emissions of fuel production pathways.

Much of existing U.S. ethanol production was exempted from meeting GHG reduction targets; new renewable fuel production was required to show reductions in estimated life-cycle GHG reductions relative to a 2005 petroleum baseline specified by EPA. Specifically, biomass-based diesel and advanced biofuels must meet a 50 percent life-cycle GHG reduction; cellulose biofuels must meet a 60 percent reduction; and conventional biofuels, such as ethanol derived from corn starch at new facilities, must meet a 20 percent reduction (EPA, n.d.-b). Life-cycle GHGs are defined under the Clean Air Act (42 U.S.C. § 7545(0)) to include direct and "significant" indirect emissions, such as land use changes related to the full fuel life cycle. Since 2010, more than 200 fuel pathways have been approved using methods of LCA (EPA, n.d.-a).

Corporate Average Fuel Economy Standards

Perhaps the most prominent federal programs regulating transportation fuel consumption in the United States are the Corporate Average Fuel Economy (CAFE) standards, regulated by the National Highway Traffic Safety Administration (Federal Register, 2020), and their companion light-duty fleet GHG emissions standards, regulated by the EPA (Federal Register, 2021). CAFE and the light-duty GHG standard do not set GHG emissions limits for fuels. But CAFE uses a placeholder for the efficiency contribution of alternative fuels, and the light-duty GHG standard treats electricity and hydrogen as zero

---

[1] Information from a presentation to the committee by V. Reed, J. Fitzgerald, and A. Haq of the Bioenergy Technologies Office of the U.S. Department of Energy, "Life-Cycle Analysis for Biofuels and Bio-Products."
[2] This program was authorized under the Energy Policy Act of 2005 and expanded under the Energy Independence and Security Act of 2007 (EPA, n.d.-b).
[3] The terms life-cycle assessment and life-cycle analysis are used interchangeably in this report, and the acronym, LCA, is used for both.

emission fuels (NASEM, 2021b). These assumptions, in effect, incentivize electric vehicle and alternative fuel adoption, and may have several effects on future vehicle GHG emissions (Gan et al., 2021; Jenn et al., 2019).

State-Level Low-Carbon Fuel Standards

At the state level, California and Oregon have adopted low-carbon fuel standards (LCFSs). Both standards are based on life-cycle carbon reduction pathways. Credits are given for fuels based on life-cycle GHG reductions.

California first adopted an LCFS in 2009, with the goal of reducing transportation fuel carbon intensity (CI) by at least 20 percent by 2030 (California Air Resources Board, 2020). LCA is used to calculate a CI score, including both the direct and indirect effects of fuel production, which in turn is used for crediting purposes. Oregon's Clean Fuels program, started in 2016, also assesses life-cycle GHGs of fuels, and it requires a 10 percent reduction in pollution from transportation fuels used in Oregon below 2015 levels by 2025 (Oregon Department of Environmental Quality, n.d.) This program also uses a credit system and applies to all fuels used in Oregon. Oregon and California's programs are aligned with regards to system boundary definitions, as well as data availability and transparency.[4] In a presentation to this committee, a representative from the Oregon program highlighted the opportunity for shared learning and information exchange regarding LCA.

There is growing interest in a national LCFS. A recent majority report from the congressional staff of the House Select Committee on the Climate Crisis calls for Congress to develop an LCFS that builds on the RFS, sets a benchmark for liquid and non-liquid fuels that is technology- and feedstock-neutral, and is tied to LCA for determining a fuel's CI (Select Committee on the Climate Crisis, 2020). Additionally, a report on behalf of a bipartisan network of former EPA career employees has called on the agency to evaluate the adoption of a federal LCFS (Environmental Protection Network, 2020).

LIFE-CYCLE ANALYSIS TO ASSESS GREENHOUSE GAS EMISSIONS

To ensure reductions in GHG emissions, metrics and accurate measurements are needed. LCA is the tool used to measure and account for the full environmental impacts of a transportation fuel, including impacts associated with feedstock production or extraction, transportation and manufacturing, and use in vehicles. LCA aims to include emissions that may be considered indirect as well as direct emissions of GHGs. A notable example is land use change stemming from increased demand for biofuels. Published LCA studies have differed in their implementation, with methodological differences affecting choices of system boundaries, quantification of market-induced effects, and allocation of emissions among co-products. Additionally, there have been questions regarding the quantity, quality, and availability of data used in LCA. If low-carbon fuel policies are to rely on LCA, the methodologies and assumptions need to be assessed, with approaches defined for how to navigate results determined by uncertain parameters, models, and assumptions.

THE COMMITTEE'S CHARGE AND APPROACH

In May 2021 Breakthrough Energy[5] requested the National Academies of Sciences, Engineering, and Medicine to appoint an ad hoc committee that would assess current methods for estimating GHG

---

[4] Information from a presentation to the committee by C. McConnaha, C-A. Wind, and K. Winans, K., "Oregon Department of Environmental Quality Clean Fuels Program Presentation to the National Academy of Sciences, Engineering, and Medicine Committee, "Current Methods for Life Cycle Analyses of Low-Carbon Transportation Fuels in the United States."

[5] Breakthrough Energy, founded by Bill Gates in 2015, is a network of entities supporting investments intended to help attain net-zero greenhouse gas emissions. For information about the organization and its mission, see https://www.breakthroughenergy.org/our-story/our-story.

emissions associated with transportation fuels (liquid and non-liquid) for potential use in a national low-carbon fuels program. The committee's statement of task is presented in Box 1-1.

Individuals appointed to the committee were chosen for their individual expertise and the relevance of their experience and knowledge to the task, not their affiliation with any institution. All committee members volunteer their time to participate in a National Academies consensus study. Areas of expertise represented on the committee include LCA, fuel production and use (including fossil fuels, biofuels, and electricity), economics, GHG emission modeling, uncertainty analysis, environmental policy decision-making, and biofuel impacts and fuel policy. For biographical sketches of the committee members, see Appendix B.

The committee organized its work by focusing on the methods of LCA and the capabilities needed for potential use in a national low-carbon fuels program. The committee examined general methodological approaches of LCA, key issues for evaluating GHG emissions, issues that arise for transportation fuels, and methodological issues that arise for characteristic types of transportation fuel.

The committee decided not to review or emphasize comparison of the numerical results of different LCAs of transportation fuels, but rather to keep the focus on the methods of GHG emission LCA for fuels. That is, the committee did not include tables compiling or comparing results from different studies, different methods, different years, or different fuels. The committee does not endorse the numerical result of any particular LCA or method. Instead, the committee focused, and the report emphasizes, what methods and approaches could be considered in order to develop reliable quantitative estimates of GHG emissions. Moreover, the committee focused on developing conclusions and recommendations for the use of LCA of transportation fuels that could be used to support an LCFS policy. That is, the committee considered that policymakers and the public would want the LCA to be able to reliably estimate the effect of a low-carbon fuel policy on reducing emissions of GHGs. To that end, the committee and the report emphasize methods to evaluate the consequences of a potential U.S. LCFS policy. The committee also, consistent with its task, identified needs for additional data, methods for data collection, standardized inputs for LCA, and model improvements that could provide a basis for strengthening the reliability and consistency of how LCA is applied for LCFSs.

BOX 1-1 Statement of Task

An ad hoc committee of the National Academies of Sciences, Engineering, and Medicine will assess current methods for estimating life-cycle greenhouse gas (GHG) emissions associated with transportation fuels (liquid and non-liquid) for potential use in a national low-carbon fuels program. In carrying out its assessment, the committee will identify the general characteristics and capabilities of GHG emissions estimation methods that would be commonly needed across various types of low-carbon fuels programs applied at a national level. The committee will include these considerations:

- Direct GHG emissions over the entire life cycle of a given transportation fuel, including feedstock generation or extraction, feedstock conversion to a finished fuel or blendstock, distribution, storage, delivery, and use of the fuel in vehicles.
- Potentially significant indirect GHG emissions, such as those associated with indirect land use changes attributed to biofuels production.
- Key assumptions, input parameters, and data quality and quantity for application of lifecycle GHG emission models for different regions of the United States.
- Needs for additional data, methods for data collection, standardized inputs for lifecycle analyses, and model improvements.

The committee deliberated and gathered information from June 2021 to February 2022, holding 10 virtual meetings. Five of these meetings included an open session at which the committee members had the opportunity to hear from and have Q&A sessions with a representative of Breakthrough Energy, the study sponsor; federal and state agency representatives; and invited speakers. The invited speakers were requested to submit recorded presentations on topics relevant to the study prior to the open sessions. The agendas for the open session are provided in Appendix C. Video recordings of the speaker presentations and speakers' slides are available on the study website.

Throughout the study, the committee also received input from interested stakeholders and the public through the study website, public comments periods in the open meetings, or by e-mail. All submitted comments and documents were added to the study's public access file, which is available on request from the National Academies' Public Access Records Office.

ORGANIZATION OF THE REPORT

The report's 10 chapters are divided in three parts. In addition to this chapter's general background for the study, Part I covers the phases and types of LCA (Chapter 2) and a discussion of LCA in an LCFS policy (Chapter 3). Part II addresses the general and specific considerations for LCA: direct and indirect effects, uncertainty and variability, and scale of production are discussed in Chapter 4; verification is discussed in Chapter 5; and specific issues and methods for LCA are discussed in Chapter 6. Part III addresses specific fuel issues for LCA: issues related to fossil and gaseous fuels for road transportation are discussed in Chapter 7; issues pertaining to aviation and maritime fuels are discussed in Chapter 8; issues related to biofuels are discussed in Chapter 9; and issues related to electricity as transportation fuel are discussed in Chapter 10.

REFERENCES

Environmental Protection Network. 2020. *Resetting the Course of EPA: Reducing Air Emissions from Mobile Sources.* https://www.environmentalprotectionnetwork.org/wp-content/uploads/2020/08/Reducing-Air-Emissions-from-Mobile-Sources.pdf.

EPA (U.S. Environmental Protection Agency). n.d.-a. *Approved Pathways for Renewable Fuel.* https://www.epa.gov/renewable-fuel-standard-program/approved-pathways-renewable-fuel.

EPA. n.d.-b. *Overview for Renewable Fuel Standard.* Overviews and Factsheets. https://www.epa.gov/renewable-fuel-standard-program/overview-renewable-fuel-standard.

Federal Register V85 n84 p24174-25024. https://www.govinfo.gov/content/pkg/FR-2020-04-30/pdf/2020-06967.pdf.

Federal Register V86 n248 p74434-74526. https://www.govinfo.gov/content/pkg/FR-2021-12-30/pdf/2021-27854.pdf.

Gan, Y., M. Wang, Z. Lu, and J. Kelly. 2021. Taking into account greenhouse gas emissions of electric vehicles for transportation de-carbonization. *Energy Policy,* 155, 112353. https://doi.org/10.1016/j.enpol.2021.112353.

House Select Committee on the Climate Crisis. 2020. *Solving the Climate Crisis: The Congressional Action Plan for a Clean Energy Economy and a Healthy, Resilient, and Just America.* https://climatecrisis.house.gov/sites/climatecrisis.house.gov/files/Climate%20Crisis%20Action%20Plan.pdf.

Jenn, A., I. L. Azevedo, and J. J. Michalek. 2019. Alternative-fuel-vehicle policy interactions increase U.S. greenhouse gas emissions. *Transportation Research Part A: Policy and Practice*, 124, 97–407.

NASEM (National Academies of Sciences, Engineering, and Medicine). 2021a. *Accelerating Decarbonization of the United States Energy System.* Washington, DC: National Academies Press. https://doi.org/10.17226/25932.

NASEM (National Academies of Sciences, Engineering, and Medicine). 2021b. *Assessment of Technologies for Improving Light-Duty Vehicle Fuel Economy—2025-2035.* Washington, DC: National Academies Press. https://www.nap.edu/read/26092.

Oregon Department of Environmental Quality. n.d. *Oregon Clean Fuels Program Overview.* https://www.oregon.gov/deq/ghgp/cfp/Pages/CFP-Overview.aspx.

U.S. Energy Information Administration. 2020. *Energy and the Environment Explained: Where Greenhouse Gases Come from.* https://www.eia.gov/energyexplained/energy-and-the-environment/where-greenhouse-gases-come-from.php.

U.S. Government Accountability Office. 2019. *Renewable Fuel Standard: Information on Likely Program Effects on Gasoline Prices and Greenhouse Gas Emissions.* GAO-19-47. https://www.gao.gov/products/gao-19-47.

2
Fundamentals of Life-Cycle Assessment

Life-cycle assessment (LCA) is an analysis technique that can be used to quantify a wide variety of environmental and social impacts that can be attributed to the provision of a good or service. This report focuses on the use of LCA to estimate greenhouse gas (GHG) emissions associated with transportation energy sources. According to the definition given by the International Standardization Organization (ISO) in the ISO 14040:2006 series standard, LCA is a "compilation and evaluation of the inputs, outputs and the potential environmental impacts of a product system throughout its life cycle." (ISO 14040:2006). This chapter provides background on how LCAs are commonly conducted, reviews the foundational concepts in LCA, and disambiguates key terminology that is used throughout this report. Although this chapter describes LCA methods in the context of transportation fuels, the fundamental LCA concepts and guidelines can apply to a wide range of goods and services.

This chapter first describes the four phases involved in conducting an LCA as defined in the ISO 14040/14044 standards. Next, two types of LCA are presented: attributional LCA (ALCA) and consequential LCA (CLCA). The selection of functional units and system boundaries are then discussed in the context of these types of analysis.

THE FOUR PHASES OF CONDUCTING A LIFE-CYCLE ASSESSMENT

The ISO 14040/14044 (ISO, 2006a,b) series is a commonly used standard for LCA. It provides principles and a framework for LCA (ISO 14040:2006) and requirements and guidelines (ISO 14044:2006) for conducting an LCA. Notably, it does not provide specific recommendations for methods, datasets, or tools for conducting an LCA, noting that "there is no single method for conducting LCA" (ISO 14040:2006). In essence, it provides a common language for conducting an LCA and basic guidelines for structuring an analysis without being prescriptive on the details of how an LCA should be performed. Although other standards for LCA exist (e.g., Guinée et al., 2018), ISO is arguably the most widely used and this chapter makes use of ISO terminology in its description of LCA.

Per the ISO standard, an LCA has four phases (see Figure 2-1). The first of these is the goal and scope definition phase. This phase lays the foundation for an LCA, specifying the goal of conducting the analysis. For transportation fuels policy, this goal could be to assign GHG emissions-intensity scores that can be compared across a range of fuel options. Another goal could be to inform and prioritize technology development or operational choices to reduce the environmental impact of a particular fuel. In some other cases, the goal of an LCA may be to conduct regulatory impact assessment to understand how a transportation fuel policy will change system-wide emissions at a national or global scale. Clarity on the goal of the study and how its results will be used helps shape all subsequent decisions for its design. The scope of the study clarifies which systems will be included. In the case of a fuel's life cycle, a single fuel production route (commonly referred to as a "fuel pathway") may be considered, or the scope may expand to include all sectors that are linked to the product or action being studied. The resulting differentiation between what sectors/activities are included versus excluded is determined by the system boundary, and any activities outside the system boundary are not included in the analysis.

The second phase of an LCA is a life-cycle inventory (LCI) analysis, which entails cataloging the material and energy flows across a fuel's life cycle. This may be done in a bottom-up manner using facility-level data, estimated through top-down approaches that rely on sector-level data, or through some combination of the two (see further discussion below). In a bottom-up analysis of fuels produced from petroleum, the LCI may involve compiling data on energy consumed and emissions produced during crude oil extraction and refining, on the transportation of intermediate and final petroleum products, and on the eventual combustion of the fuel in a vehicle. These activities make up the primary supply chain for a petroleum-derived fuel and, when combined, are considered to be a fuel pathway. Each stage in the fuel pathway also has its own supply chain, with corresponding upstream resource consumption and emissions. For example, petroleum refineries consume electricity, which is generated using multiple energy sources. Building materials to construct facilities and purchases from service sectors such as insurance are all part of this extended supply chain and can be important to include. As discussed further in the CLCA section, an LCI may also be used to quantify economy-wide changes in material and energy flows associated with the implementation of a policy or a change in production of an individual fuel.

In systems with mature technologies, these data may be obtained from government or industry reports. In emerging systems that are not yet well established at a commercial scale, LCI data may be generated through a combination of engineering models and empirical data from small-scale pilot or demonstration operations. For analyses seeking to understand economy-wide impacts of an action or policy, general equilibrium models or other economic models and data may also be used to capture market-mediated effects. Examples of market-mediated effects include the land use change (LUC) from production of biofuels and shifts in total market demand for fuels and other co-products as a result of a change in supply. Data sources can vary across an LCI; reporting one's data sources transparently can increase confidence in LCA results and enable reproducibility. Data quality in an LCI directly influences the quality of LCA results. For studies that are focused on GHG emissions, carbon dioxide (CO_2) from combustion can usually be approximated by using information about the fuel type(s) combusted for different activities across the supply chain and the stoichiometry of complete combustion. However, non-combustion emissions also have to be accounted for in the inventory phase and these emission factors often rely on field measurements, satellite data, or self-reported data from industry. For example, natural gas systems emit fugitive methane emissions, and agricultural systems emit nitrous oxide (N_2O) and methane (CH_4).

Life-cycle impact assessment is the third phase of conducting an LCA. In a life-cycle impact assessment, the data from the inventory phase, usually reported in physical units (e.g., kg of pollutant emitted or MJ [megajoule] of a fuel consumed), are used to calculate impact results in terms of multiple so-called indicators, which capture a wide range of human health, climate, and ecological impacts. There are numerous calculation methods available to convert LCI data to indicators, including but not limited to ReCiPe, USEtox, TRACI, and IMPACT 2002+ (Wang et al., 2020). For example, life-cycle GHG emissions may be calculated on an individual basis (separate inventories for CO_2, CH_4, N_2O, refrigerants, and any other relevant GHGs) based on the amounts and types of energy combusted, by process (non-combustion) emissions, and measured or simulated levels of fugitive emissions. These individual emissions totals can then be combined based on their relative climate impact and reported as global warming potential (GWP) in the form of CO_2-equivalents (CO_{2e}) (Peters, 2010). This reported CO_{2e} is often referred to as "carbon intensity" (CI), "carbon footprint," or "GHG footprint", despite the fact that not all emissions commonly included in the footprint (namely N_2O) contain carbon. Impacts can also be converted to costs (or net benefits) by estimating the monetized damages to society associated with each impact. GHG emissions are typically translated into monetized damages by using a value known as the social cost of carbon, which can be helpful in conducting cost–benefit analyses for emissions mitigation efforts, although social cost of carbon estimates may be incomplete in their accounting of potential damages (Bressler 2021; IWG, 2021).

The fourth phase of an LCA is the interpretation phase, in which impact assessment results are translated into meaningful information and guidance. In this phase, LCA practitioners interpret the impact assessment to inform policy or advice on fruitful directions for research and development to reduce system-wide effects. As noted in Figure 2-1, the process is often iterative, and the interpretation phase may highlight the need for collection of additional inventory data to address key sources of uncertainty or even a revision in the study's goal and scope. Interpretation is not the last phase in LCA but rather part of an iterative process concurrently with the other phases so as to inform LCA design.

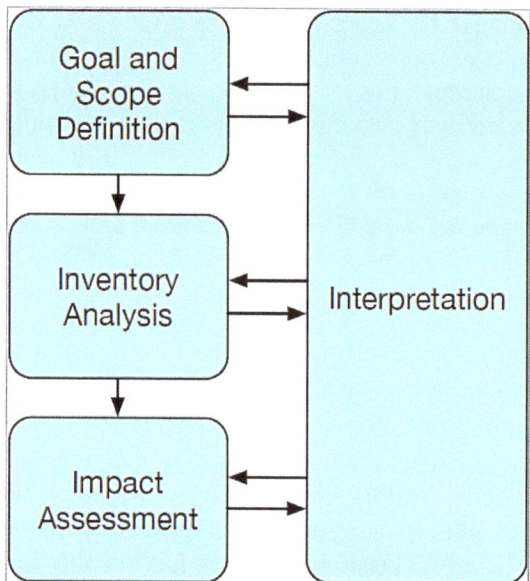

FIGURE 2-1 Four phases of life-cycle assessment. SOURCE: Adapted from ISO 14040:2006. ©ISO. This material is reproduced from ISO 14040:2006 with permission of the American National Standards Institute (ANSI) on behalf of the International Organization for Standardization. All rights reserved.

TWO BROAD CATEGORIES OF LIFE-CYCLE ASSESSMENT

In defining the goal and scope of an LCA, a practitioner may select from a wide variety of desired outcomes: perhaps two different products are being compared to inform the selection of one based on its environmental impact or an industrial production process is being assessed to identify opportunities for reducing its life-cycle impacts. However, there are two broad categories of LCA that are relevant to this report and require such fundamentally different approaches that they are important to discuss in greater detail: attributional LCA and consequential LCA. **ALCA** is defined by "environmentally relevant physical flows to and from a life cycle and its subsystem" (Finnveden et al., 2009). ALCA seeks to *attribute* a portion of total observed environmental impacts from human activities to the provision of a specific good or service. In contrast, **CLCA** is defined by its aim to describe "how environmentally relevant flows will change in response to possible decisions" (Finnveden et al., 2009). In other words, CLCA captures the *consequences* of some change in the provision of goods or services. Table 2-1 provides a list of definitions of attributional and consequential LCA from the research literature. Note that one of the examples in Table 2-1 uses the term "hybrid" LCA to refer to a mix of ALCA and CLCA; in this report, the term "hybrid" is used to refer to a combination of process-based and economic input-output (EIO) LCA, as discussed further in the ALCA section. While these definitions of ALCA and CLCA vary slightly, the common thread is that ALCA estimates emissions as they are or could be in some projected future state (among other things, requiring choices about how to assign emissions to co-products), and CLCA estimates how emissions will change in response to a decision or action. Both ALCA and CLCA can be useful in research, analysis, and policy design, but they answer different questions and will produce different results (see Figure 2-2). ALCA and CLCA can be applied to quantify a wide variety of impacts well beyond GHG emissions. Although not discussed in detail here, ALCA and CLCA can be applied to quantify a wide variety of impacts well beyond GHG emissions. Social LCA, for example, estimates social and socio-economic impacts (UNEP-SETAC, 2020) and can be useful in regulatory impact assessment.

TABLE 2-1 Definitions of Attributional and Consequential LCA from the Literature

Source	Definitions of Attributional and Consequential LCA
Matthews et al. (2014)	"**Attributional** LCAs seek to determine the effects now, or in the past, which inevitably means that our concerns are restricted to average effects. However, emerging practice and need in LCA often seeks to consider the consequences of product systems or changes to them. In **consequential** LCA studies, marginal, instead of average, effects are considered (Finnveden et al. 2009). Marginal effects are those effects that happen 'at the margin', and in economics refer to effects associated with the next additional unit of production. Furthermore, consequential analyses seek to determine what would change or need to change given the influence of changing product systems on markets."
NRC (2012)	"**Attributional** LCA, the more traditional form, traces the material and energy flows of a biofuel supply chain and seeks to attribute environmental impact to a biofuel based upon these flows. **Consequential** LCA, on the other hand, considers the environmental effects of the cascade of events that occur as a result of a decision to produce or not to produce a given biofuel."
RFS2 Regulatory Impact Assessment (EPA, 2010)[a]	"Lifecycle assessments can be divided into two major methodological categories: attributional and consequential. An **attributional** approach to GHG emissions accounting in products provides information about the GHG emitted directly by a product and its life cycle. The product system includes processes that are directly linked to the product by material, energy flows or services following a supply-chain logic. A **consequential** approach to GHG emissions accounting in products provides information about the GHG emitted, directly or indirectly, as a consequence of changes in demand for the product. This approach typically describes changes in GHG emissions levels from affected processes, which are identified by linking causes with effects. The definition of lifecycle greenhouse gas emissions established by **Congress** states that: "The term 'lifecycle greenhouse gas emissions' means the aggregate quantity of greenhouse gas emissions (including direct emissions and significant indirect emissions such as significant emissions from land use changes), as determined by the Administrator, related to the full fuel lifecycle, including all stages of fuel and feedstock production and distribution, from feedstock generation or extraction through the distribution and delivery and use of the finished fuel to the ultimate consumer, where the mass values for all greenhouse gases are adjusted to account for their relative global warming potential." This definition and specifically the clause "(including direct emissions and significant indirect emissions such as significant emissions from land use changes)" requires the Agency to consider a consequential lifecycle analyses and to develop a methodology that accounts for all of the important factors that may significantly influence this assessment, including the secondary or indirect impacts of expanded biofuels use.
British Columbia Low Carbon Fuel Standard Avoided Emissions Policy: Intentions Paper for Consultation[b]	"An **attributional** LCA accounts for only the direct emissions associated with the fuel lifecycle, including the emissions from production of energy and material inputs to the fuel life cycle. Emissions are allocated between co-products based on a physical quantity and indirect impacts are not considered...A **consequential** LCA determines the comprehensive greenhouse gas (GHG) emissions of a product by assessing the direct and indirect impacts of the fuel on external markets. A consequential LCA considers the market effects of a change in production, expands the system boundary to include non-fuel system impacts, and includes the indirect effects of the fuel production on the environment (e.g. indirect land use change)... The consequential approach to LCA essentially compares a scenario without the fuel to one with the fuel and attributes the resulting changes in affected markets to the fuel. **Hybrid** LCA: Hybrid LCA is a combination of attributional and consequential LCA."

continued

TABLE 2-1 continued

Source	Definitions of Attributional and Consequential LCA
Finnveden et al. (2009)	**Attributional LCA** is defined by its focus on describing the environmentally relevant physical flows to and from a life cycle and its subsystems. **Consequential LCA** is defined by its aim to describe how environmentally relevant flows will change in response to possible decisions (Curran et al., 2005). Similar distinctions have been made in several other publications (Ekvall, 1999), but often using other terms to denote the two types of LCA (such as descriptive versus change-oriented) and sometimes including further distinctions of subcategories within the two main types of LCA (Guineé et al., 2002).
Ekvall et al. (2016)	**Attributional LCI** considers the flows in the environment within a chosen temporal window. **Consequential LCI** considers how the flows may change in response to decisions.
Ekvall (2019)	**Attributional LCA:** LCA aiming to describe the environmentally relevant physical flows to and from a life cycle and its subsystems. **Consequential LCA:** LCA aiming to describe how environmentally relevant flows will change in response to possible decisions. Ekvall has developed these definitions based on Finnveden et al. (2009) and argued that: "These definitions clearly connect ALCA/CLCA not only to methodological choices but also to the goal of the study, because they respond to different questions" described in Figure 2-2.
UNEP-SETAC (2011)	The **attributional** approach attempts to provide information on what portion of global burdens can be associated with a product (and its life cycle). In theory, if one were to conduct attributional LCAs of all final products, one would end up with the total observed environmental burdens worldwide. The **consequential** approach attempts to provide information on the environmental burdens that occur, directly or indirectly, as a consequence of a decision (usually represented by changes in demand for a product). In theory, the systems analyzed in these LCAs are made up only of processes that are actually affected by the decision.
EUCAR (2020)[c]	**Attributional** LCA: It depicts the potential environmental impacts that can be attributed to a system (e.g. a product) over its life cycle, i.e. upstream along the supply-chain and downstream following the system's use and end-of-life value chain. **Consequential** LCA: It aims at identifying the consequences that a decision in the foreground system has for other processes and systems of the economy, both in the analyzed system's background system and on other systems. It models the analyzed system around these consequences.

[a] Renewable Fuel Standard Program (RFS2) Regulatory Impact Assessment. Report number: EPA-420-R-10-006; Date published: February 2010. URL: https://nepis.epa.gov/Exe/ZyPURL.cgi?Dockey=P1006DXP.TXT.
[b] See https://www2.gov.bc.ca/assets/gov/farming-natural-resources-and-industry/electricity-alternative-energy/transportation/renewable-low-carbon-fuels/bc_low_carbon_fuel_standard_avoided_emission_policy_-_intentions_paper_for_consultation.pdf.
[c] European Council for Automotive R&D; see https://www.eucar.be/lca-in-wtt-and-wtw-review-and-recommendations/.

ATTRIBUTIONAL LIFE-CYCLE ASSESSMENT

In an ALCA, an inventory of emissions or impacts that occur along each stage of a supply chain are assigned or attributed to a functional unit. A functional unit is a core characteristic of ALCA (although not exclusive to ALCA) and it is the common basis on which environmental effects are evaluated and reported. Functional units serve as the denominator in LCA results (impact per functional unit), so they must be defined in a manner that captures the value or function provided by a product. In the case of transportation fuels, common functional units are MJ of fuel and vehicle-mile traveled (Chapter 6 discusses in

Fundamentals of Life-Cycle Assessment

more detail vehicle-fuel combinations and the types of functional units that are appropriate for drawing comparisons across multiple fuels and vehicle technologies). Other functional units may be more appropriate when different modes of transportation are being compared for the movement of freight or people, such as ton-mile or passenger-mile traveled. Use of common functional units is one important step to enable comparison across different ALCAs.

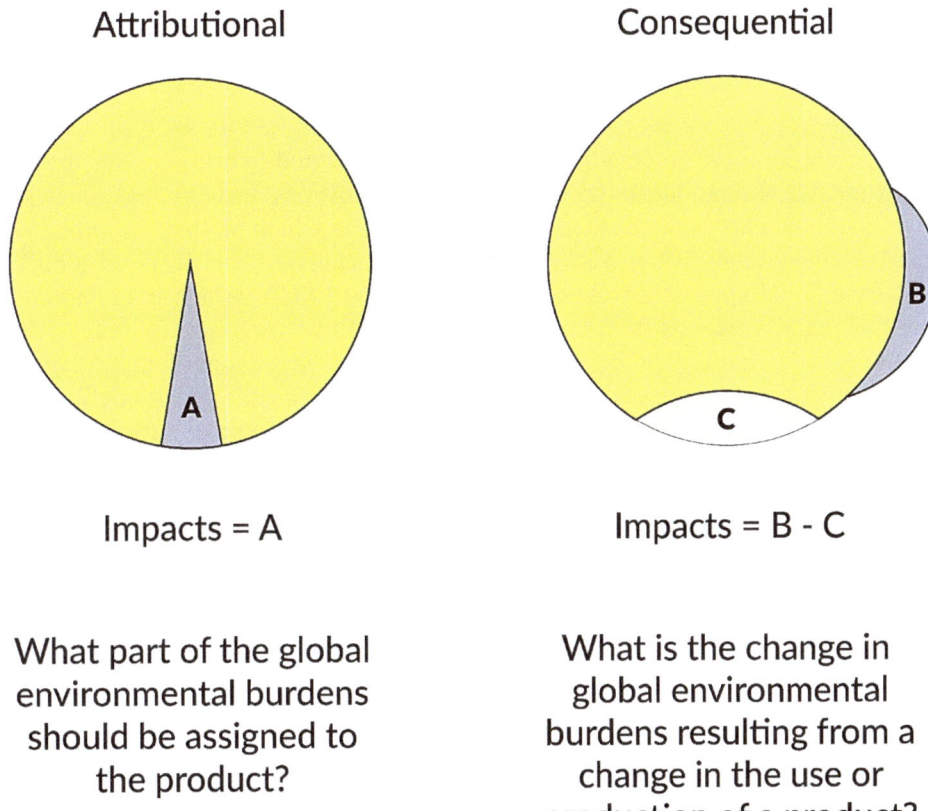

FIGURE 2-2 Illustration of how attributional (left) and consequential (right) LCA address different questions. NOTE: The yellow circles refer to global environmental burdens. SOURCE: Weidema (2003). Reprinted with permission from Copenhagen: Danish Environmental Protection Agency (Environmental Project no. 863).

The system boundary is a second core characteristic of ALCA. Establishing an appropriate system boundary requires several steps. First, the primary stages of the product's supply chain has to be selected for analysis. In the case of transportation fuels, it is common to include material extraction, transportation of raw material to point of processing, raw material conversion into fuel, transportation of fuel to points of distribution, and combustion of the fuel. In a system boundary that encompasses multiple fuel systems, this consideration will be more complex.

A second consideration is the time scale of the study. An ALCA may consider an existing technology operating in the context of current infrastructure systems or it may be focused on some future state in which the technology or infrastructure systems have evolved. Finally, a geographic scope has to be established. An ALCA may limit the system boundary to a production and use occurring in one or more specific geographic regions.

There are three main techniques for carrying out ALCA: process-based LCA, EIO LCA, and hybrid LCA, which combines elements of the first two.

- Process-based LCA uses a bottom-up emissions accounting approach, measuring or estimating emissions from each activity within the chosen system boundary.
- EIO LCA uses a top-down emissions accounting approach, leveraging data on economic trade and emissions from each sector of the economy to estimate emissions associated with economic activity in particular sectors.[1]
- Hybrid Process/EIO LCA combines bottom-up process-based LCA estimates with EIO LCA estimates.

These three approaches (process-based LCA, EIO LCA, and hybrid) answer the question "what emissions are attributable to a product or process?" based on decisions made by the modeler about what to include in the system boundary and which emissions to assign to which products or processes when there are co-products (Matthews et al., 2014). Each of the approaches answers this question at differing levels of detail. The most traditional approach to ALCA is process-based LCA, which uses bottom-up emissions accounting to estimate the emissions from material and energy flows for producing a fuel, including a portion of or all of its supply chain. A process-based LCA, by necessity, cannot include every supply chain activity, so system boundaries are drawn to prevent the analysis from continuing indefinitely. The level of detail used for modeling each supply change stage may vary. For example, national-level material and energy flows may be used in some cases (e.g., crude oil extraction, corn agriculture, average petroleum refinery energy and material consumption), facility-specific data may be used in other cases (e.g., from a specific refinery), and, in the absence of such data, a process model may be built to estimate anticipated

---

[1] Committee member Farzad Taheripour wishes to clarify the following: This report classifies the EIO method as an ALCA approach. This classification is at least inconsistent with some common ALCA definitions provided in Table 2-1 of this report. This table asserts that:

- "An attributional approach to GHG emissions accounting in products provides information about the GHG emitted directly by a product and its life cycle. The product system includes processes that are directly linked to the product by material, energy flows or services following a supply-chain logic."
- "An attributional LCA accounts for only the direct emissions associated with the fuel lifecycle, including the emissions from production of energy and material inputs to the fuel lifecycle."
- "The attributional approach attempts to provide information on what portion of global burdens can be associated with a product (and its life cycle)."

Table 2-1 also declares that:

- "A consequential approach to GHG emissions accounting in products provides information about the GHG emitted, directly or indirectly, as a consequence of changes in demand for the product. This approach typically describes changes in GHG emissions levels from affected processes, which are identified by linking causes with effects."
- "A consequential LCA determines the comprehensive greenhouse gas (GHG) emissions of a product by assessing the direct and indirect impacts of the fuel on external markets. A consequential LCA considers the market effects of a change in production, expands the system boundary to include non-fuel system impacts, and includes the indirect effects of the fuel production on the environment."
- "The consequential approach attempts to provide information on the environmental burdens that occur, directly or indirectly, as a consequence of a decision (usually represented by changes in demand for a product)."

An EIO analysis, ignoring its limitations and deficiencies, quantifies changes in direct and indirect emissions induced by changes (usually increases) in sectoral demands and or supplies. This approach does not attribute a portion of global burden to a product. It calculates direct and indirect emissions induced by changes in sectoral demands or supplies. Therefore, this method follows a consequential approach. Figure 2-2 of this report also suggests that the EIO method is consequential. Taheripour et al. (2022) have outlined how a typical input-output analysis calculates direct and indirect induced emissions due to changes in sectoral demands or supplies.

flows. Process-based LCAs can provide insight into which steps in a process are responsible for a substantial fraction of total energy consumption or emissions and therefore merit attention from engineers and designers who want to reduce environmental burdens. A limitation of process-based LCAs is their reliance on a wide variety of data sources that may vary in accuracy and representativeness. Public and commercial databases intended for use in process-based LCA can be poorly documented and may convey false precision by reporting values with significant figures well beyond what is appropriate. Another limitation of process-based LCAs is that they do not account for effects on material or energy consumption and corresponding GHG emissions outside the system boundary. Therefore process-based models are subject to truncation error, meaning they do not capture the full extent of economy-level effects and thus will underestimate these effects (Lave et al., 1995; Matthews et al., 2008). It is because of this last point that the approaches of EIO LCA, environmentally extended input–output LCA (EEIO LCA), and hybrid LCA were developed.

The second ALCA approach, EIO LCA or EEIO LCA, uses information about how much each economic sector directly purchases from other economic sectors, assembled in a matrix (input–output table) that can be used to calculate the monetary sum of all inputs that a sector requires directly or indirectly to produce its output. In the United States, this information is published regularly by the Bureau of Economic Analysis in the form of an input–output table, with 71-sector input-output data updated each year and the 405-sector input–output data updated every 5 years. Impact vectors are assembled as a set of linear multipliers that translate dollars of economic activity in each sector to a given environmental metric (e.g., CO_2 emissions) (Matthews et al., 2014). Impact vectors (e.g., emissions intensities per dollar) are usually developed by dividing sector-specific emissions totals (or other metrics, such as freshwater withdrawals) by total economic output from that sector to establish direct emissions or other metrics per dollar of economic activity in each sector. EIO LCA models are linear in nature, so each dollar of economic activity within a sector is assigned exactly the same set of impacts. A commonly-used U.S.-based EIO LCA model is national in scope (Matthews et al., 2014, Ch. 8), although multi-regional models are also available (Cicas et al., 2007; Stadler et al., 2018). Together, these data can estimate broader supply chain relationships of economic activity, and corresponding emissions. For example, production of automobiles requires production of steel, which requires production of iron ore, and so forth (Hendrickson et al., 2006).

EIO LCA models lack the technological granularity of process-based models, but can be used to screen for likely hot spots of high environmental impact across a broader system boundary. Another challenge is that flows are typically linked to environmental effects based on monetary value of materials or energy carriers. This linking requires translation of monetary values into mass or energy flows based on an assumed market value, and it does not differentiate different products or activities within an individual economic sector. Market values for any given energy carrier or material fluctuate with time, so EIO LCA results may become unrepresentative of a system when major market value shifts. If major technological advancements occur in one or more sectors, this will only be reflected once updated input–output tables and impact vectors are in place. In EIO LCA models, emissions from each sector are based on average, rather than marginal, emissions in the sector, so these analyses generally do not estimate net emissions implications of changes in fuel use unless marginal emissions are similar to average emissions in the relevant economic sectors (e.g., if emissions are linear with economic output in the relevant sectors).[2] The U.S.

---

[2] EIO LCA and process-based LCA are commonly held to be ALCA. For example, Plevin et al. (2014) state, "Conceptually and structurally, EIO is a version of ALCA, with an expanded, more interconnected set of processes than in what might be called 'traditional' once-through process-based LCA." Such a view is supported by Finnveden et al. (2009), who conclude, "With regard to the discussion on attributional and consequential LCA, it can be noted that the average data contained in an IOA [EIO LCA] are adequate for attributional LCA but less so for consequential LCA. They typically do not describe how the resource uses and emissions of a sector are affected by possible decisions." Indeed, the early developers of the EIO LCA approach note that it has "the advantage of including effects *attributable* [emphasis added] to the influences of many indirect suppliers, which can be overlooked in process models" (Hendrickson et al. 1998). Nevertheless, because EIO LCA models emissions throughout the economy, some researchers think of it as a type of CLCA. However, as noted above, because EIO LCA models average, rather than marginal, emissions from each sector, it does not meet the definitions in Table 2-1 that require CLCA to estimate the *change* in emissions resulting from a decision or action.

Environmental Protection Agency (EPA) has established a set of tools based on U.S. EEIO data which is updated on an ongoing basis (Yang et al., 2017a). Recent research has used EEIO methodology to evaluate environmental and socio-economic effects of biofuels and related technologies (Lamers et al., 2021).

The third approach, hybrid process/EIO LCA, is an attributional approach in which process-based LCA modeling for specific processes of interest is combined with economy-wide process modeling from the EIO LCA approach (Heijungs and Suh, 2002). In doing so, it attempts to extend what is known about supply chains beyond the specific process under examination (Suh et al., 2004). The method and extent of integration can vary depending on the study, and methods include tiered hybrid analysis, input–output based hybrid analysis, and integrated hybrid models (Suh et al., 2004). For example, one study may represent only the conversion stage of a fuel's life cycle with a detailed process model and use an EIO LCA model to estimate effects that occur upstream and downstream of the conversion stage. Another might use detailed process models for both the upstream and conversion stages and turn to EIO LCA to estimate emission from the downstream portions of the supply chain. As an example, EIO LCA could be used to handle the effects of co-products on the broader economy. Although different approaches may be well justified given the goals of each individual study, these inconsistencies in hybrid LCA can complicate cross-study comparisons. The main advantage in pursuing a hybrid approach is to combine the insights available at the process-level from process-based LCA with the broader reach of EIO LCAs to cover a larger swath of the economy and associated environmental effects. Although the limitations of EIO LCA still apply, the hybrid method may help to reduce systematic biases that result from truncation error in purely process-based LCAs. Recent work on hybrid LCA databases is improving the databases and methodology of hybrid LCA (Agez et al., 2021).

It is important to consider the applicability of these types of ALCA in the context of transportation fuels. Process-based, EIO, and hybrid process/EIO LCA all provide approaches to track environmental effects across fuel supply chains. Process-based LCA, for example, can be useful in informing the development of a new fuel production process (e.g., converting lignin to a hydrocarbon fuel). An analysis that is focused on understanding and improving the life-cycle GHG footprint of an industrial process may benefit most from focusing detailed analysis on emissions sources that will be directly affected by changes to the process. Some sectors that are captured in EIO LCA, such as those associated with office workers in the insurance or finance sector, may not be as directly affected by process-level details at the facility. Conversely, an analysis seeking to capture the most comprehensive picture of life-cycle environmental effects would benefit from development of a hybrid process/EIO LCA to capture effects across a broader system boundary. These three types of ALCA can all have a role to play in decision-making, depending on the types of insights sought from the analysis. However, when the question of interest is how emissions will change as a result of a policy action or a change in fuel consumption, an attributional analysis will not provide an answer; a consequential analysis is the only type of analysis targeted to answering that question.

CONSEQUENTIAL LIFE-CYCLE ASSESSMENT

As noted above, CLCA asks a different question from ALCA, focused on how emissions or impacts will *change* in response to a decision or action (see Table 2-2; Ekvall, 2019; Schaubroeck et al., 2021). It estimates the difference in total emissions or environmental effect between one or more scenarios, in which some action is taken, and one or more counterfactuals in which no action is taken. A CLCA may report results on the basis of a functional unit that corresponds to a given quantity of some product or service (e.g., a gallon of fuel). However, the application of CLCA is not limited to this type of analysis. In the context of this report, the change to be captured by CLCA may be increased fuel supply or implementation of a policy. CLCA can include cascades of effects throughout the economy, as in EIO LCA, as well as other market-mediated effects, such as the effect that increasing fuel supply has on fuel prices and ultimately on demand and emissions (Earles and Halog, 2011; Ekvall, 2019). The defining feature of CLCA that differentiates it from ALCA is that it estimates the *change* in emissions induced by a decision or action (Figure 2.2). Often the scope of CLCA studies is broad, estimating economy-wide changes induced by a decision or action, but this is not always the case, and CLCA/ALCA can both use either broad or narrow system boundaries, depending on the LCA goal and scope (Table 2-2).

Methodologies used in CLCA may include equilibrium, input–output, and dynamic models (Le Luu et al., 2020) as well as process-based models that estimate changes, rather than averages. For example, while an ALCA approach may assign the average electricity grid mix and emission factors to a process that consumes electricity, a CLCA would attempt to estimate which types of power plants are most likely to increase generation to meet the increase in power demand, or how the power sector infrastructure itself may change (Chapter 10). For biofuel LCAs, LUC is often modeled consequentially (see Chapter 9), even as the biofuel supply chains themselves are modeled using a predominantly ALCA approach. This practice of mixing ALCA and CLCA is discussed in Chapter 3.

The approach for conducting a CLCA will vary depending on the scale of the change being evaluated. When the change in question is small in comparison with an overall market for a fuel or other product, consequential emissions can be estimated using models that capture the effects at the margin. Returning to the electricity grid example, the overall generation mix in any particular region may include substantial quantities of nuclear, hydroelectricity, coal, and renewable energy sources, but depending on the time of day, the marginal grid mix may be mostly or entirely natural gas if those power plants are responsible for meeting marginal increases in demand. Actions or policies that result in larger relative differences in production require different modeling approaches to predict the structural changes needed to accommodate the change. Because the cascade of changes induced by a technology or policy change can be wide reaching and complicated, answering a consequential question may involve high uncertainty. CLCAs that seek to predict net changes of a policy or other action years into the future run the risk of failing to predict other changes that are unrelated to the policy or action but occur in parallel. This possible problem does not necessarily imply that all CLCAs result in greater uncertainty relative to ALCAs, as results will vary on a case-by-case basis. As with ALCA, there is no single approach to conducting a CLCA and the selection of models and datasets needs to be guided by the goal of the study. Table 2-3 summarizes the three types of ALCA and the category of CLCA.

COMPARISON OF ATTRIBUTIONAL AND CONSEQUENTIAL LIFE-CYCLE ANALYSIS

As discussed in the preceding sections, an important difference between ALCA and CLCA pertains to the concept of average versus marginal emissions. Equally important is the fact that CLCA tends to, in many cases, include a larger system boundary because of the need to incorporate market-mediated effects that are not captured in ALCA. Box 2-1 provides examples of system boundaries for both ALCA and CLCA in a particular study, although the presentation of this approach does not imply endorsement by this committee.

TABLE 2-2 Relationship between System Boundary and LCA Type

	System Boundary	
LCA Type	**Process and Supply Chain**	**Economy-wide**
ALCA: Average emissions attributed to products or services	Process-based ALCA	Economic input-output-LCA
CLCA: Change in emissions due to a decision or action	Process-based CLCA	Equilibrium models

BOX 2-1 Example of System Boundaries in ALCA and CLCA

A boundary map is a visual means of conveying what is included and excluded in an LCA. Martin et al. (2015) present two boundary maps showing how the system boundaries might differ between an ALCA model and a CLCA model of a municipal waste-to-ethanol production system. Such diagrams can aid in the comparison of results from different LCA studies. This paper presents CLCA as starting with the ALCA and expanding the system boundary to encompass additional background systems, namely the gasoline market and market for biofuels from purpose-grown feedstocks. The system boundary for CLCA studies are often, but not always, larger than the system boundary for ALCA studies.

ALCA:

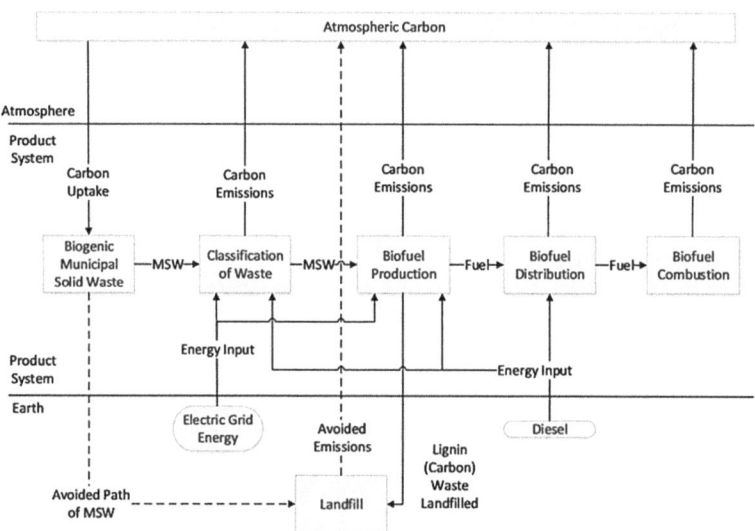

CLCA:

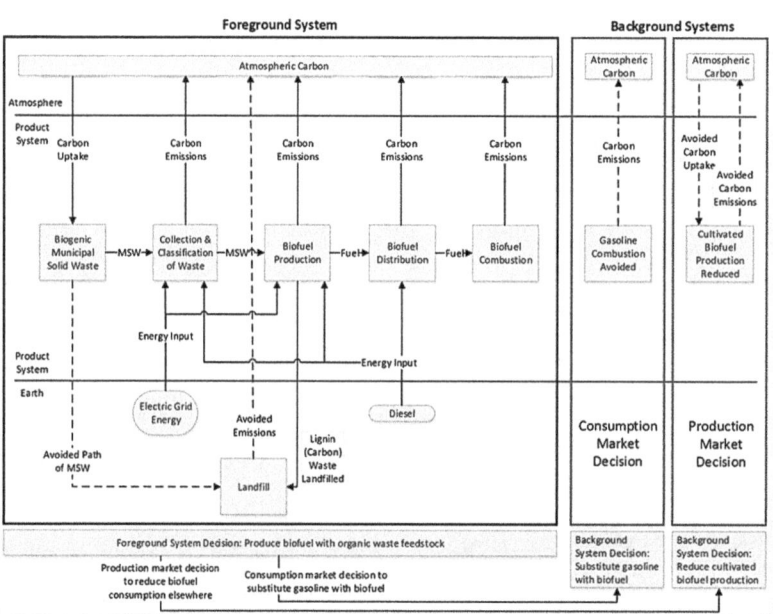

SOURCE: Martin et al. (2015). Reprinted by permission from Springer Nature: Current Sustainable/Renewable Energy Reports.

Fundamentals of Life-Cycle Assessment

Figure 2-3 shows how attributional and consequential LCA relate to the research question of interest. When an analysis is addressing a question of how emissions will change in response to a decision or action, CLCA is appropriate (see Table 2-3). If the resulting change in a fuel's consumption is small, marginal emission factors can estimate consequential emissions, but if the change is large, consequential emissions may be non-marginal. In contrast, when addressing the question of how existing emissions can be attributed to fuels, the researcher has a choice: a consequential approach may be appropriate if the primary goal is to attribute to fuels the emissions associated with changes in use of those fuels; an attributional approach may be appropriate if the primary goal is to assign a share of existing emissions to fuels and other co-products. Attributional approaches often, but not always, assign average emission rates to products.

TABLE 2-3 Summary of Major Approaches to LCA

Method	Approach	Question Addressed	System Boundary
Attributional LCA: Process-Based	Bottom-up emissions accounting	What emissions are attributable to a process or product, as approximated by a supply chain, within the system boundaries?	Typically the process in question; potentially including portions of its supply chain
Attributional LCA: Economic Input-Output	Top-down emissions accounting	What emissions are attributable to a process or product, as approximated by a sector, within the system boundaries?	National, multi-regional or global economy
Attributional LCA: Hybrid Process/EIO	Both bottom-up and top-down emissions accounting	What emissions are attributable to a process or product within the system boundaries, as approximated by a combination of supply chain and economic sector information?	National, multi-regional or global economy
Consequential LCA	Counterfactual emissions comparison	How will emissions change in response to a decision or action?	Varies, but ideally as comprehensive as possible, including global effects

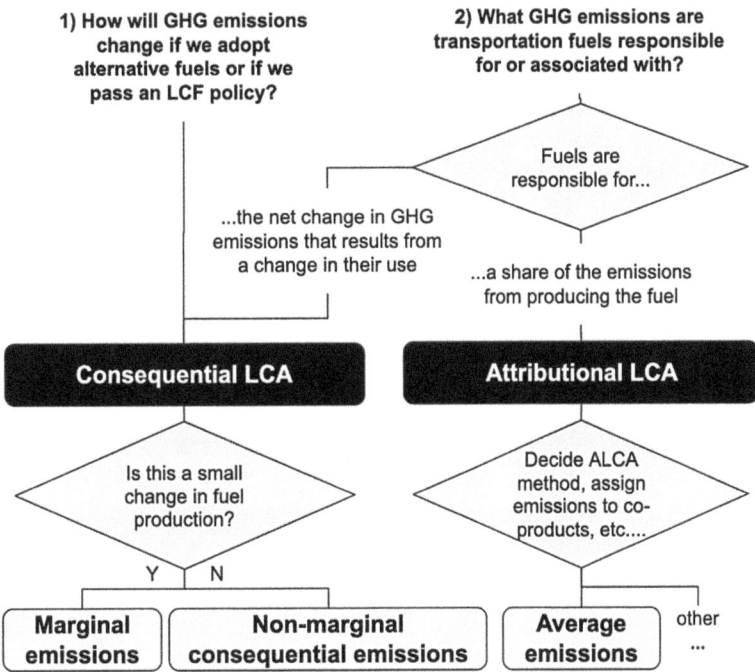

FIGURE 2-3 LCA approaches by research question and relationship to average and marginal emission.

Figure 2-4 illustrates consequential GHG emissions from increased production of a fuel for which GHG emissions increase non-linearly with production. The first scenario (current) has no change, decision, or action that would influence fuel production. The second scenario (result) includes such a change, decision, or action. Figure 2-4a shows the difference in emissions between the two scenarios. Figure 2-4b illustrates that using average emissions estimates from an attributional LCA in this situation can produce poor estimates of the emissions consequences of the change unless emissions rise linearly with volume. Marginal emissions are shown in Figure 2-4c and represent the slope of the emissions curve at current levels. Marginal emissions can produce good estimates of consequential emissions when changes are small or when or the emissions curve is linear but can produce poor estimates when changes are large and the emissions curve is nonlinear, as in the illustration. Non-marginal consequential analysis can estimate the consequential change directly (Figure 2-4a) by estimating the difference between emission levels with and without a proposed change, decision or action.

Importantly, in practice, many LCAs draw on elements from more than one of these approaches. For example, many CLCAs make use of average estimates for particular products or processes when consequential estimates are unavailable. Additionally, some largely process-based LCAs attempt to account for some consequential effects, such as LUC, credits for avoided burdens from co-product displacement or substitution, and other estimated changes relative to the counterfactual (baseline).

As a consequence of the variety of approaches that an LCA practitioner can adopt, it is unsurprising that LCA studies of the same product or system can produce conflicting results. At a high-level, these differences can result from (1) different questions being asked, (2) different methods being used to answer these different questions, (3) different underlying data or (4) different scope and assumptions.[3]

Conclusion 2-1: The approach to LCA needs to be guided on the basis of the question the analysis is trying to answer. Different types of LCA are better suited for answering different questions or achieving different objectives, from fine tuning a well-defined supply chain to reduce emissions, to understanding the global, economy-level effect of a technology or policy change.

Conclusion 2-2: Process-based ALCAs entail bottom-up accounting where emissions are assigned to products or processes based on modeling approach of a static world. Process-based ALCA can identify major sources of emissions in well-defined supply chains and identify opportunities to reduce supply chain carbon intensity, especially when case-specific process-data can be used instead of generic data. Economic input-output life-cycle assessment (EIO LCA) identifies implications of interactions across broad sectors of the economy. It can capture emissions that may not be immediately apparent if only a well-defined supply chain is evaluated. It also is helpful in flagging emissions sources that are far-removed from the foreground system but are major contributors to total environmental effects. Hybrid Process/EIO ALCA identifies major sources of emissions beyond well-defined supply chains to include economy-wide effects. CLCA assesses the net effect of a decision or action, such as a change in fuel use or a change in policy, on total GHG emissions.

Conclusion 2-3: LCA results can vary depending on which methods are used, which data are used, which assumptions are made, what scope is defined, and what question is asked.

Recommendation 2-1: When emissions are to be assigned to products or processes based on modeling choices including functional unit, method of allocating emissions among co-products, and system boundary, ALCA is appropriate. Modelers should provide transparency, justification, and sensitivity or robustness analysis for modeling choices.

---

[3] For a detailed description of this effect, see Box 2-2 and Chapter 5 in National Research Council (2011).

Fundamentals of Life-Cycle Assessment

Recommendation 2-2: When a decision-maker wishes to understand the consequences of a proposed decision or action on net GHG emissions, CLCA is appropriate. Modelers should provide transparency, justification, and sensitivity/robustness analysis for modeling choices for the scenarios modeled with and without the proposed decision or action.

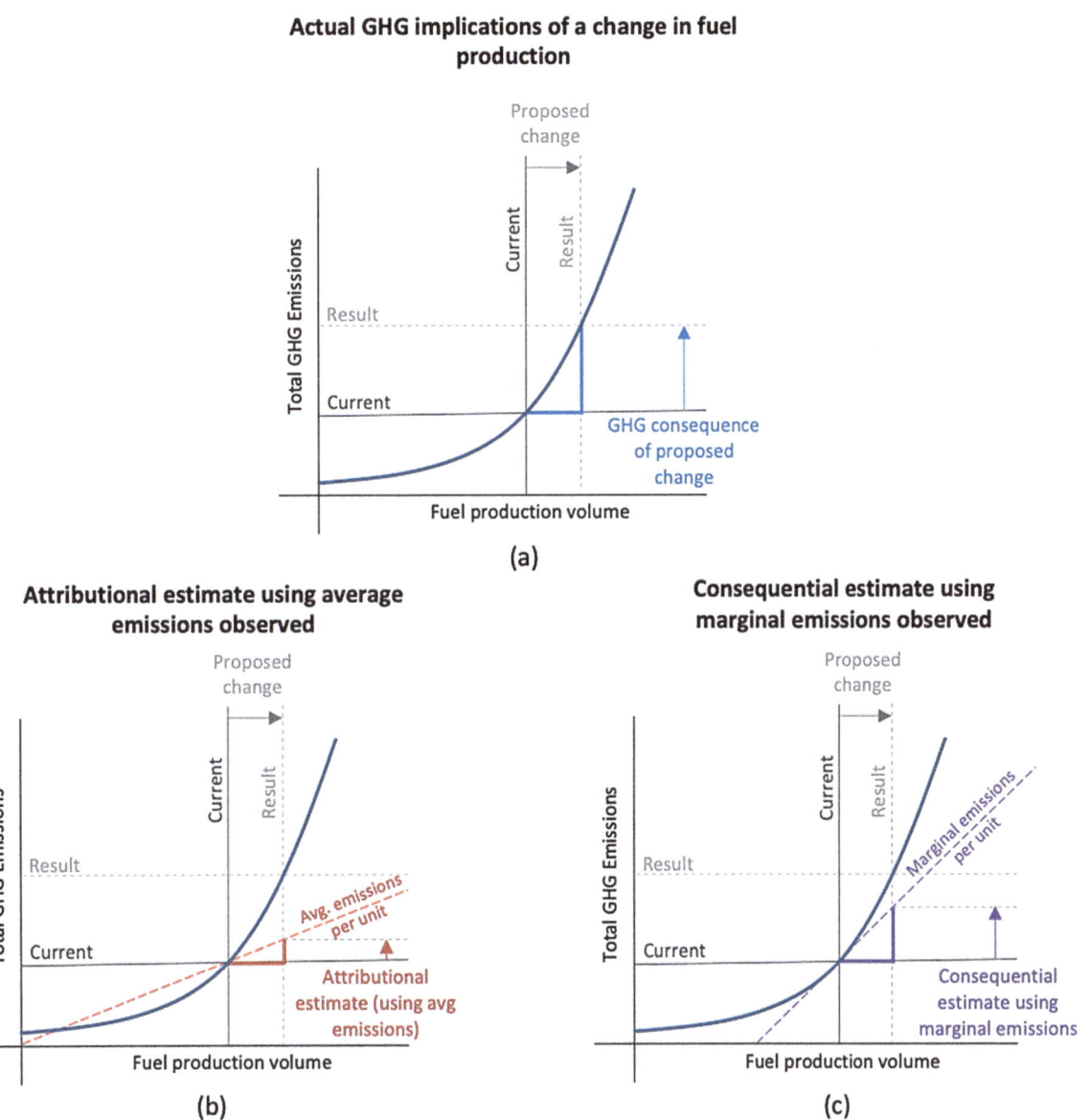

FIGURE 2-4 Illustration of the hypothetical relationships between attributional and consequential emissions and average and marginal emission factors for a single fuel, shown for one possible case when GHG emissions from increased production of this fuel are convex with fuel production volume. The figure is not intended to imply that GHG emissions are always convex with production volume. Emissions' response to production volumes, particularly when large shifts in production occur, may take other shapes, including concave, linear, or nonconcave/nonconvex. The implementation of these models may or may not translate analytically in results depicted by these figures, as each model differs in how it represents emissions and production volumes, as well as their relationship.

REFERENCES

Agez, M., E. Muller, L. Patouillard, C-J. H. Södersten, A. Arvesen, M. Margni, R. Samson, and G. Majeau-Bettez. 2021. Correcting remaining truncations in hybrid life cycle assessment database compilation. *Journal of Industrial Ecology*. https://doi.org/10.1111/jiec.13132.

Bressler, R. D. 2021. The mortality cost of carbon. *Nature Communications* 12.1:1-12. https://doi.org/10.1038/s41467-021-24487-w.

Cicas, G., C. T. Hendrickson, and A. Horvath, H.S. Matthews. 2007. A regional version of a US economic input-output life-cycle assessment model. *International Journal of Life Cycle Assessment* 12: 365–372. https://doi.org/10.1065/lca2007.04.318.

Curran, M. A., M. Mann, and G. Norris. 2005. "The international workshop on electricity data for life cycle inventories." *Journal of Cleaner Production* 13(8):853-862. https://doi.org/10.1016/j.jclepro.2002.03.001.

Earles, J. M., and A. Halog. 2011. Consequential life cycle assessment: A review. *International Journal of Life Cycle Assessment*, 16, 445–453. https://doi.org/10.1007/s11367-011-0275-9.

Ekvall, T. 1999. *System expansion and allocation in life cycle assessment: With implications for wastepaper management*. Chalmers University of Technology.

Ekvall, T., A. Azapagic, G. Finnveden, T. Rydberg, B. P. Weidema, and A. Zamagni. 2016. Attributional and consequential LCA in the ILCD handbook. *The International Journal of Life Cycle Assessment* 21(3), 293-296. https://doi.org/10.1007/s11367-015-1026-0.

Ekvall, T. 2019. Attributional and consequential life cycle assessment. In *Sustainability Assessment at the 21st Century*, M. J. Bastante-Ceca, J. L. Luis Fuentes-Bargues, L. Hufnagel, F-C. Mihai, and C. Iatu, eds. London: IntechOpen. https://www.intechopen.com/chapters/69212.

Finnveden, G., M. Z. Hauschild, T. Ekvall, J. Guinée, R. Heijungs, S. Hellweg, A. Koehler, D. Pennington, and S. Suh. 2009. Recent developments in life cycle assessment. *Journal of Environmental Management* 91(1):1-21.

Guinée, J. B., S. Cucurachi, P. J. Henriksson and R. Heijungs. 2018. Digesting the alphabet soup of LCA. *International Journal of Life Cycle Assessment* 23(7):1507–1511.

Guinée J. B., M. Gorreé, R. Heijungs, G. Huppes, R. Kleijn, A. de Koning, L. van Oers, A. Wegener Sleeswijk, S. Suh, H. A. Udo de Haes, J. A. de Bruijn, R. van Duin, and M .A. J. Huijbregts. 2002. Handbook on Life Cycle Assessment: Operational Guide to the ISO Standards. Series: Eco-efficiency in Industry and Science. Kluwer Academic Publishers, Dordrecht.

Heijungs, R. and S. Suh. 2002. The Computational Structure of Life Cycle Assessment. Vol. 11. Springer Science & Business Media. https://www.semanticscholar.org/paper/The-computational-structure-of-life-cycle-Heijungs-Sun/4b80bf5f6254d25f1e0186f54aadf233798668b2.

Hendrickson, C. T., A. Horvath, S. Joshi, and L. B. Lave. 1998. "Peer Reviewed: Economic Input–Output Models for Environmental Life-Cycle Assessment." *Environmental Science & Technology* 32(7):184A-191A. https://doi.org/10.1021/es983471i.

Hendrickson, C.T., L. B. Lave, and H. S. Matthews. 2006. *Environmental Life Cycle Assessment of Goods and Services: An Input-Output Approach* (1st ed.). Routledge. https://doi.org/10.4324/9781936331383.

ISO (International Organization for Standardization). 2006a. ISO 14040: 2006. Environmental management–Life cycle assessment–Principles and framework.

ISO. 2006b. ISO 14044: 2006. Environmental management. Life cycle assessment. Requirements and guidelines.

IWG (Interagency Working Group on the Social Cost of Greenhouse Gases. 2021. Technical Support Document: Social Cost of Carbon, Methane, and Nitrous Oxide: Interim Estimates under Executive Order 13990. https://www.whitehouse.gov/wp-content/uploads/2021/02/TechnicalSupportDocument_SocialCostofCarbonMethaneNitrousOxide.pdf.

Lamers, P., A. F. T. Avelino, Y. Zhang, E. C. D. Tan, B. Young, J. Vendries and H. Chum. 2021. Potential socioeconomic and environmental effects of an expanding U.S. Bioeconomy: An assessment of near-commercial cellulosic biofuel pathways. *Environmental Science & Technology* 55(8):5496–5505. https://pubmed.ncbi.nlm.nih.gov/33764760/.

Lave, L. B., E. Cobas-Flores, C. T. Hendrickson and F. McMichael. 1995. Life cycle assessment: Using input-output analysis to estimate economy-wide discharges. *Environmental Science & Technology* 29:420A–426A.

Le Luu, Q., S. Longo, M. Cellura, E. R. Sanseverino, M. A. Cusenza and V. Franzitta. 2020. A conceptual review on using consequential life cycle assessment methodology for the energy sector. *Energies* 13(12):3076.

Martin, E. S., M. V. Chester and S. E. Vergara. 2015. Attributional and consequential life-cycle assessment in biofuels: A review of recent literature in the context of system boundaries. *Current Sustainable/Renewable Energy Reports* 2:82–89. https://doi.org/10.1007/s40518-015-0034-9.

Matthews, H. S., C. T. Hendrickson and D. Matthews. 2014. *Life Cycle Assessment: Quantitative Approaches for Decisions that Matter*. https://www.lcatextbook.com/.

Matthews, H. S., C. Weber and C. T. Hendrickson. 2008. Estimating carbon footprints with input-output models. *Proceedings of the International Input-Output Meeting on Managing the Environment*. http://iioa.org/conferences/intermediate-2008/pdf/1c1_Matthews.pdf.

National Research Council. 2011. *Renewable Fuel Standard: Potential Economic and Environmental Effects of U.S. Biofuel Policy*. National Academies Press: Washington, DC.

Peters, G. P. 2010. Carbon footprints and embodied carbon at multiple scales. *Current Opinion in Environmental Sustainability* 2(4):245-250.

Plevin, R., M. Delucchi and F. Creutzig. 2014. Response to Comments on 'Using Attributional Life Cycle Assessment to Estimate Climate-Change Mitigation .....*Journal of Industrial Ecology* 18(3):468–70. https://doi.org/10.1111/jiec.12153.

Schaubroeck, T., S. Schaubroeck, R. Heijungs, A. Zamagni, M. Brandão and E. Benetto. 2021. Attributional & consequential life cycle assessment: Definitions, conceptual characteristics and modelling restrictions. *Sustainability* 13(13):7386.

Stadler, K., R. Wood, T. Bulavskaya, C. J. Södersten, M. Simas, S. Schmidt, A. Usubiaga, J. Acosta-Fernández, J. Kuenen, M. Bruckner and S. Giljum. 2018. EXIOBASE 3: Developing a time series of detailed environmentally extended multi-regional input-output tables. *Journal of Industrial Ecology* 22(3):502-515. https://doi.org/10.1111/jiec.12715.

Suh, S., M. Lenzen, G. J. Treloar, H. Hondo, A. Horvath, G. Huppes, O. Jolliet, U. Klann, W. Krewitt, Y. Moriguchi and J. Munksgaard. 2004. System boundary selection in life-cycle inventories using hybrid approaches. *Environmental Science & Technology* 38(3):657-664.

Taheripour, F., M. Chepeliev, R. Damania, T. Farole, N. L. Gracia and J. D. Russ. 2022. Putting the Green Back in Greenbacks: Opportunities for a Truly Green Stimulus. *Environmental Research Letters* 17(4):044067. https://doi.org/10.1088/1748-9326/ac6003.

UNEP-SETAC (U.S. Environment Program-The Society of Environmental Toxicology and Chemistry). 2011. *Global Guidance Principles for Life Cycle Assessment Databases – A Basis for Greener Processes and Products*. Shonan Guidance Principles. UNEP-SETAC: Paris, France. https://www.lifecycleinitiative.org/wp-content/uploads/2012/12/2011%20-%20Global%20Guidance%20Principles.pdf.

UNEP-SETAC. 2020. Guidelines for Social Life Cycle Assessment of Products and Organisations 2020. https://wedocs.unep.org/handle/20.500.11822/34554;jsessionid=7ED66B6DC7485618EBFE17904FF3467A (accessed July 25, 2022).

Wang, Z., P. Osseweijer and J. A. Posada. 2020. Human health impacts of aviation biofuel production: Exploring the application of different life cycle impact assessment (LCIA) methods for biofuel supply chains. *Processes* 8(2):58.

Weidema, B. P. 2003. Market Information in Life Cycle Assessment. Environmental Project no. 863. Danish Environmental Protection Agency; Copenhagen, p. 147. https://www2.mst.dk/Udgiv/publications/2003/87-7972-991-6/pdf/87-7972-992-4.pdf.

Yang, Y., R. Heijungs and M. Brandão. 2017b. Hybrid life cycle assessment (LCA) does not necessarily yield more accurate results than process-based LCA. *Journal of Cleaner Production* 150:237-242.

Yang, Y., W. W. Ingwersen, T. R. Hawkins, M. Srocka and D. E. Meyer. 2017a. USEEIO: A new and transparent United States environmentally-extended input-output model. *Journal of Cleaner Production* 158:308−318.

3
Life-Cycle Assessment in a Low-Carbon Fuel Standard Policy

A low-carbon fuel standard (LCFS) refers to a policy intended to reduce the overall climate forcing that results from the production and use of transportation fuels by incentivizing the use of fuels with reduced emissions intensities. In this chapter, the term LCFS does not refer to any specific existing policy. Individual policies are specified differently; for example, the California LCFS is referred to as CA-LCFS. In general, an LCFS is predicated on three ideas:

1) There are multiple energy sources (fuels) for accomplishing the same goal (transportation),
2) The production and use of different transportation fuels will result in different climate forcing effects, driven by greenhouse gas (GHG) emissions and other climate forcers, and
3) There is a mechanism for increasing the use of fuels with reduced contributions to climate change (referred to as low-carbon fuels) over others by regulating and incentivizing individual fuels based on their estimated ability to reduce GHG emissions and other climate forcing effects relative to an assumed baseline fuel.

In this context, the term "low-carbon" in "low-carbon fuel" refers not to the chemical composition of a fuel itself, but rather to the climate forcing, as estimated for use in the LCFS, associated with the production and use of a fuel relative to the impact of the use of other options. Climate forcing is typically accounted for based on the global warming potential of GHGs including carbon dioxide (CO_2), methane (CH_4), nitrous oxide (N_2O), fluorinated gases, and other climate forcers–some of which may not be GHGs (i.e., aerosols, albedo changes, and other biophysical impacts). Because identical fuel molecules can be produced from different raw materials using different conversion methods, a specific set of material inputs (feedstocks) and conversion processes is referred to in many transportation fuel policies as a "pathway."

As highlighted in Chapter 2, there is no single life-cycle assessment (LCA) method capable of answering all possible questions related to the climate impacts of a transportation fuel. Attributional LCA (ALCA) and consequential LCA (CLCA) both have important roles to play in understanding the climate impacts of transportation energy use, and each answers fundamentally different sets of questions. For each approach, there are methodological choices that will result in different outcomes. This chapter introduces some of the practical considerations in applying LCA methods to LCFS, both in terms of its direct integration into policy as well as its use in regulatory impact assessment (RIA). It summarizes the use of LCA in LCFS to date and discusses the appropriateness and tradeoffs of different approaches and methods for estimating life-cycle GHG emissions associated with transportation fuels for potential use in a national LCFS. The chapters in Parts II and III of this report explore technology-specific challenges and specific methodological issues in greater depth.

HISTORICAL CONTEXT

The use for LCA in regulating transportation-sector GHG emissions stems from the fact that emissions do not only occur at vehicle tailpipes. Raw material production, conversion to fuel, and the transportation of fuels to consumers can be important contributors to the total carbon footprint. Furthermore, many of those upstream emissions sources are not directly regulated. Regulating only vehicle tailpipe emissions, for example, would unfairly advantage technologies that have upstream impacts but zero tailpipe emissions, such as battery electric or hydrogen fuel cell vehicles. Biofuels were, and continue to be, a focus in the development of LCA methods for LCFS because there are GHG emission sources along biofuel supply chains (e.g., N_2O from farmland and fossil CO_2 emissions from combustion of diesel fuel and natural gas in farm equipment and biorefineries). The LCFS concept was also conceived as a way to track and disincentivize the consumption of more emissions-intensive fossil resources.

In 2007, when federal and California policies were beginning to incorporate life-cycle GHG emissions, oil prices were increasing, stimulating increased production of tar sands, oil shale, and consideration of other emissions-intensive alternative production routes, such as coal-to-liquids. Some held the belief that this trend was unlikely to reverse and global "peak oil" (referring to conventional oil production) would soon be reached, implying that prevention of a rise in unconventional fossil fuel production was key to avoiding catastrophic climate change (Kharecha and Hansen, 2008). However, these alternative production routes were not being penalized for their emissions-intensive upstream activities and contributions to the GHG-intensity of transportation. As Farrell et al. noted in their 2007 policy analysis of the CA-LCFS, "[T]he LCFS is a response to this recarbonization of transportation fuels, as well as the many market failures blocking innovation and investments in low-carbon alternatives to petroleum" (Farrell et al., 2007, p. 3). In short, an LCFS attempts to capture both combustion and supply chain emissions by incorporating upstream GHG emissions sources that are otherwise not adequately regulated directly, such as power plants, farms, biorefineries, and hydrogen production facilities. This practice of regulating upstream emission sources can improve the comprehensiveness of an LCFS, but it also introduces additional complexity and uncertainty.

Challenges in the application of LCA for regulatory purposes have long been recognized (Hunt et al., 1996). For example, Lave et al. (1995, p. 420) stated with respect to one LCA method commonly used today (process-based), "Equally credible analyses can produce qualitatively different results, so the results of any particular life-cycle analysis cannot be defended scientifically." More recently, Ekvall (2019, p. 1), noted:

"[I]t was clear almost from the start that results from different LCAs can contradict each other. This is still true, despite many attempts to harmonize, standardize, and regulate LCA. From history, we learn that it is not realistic to expect LCA to deliver a unique and objective result. It should not be regarded as a single unique method; it is more fruitful to consider it a family of methods." Meta-analyses and harmonization efforts can adjust for differences in underlying assumptions and methodological choices that drive many of the contradictions across different LCAs, particularly for ALCAs (Heath and Mann, 2012). Despite the complexities of using LCA in a regulatory context, a desire to design policies that achieve GHG emissions reductions has motivated an increase in the use of LCA for policy development, particularly in the context of fuels

LOW-CARBON FUEL POLICIES IN THE UNITED STATES, EUROPE, AND BRAZIL

Although a comprehensive overview of international LCFS is outside the scope of this report, there are several examples of transportation fuel policies across the world that use LCA to varying degrees. This section summarizes a set of policies selected to provide a representative range of approaches to LCA implementation.

In the United States, there have been federal- and state-level policies that aim to reduce GHG emissions from transportation fuels in which LCA has been used. For the United States, the focus is on two groups of policies. The first is the Renewable Fuel Standard (RFS) (75 Fed. Reg. 14863), which was created by the Energy Policy Act of 2005; the life-cycle provisions were added when the program was amended by the Energy Independence and Security Act of 2007 (EISA). The U.S. Environmental Protection Agency (EPA) issued the final rule implementing EISA in 2010. A second set of policies targeting transportation fuels includes the CA-LCFS, which took effect in 2011, and a related policy called the Clean Fuel Program in Oregon, which took effect in 2016. Washington State also passed legislation authorizing a Clean Fuel Standard in 2021, which will go into effect in 2023. The California, Oregon and Washington programs are closely related and for simplicity, this chapter focuses primarily on the CA-LCFS because of its longer history and its similarity to the other more recently-crafted state policies.

Outside of the United States, one policy in the European Union (EU) and another in Brazil are explored. The EU's 2009 Renewable Energy Directive (RED) and the updated 2018 Renewable Energy Directive (RED II) cover multiple sectors but have specific targets for the use of low-carbon fuels in transportation sector. The RenovaBio policy introduced in Brazil in 2017 sets carbon intensity (CI) targets for road transportation and assigns credits based on individual fuels' contributions to achieving those targets.

The U. S. Renewable Fuel Standard

The RFS sets threshold GHG emissions reductions compared to a fixed baseline for petroleum-derived fuels that a renewable fuel must achieve to be eligible for the program. In addition to transportation fuels, liquid fuels intended for use as heating oil are also eligible. The legislation contains definitions for what types of feedstocks make a fuel eligible, independent of their GHG footprint. Generally speaking, biofuels, electricity generated using biomass as the fuel, and renewable compressed natural gas (biogas that has been cleaned and upgraded to pipeline quality methane) are currently, or could be, qualifying fuels in the RFS. Notably, the RFS excludes non-biomass sources of renewable electricity, such as solar and wind. Qualifying fuels are then awarded Renewable Identification Numbers (RINs), which serve as the "currency" of the RFS program. There are four categories of RINs, each with different GHG reduction thresholds that fuels must meet according to the specific LCA methods used by EPA:

- Conventional renewable fuel (RIN category D6), 20 percent;
- Advanced biofuel (RIN category D5), 50 percent;
- Biomass-based diesel (RIN category D4); and
- Cellulosic biofuel (RIN categories D3 andr D7), 60 percent

Fuels must be, according to the specific LCA methods used by EPA, 20 percent, 50 percent, 50 percent, and 60 percent less GHG intensive than petroleum-derived fuels, respectively, as assessed in the last year of statutory volumes (2022) (as shown in Figure 9-1, Chapter 9 of this report). The minimum 20 percent GHG reduction for D6 fuels was not strictly required for grandfathered facilities, including biorefineries that were existing, idled, or under-construction as of the November 2009 industry assessment, as well as coal-burning ethanol plants whose construction commenced before December 19, 2007. As of 2010, the EPA calculated that 14.8 billion gallons of ethanol production capacity could be grandfathered in regardless of whether they met the GHG reduction requirement, just short of the total 15 billion gallon cap for corn ethanol (EPA, 2010). There are additional eligibility criteria limiting the feedstocks from which the different fuel categories can be produced, including some safeguards against specific forms of domestic land conversion (EPA, 2010b).

As noted above, the RFS uses a combination of life-cycle emissions estimates and feedstock categorization to assign different alternative fuels to different compliance categories. Therefore, eligibility is based on both the definitions of qualifying fuels set forth in EISA and a fuel's partial life-cycle GHG emissions relative to the baseline, fossil fuel life-cycle CI determined by EPA. As part of the RFS implementation, EPA has assessed the LCA of a variety of fuel pathways; new fuel pathways may be eligible for the RFS, but producers must first petition and submit data for EPA to conduct an LCA for final approval. EPA (2010) uses a combination of CLCA and ALCA (see Chapter 2) in its assessment of the life-cycle emissions for each fuel pathway. EPA (2010a) assessed the consequential impacts of RFS-induced biofuel demand on agricultural emissions and global land use, but not on fuel markets, for the regulatory impact analysis of the policy, supplemented with attributional emissions to estimate the impacts of biofuel production for four biofuel pathways.[1]

---

[1] For a high-level summary of the regulatory impact assessment, see RIAs https://nepis.epa.gov/Exe/ZyPURL.cgi?Dockey=P1006DXP.txt.

An initial RIA (EPA, 2020) is supplemented by additional analysis for new pathways prior to their approval, based on a predominantly attributional analysis based on the energy and emissions associated with feedstock production and conversion. The hybrid ALCA and CLCA-based GHG footprints are used to determine, along with the feedstock type, in which RIN category a given fuel will be placed. RINs are generated when a qualifying fuel is produced and are retired when the fuel is blended. As can be seen above, the conventional renewable fuel (D6) category is the broadest and requires only a 20 percent life-cycle GHG reduction as assessed in 2022, in contrast; the cellulosic biofuel (D3 RIN and, for cellulosic diesel, D7 RIN) requires both a cellulosic feedstock and a 60 percent GHG reduction (although some types of biogas including landfill gas, once upgraded to renewable compressed natural gas, also qualify). EPA annually sets volume requirements based on statutory targets set out in EISA, with periodic adjustments based on various market factors. EPA mandates the blending of an overall quantity of renewable fuel annually, with separate volumetric sub-targets for certain compliance categories. RINs can be traded and carried over time. The four RIN categories are nested, meaning that some types of RINs can be used in compliance for other RINs. For example, biomass-based diesel (D4 RINs) can be used to satisfy advanced biofuel (D5) mandates for advanced biofuels or conventional renewable fuel (D6) for renewable fuels; see Figure 3-1.

Notably, the decision to incorporate CLCA in RFS decision-making was based on guidance in EISA, which included the following definition:

> "LIFECYCLE GREENHOUSE GAS EMISSIONS.—The term 'lifecycle greenhouse gas emissions' means the aggregate quantity of greenhouse gas emissions (including direct emissions and significant indirect emissions such as significant emissions from land use changes), as determined by the Administrator, related to the full fuel lifecycle, including all stages of fuel and feedstock production and distribution, from feedstock generation or extraction through the distribution and delivery and use of the finished fuel to the ultimate consumer, where the mass values for all greenhouse gases are adjusted to account for their relative global warming potential."

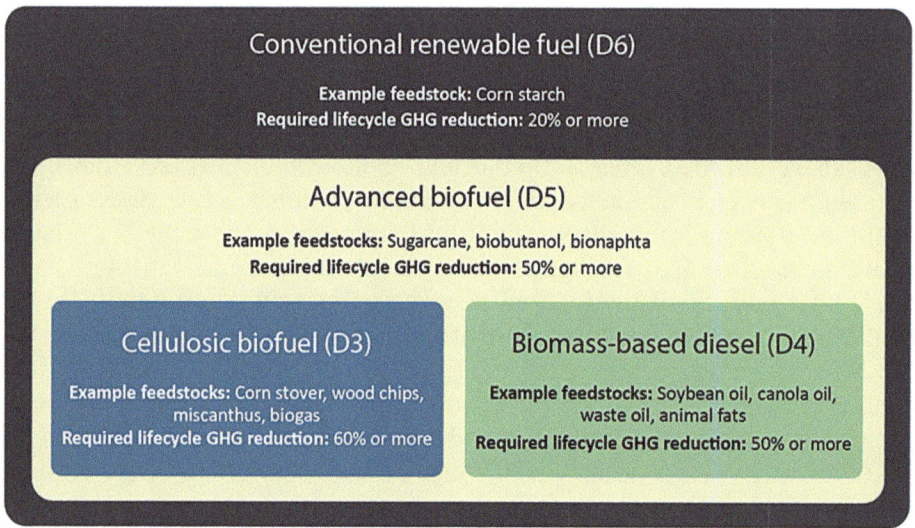

FIGURE 3-1 Nested compliance categories in the EPA Renewable Fuel Standard (RFS2). SOURCE: EPA (n.d.). NOTE: D7 is not shown, and refers to cellulosic diesel; it is subject to the same requirements as D3 fuels.

The EPA 2010 final rule acknowledged the uncertainty associated with estimating land use change (LUC) and presented results for emissions from domestic and international LUC as a mean value bracketed by low and high values. Because the RFS uses thresholds for GHG emissions reductions rather than directly tying incentives to CI scores, fuel producers are somewhat insulated from the uncertainty around the impacts of LUC. Once it is demonstrated that a fuel meets this threshold, more precise estimates of the life-cycle GHG emissions of the fuel have no impact on the administration of the policy. Furthermore, while EPA carried out a CLCA including indirect land use change (ILUC, also referred to as market-mediated LUC) for its RIA, it did not ultimately include other market-mediated effects such as potential changes in emissions from changes in fuel markets.

The California Low-Carbon Fuel Standard

The CA-LCFS is a performance-based standard that requires obligated parties—in this case, petroleum refiners and importers—to reduce the CI of the fuel mix intended for road transportation supplied to the road sector, as assessed and approved by the California Air Resources Board (CARB). Amended in 2018 to extend beyond 2020, it mandates a reduction of 20 percent by 2030 of the road sector fuel mix relative to 2010 levels (17 CCR § 95482), using CARB's accounting methods. Deficits are generated from the production and import of fossil-derived gasoline and diesel fuel; credits may be generated through the use of alternative fuels based on their CI as approved by CARB relative to the benchmark CI assigned to each fuel type, which CARB updates annually. "Opt-in" credits may be generated for pathways such as aviation fuels and electricity, whose producers are not obligated parties under the CA-LCFS. In other words, petroleum-derived aviation fuels do not generate deficits, but alternative aviation fuels with a CI below the baseline jet aviation CI can generate credits.

Unlike the volumetric mandate of the federal RFS, credit generation in the CA-LCFS is based on the quantity of estimated GHG reductions from blended fuel rather than the volumes. Also in contrast with the RFS, non-biomass fuels such as electricity and hydrogen generated from wind and solar can be used for compliance. Similar CI-based standards have been implemented in British Columbia and Oregon and are in the process of implementation in the state of Washington and other provinces in Canada (British Columbia, 2008; Government of Canada, 2020; Oregon DEQ, 2021; Reducing Greenhouse Gas Emissions by Reducing the Carbon Intensity of Transportation Fuel, 2021).[2]

To be eligible under the CA-LCFS, every fuel producer must develop an LCA for each specific pathway and submit it for certification with CARB. The program includes pathway-specific life-cycle emissions estimates calculated using an ALCA approach, as well as a set of ILUC factors estimated for a selection of crop-based fuels using the Global Trade Analysis Project (GTAP)-BIO model (CARB, 2015). The ALCA portion of the assessment uses a modified version of Argonne National Laboratory's Greenhouse Gases, Regulated Emissions, and Energy Use in Technologies (GREET) model (referred to as CA-GREET). For biofuel pathways that are well-characterized in the program (e.g., corn ethanol) users may submit a simplified (Tier 1) application that requires inputting facility-specific parameters into a calculator; novel pathways with unique operating considerations require a Tier 2 application that may include a modified CA-GREET model for certification. Pathway applications for new fuels require a combination of validation and verification by third-party verifiers as well as review by CARB.

In contrast with the RFS, the CA-LCFS is performance-based, so the precise CI value assigned to each fuel is integral to the implementation of the policy. The credits a fuel can earn under this policy are proportional to its estimated life-cycle GHG emissions. Accordingly, there is a strong incentive to demonstrate estimated life-cycle GHG emissions that are as low as possible. As noted previously, life-cycle GHG emissions of fuels under the CA-LCFS, excluding ILUC, are generally estimated using ALCA. However, CARB uses a CLCA approach to evaluate GHG emissions of ILUCs.

The policy paper developing the CA-LCFS concept in response to the California governor's executive order S-01-07 recommended including some non-zero representation of LUC emissions, which

---

[2] See Washington State's 2021 Senate Bill 5231 at https://www.washingtonvotes.org/2021-SB-5231.

inherently requires a CLCA approach. However, the executive order acknowledged that these estimates were likely subject to change while methodologies were further developed. The executive order included the following recommendation (California Governor's Executive Order S-01-07):

> "Develop a non-zero estimate of the global warming impact of direct and indirect land use change for crop-based biofuels, and use this value for the first several years of the LCFS implementation. Participate in the development of an internationally accepted methodology for accounting for land use change, and adopt this methodology following an appropriate review."

Methodologies to quantify the emissions from ILUCs caused by increased biofuels production and use have been developed and refined, as further described in Chapter 9, but no firm consensus methodology has emerged that has been universally adopted by regulators.

The European Union Renewable Energy Directive

The EU's Renewable Energy Directive (RED) is a broad, cross-sectoral energy mandate with a sub-target for energy supplied to the road and rail sectors (Directive 2009/30/EC). As a directive, it provides high-level targets that shall be implemented with some flexibility by member states. RED II, updated in 2018, mandates a 14 percent transport sector energy target (Directive 2018/2001/EU). There are different GHG reduction requirements for fuels to qualify for RED II, depending on fuel type and what year a facility entered operation; fuels from facilities operating prior to October 2015 must generate at least 50 percent GHG savings as estimated by RED relative to the fossil fuel comparator; fuels from facilities entering operation in 2021 must generate at least 70 percent GHG savings. These emissions reduction thresholds are based only on the GHG emissions estimated with ALCA for each pathway they do not include emissions from ILUCs. Separate from the life-cycle GHG footprint estimates, fuels must comply with a set of sustainability criteria, including a prohibition on biofuels made from crops grown on high carbon stock land converted after 2008 to crop production. To verify that fuels counted toward the RED II are in compliance with the RED II, qualifying third-party verification schemes verify the supply chains of those fuels. These third-party certifiers at a minimum evaluate the feedstock source, land conversion, and life-cycle GHG savings from biofuels, as assessed by RED II, to ensure alignment with the sustainability criteria, and some schemes go further and also assess other environmental and social impacts. These schemes must be approved by the European Commission in order to be recognized.

RED II includes a variety of additional sub-targets, caps, and incentives for specific fuel categories, as well as consideing ILUC elements. The contribution of food-based biofuels is capped at 7 percent of transport sector energy or each member state's 2020 production, whichever is lower. The cap drops to 3.8 percent in future years. RED II also includes a sub-target of 3.5 percent for advanced biofuels, which includes primarily lignocellulosic feedstocks, wastes, and residues — these fuels may be double-counted towards both the 3.5 percent sub-target and the overall 14 percent transport target. Other fuels, primarily waste oils, fats, and greases, also double count towards the 14 percent target, but their overall contribution is capped at 1.7 percent. The contribution of non-food sustainable aviation fuels and maritime fuels is adjusted to 1.2 times their energy value towards the 14 percent target.

Brazil's RenovaBio Policy

Brazil's RenovaBio policy, introduced in 2017, is a CI standard for biofuels (Presidência da República Secretaria-Geral Subchefia para Assuntos Jurídicos, 2017).[3] RenovaBio establishes a 10-year time horizon for the reduction in the CI of road transport fuels consumed in Brazil, mandating a 10.2 percent CI reduction from 73.6 gCO$_2$e/MJ to 66.1 gCO$_2$e/MJ (USDA GAIN, 2021). Through the program's biofuel

---

[3] See the National Biofuel Policy (Law 13.576/2017 - RenovaBio). https://www.gov.br/mme/pt-br/assuntos/secretarias/petroleo-gas-natural-e-biocombustiveis/renovabio-1/renovabio-ingles.

certification scheme, individual CI scores will be attributed to each biofuel producer or importer based on the CI of fuel supplied. Fuels generate compliance credits on the basis of their CI relative to the fossil fuel baseline, as estimated using the policy's RenovaCalc LCA model. Biofuel producers must hire inspection firms accredited by the National Petroleum, Natural Gas and Biofuels Agency in order to achieve certification and become eligible to generate compliance credits.

To summarize, the federal RFS and CA-LCFS policies both estimate fuel CIs and include emissions from ILUCs in the calculation of CI for biofuels. The EU RED II does not include those changes in its assessment of GHG savings for specific fuel pathways, but it does include limits on fuels with high emissions from ILUCs. The RenovaBio policy implemented in Brazil has defined CI standards, but it does not include emissions from ILUCs. None of these policies include emissions from changes in fuel markets that arise from changes in fuel production.

CONSIDERATIONS IN APPLYING ATTRIBUTIONAL LIFE-CYCLE ASSESSMENT AND CONSEQUENTIAL LIFE-CYCLE ASSESSMENT IN LOW-CARBON FUEL STANDARDS

Over the years, the use of LCA in the development of the RFS, CA-LCFS, and other policies across the United States and globally have been the subject of much discussion. The charge for this report is to "assess current methods for estimating lifecycle GHG emissions associated with transportation fuels (liquid and non-liquid) for potential use in a national low-carbon fuels program." This section outlines three different specific applications of LCA in an LCFS and comments on the high-level methodological choices that must be made in each application, namely choosing between ALCA, CLCA, or some combination of both.

The process of estimating life-cycle emissions may be used for several purposes in an LCFS:

1. Policy implementation: LCA methods have been used to assign CI scores to individual fuel pathways (production routes). Historically, this process has relied primarily on ALCA methods with a CLCA element added to account for emissions from LUCs, although this is not by any means the only viable approach to developing CI scores.
2. Policy design: In addition to the use of CI scores for purposes of calculating credits and deficits used for LCFS policy implementation, LCA methods can also be used to guide other aspects of policy design. For example, the EU's RED relies heavily on LUC modeling, and the results are used to establish limits and eligibility for particular feedstocks rather than modifying CI scores for a performance-based policy. This approach is meant to promote a policy that does not have market effects that decision-makers deem unacceptable.
3. Regulatory impact assessment: Like policy design applications, RIAs are usually undertaken to understand the impacts of implementing or changing a policy and to avoid unintended negative impacts. A RIA would likely ask, among other things, how GHG emissions and other factors will change if the policy in question is enacted (relative to what would happen if the policy is not enacted), so a consequential LCA framing, with appropriate characterization of uncertainty, is well suited for this purpose: see Figure 3-2. The scope of a regulatory impact assessment may be much broader than the methods discussed at length in this report. For example, social LCA may be used to quantify the social and socio-economic impacts of implementing or altering an LCFS.

Of the three low-carbon fuel policies discussed previously, the CA-LCFS serves as a particularly interesting case study in applying LCA to policy. The CA-LCFS is performance-based and uses LCA in policy implementation through its use of CIs. California's policy defines "life cycle greenhouse gas emissions" as:[4]

---

[4] See https://ww2.arb.ca.gov/sites/default/files/2020-07/2020_lcfs_fro_oal-approved_unofficial_06302020.pdf.

"the aggregate quantity of GHG emissions (including direct emissions and significant indirect emissions, such as significant emissions from land use changes), as determined by the Executive Officer, related to the full fuel life cycle, including all stages of fuel and feedstock production and distribution, from feedstock generation or extraction through the distribution and delivery and use of the finished fuel to the ultimate consumer, where the mass values for all greenhouse gases are adjusted to account for their relative global warming potential."

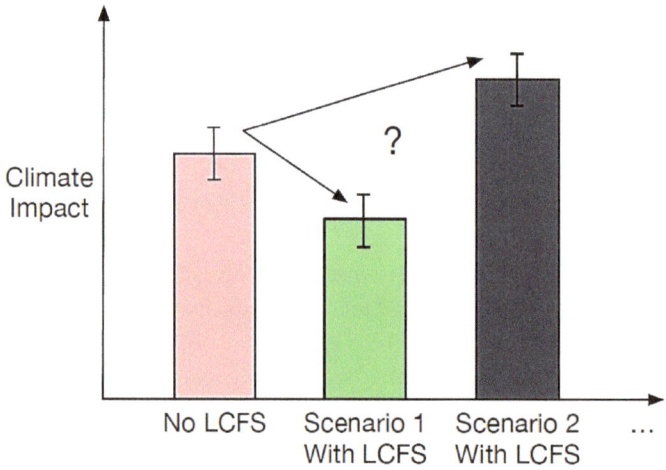

FIGURE 3-2 Illustration of potential consequential outcomes in regulatory impact assessment for a low-carbon fuel standard (LCFS). NOTE: No specific outcome is implied; rather, the figure shows how an LCFS could decrease or increase emissions in total, depending on its design and other factors.

Although this wording does not explicitly require a particular approach, the phrase "related to the full fuel life cycle" has so far been interpreted as requesting a mostly ALCA: the task is to estimate emissions of processes "related to" fuel production and decide which of these emissions to allocate to fuel production and which to attribute to other co-products produced by the supply chain. There are a few caveats worth highlighting: CA-LCFS does assign offset/avoidance credits that are arguably more consistent with a CLCA approach. For example, fuel pathways that use dairy and swine manure as a feedstock for biogas production are awarded emissions avoidance credits relative to a counterfactual in which manure is stored in lagoons that emit methane to the atmosphere (CARB, 2022). It should be noted that these methane avoidance credits are the subject of debate, including a recent petition to exclude dairy and swine manure digesters from the LCFS; in January 2022 CARB declined to take immediate action on this petition but it promised to consider the issue in an LCFS rulemaking process in 2023 or sooner (Corey, 2022). The assessed CI of dairy biogas depends on the assumed counterfactual behavior in the absence of the LCFS; with binding methane regulations in place, it may no longer be appropriate to assume that the methane would have been released into the atmosphere if it were not converted into fuel.[5] An additional example of offset credits in the CA-LCFS is the use of a co-product accounting method called system expansion, in which co-products are credited on the basis of the conventional product they displace in their respective markets. For example, dried distillers grains and solids are accounted for by subtracting the impacts of producing a functionally equivalent quantity of protein-containing animal feed (e.g., soybean meal). Similarly, electricity exported from fuel production facilities is accounted for by assigning an emissions offset credit consistent with the average grid mix in that region. Arguably, the application of system expansion to account for electricity exports is consistent with CLCA, but the use of average grid mixes for those credits is consistent with ALCA. CARB does not provide clear guidance for when to use ALCA or CLCA. An overview of ALCA models applied to transportation policies is provided in Box 3-1.

---

[5] See https://ww2.arb.ca.gov/sites/default/files/classic//fuels/lcfs/lcfs_meetings/041717discussionpaper_livestock.pdf.

> **BOX 3-1** Attributional LCA Models Used for Transportation Fuel Policies
>
> There are numerous attributional models used for transportation fuels, some of which are built for a specific study or a very narrow user base, while others are designed for wider dissemination and use, including use for informing and developing policy. The Greenhouse gases, Regulated Emissions and Energy use in Technologies ™ (GREET) model is an example of a widely-used U.S.-focused ALCA model that has been used in the development of policy (Argonne National Laboratory, 2021). Parts of GREET such as GHG emission factors for fertilizer production were used as the national RFS was being developed and biofuel LCAs were conducted by EPA. The California Air Resources Board (CARB) supported the development of a CARB-specific version, called CA-GREET, for use in California's Low Carbon Fuel Standard. The Oil Production Greenhouse gas Emissions Estimator (OPGEE) developed at Stanford (see https://eao.stanford.edu/opgee-oil-production-greenhouse-gas-emissions-estimator) was also adopted by CARB for use in the CA-LCFS. Another example of a petroleum-focused model is the Petroleum Life Cycle Inventory model (PRELIM) (see https://www.ucalgary.ca/energy-technology-assessment/open-source-models/prelim), which is also an open-source model designed for use in policy that explores energy consumption and GHG emissions in refineries with different configurations.

In contrast to the ALCA framing relied on in most of the CA-LCFS LCA, ILUC is modeled with a CLCA approach (the word "change" in ILUC itself implies a consequential framing). These analyses estimate how land use will change if a fuel is produced in higher quantities, and then translate that change in land use to a net change in GHG emissions. This type of CLCA, in which market-mediated effects are captured, is not applied for any other element of the analysis underpinning the CA-LCFS. For example, the potential rebound effect associated with increased supply of alternative fuels is not modeled or accounted for (CARB, n.d.).

There are some drawbacks to mixing ALCA and CLCA in the manner that CARB has done in the CA-LCFS. Combining ALCA and CLCA approaches limits the interpretability of the CIs because some parts of the underlying analysis answer one question and other parts answer a different question. In other words, CIs in the CA-LCFS do not reflect the average emissions associated with a fuel nor the emissions implications of changing fuel use, per the modeling methods described by CARB (CARB, n.d.). However, it is worth acknowledging that there are practical reasons why CARB chose not to employ a purely CLCA approach. The average behavior of a system may change relatively slowly (e.g., the electricity grid will not double or halve its carbon footprint in a single year), but the behavior at the margin can evolve more quickly depending on a host of economic, geopolitical, climate, and other factors well outside the control of any entity being directly regulated. Cross-sectoral
policy coordination to reduce the chances or understand the implications of policy interaction and the risk of double-booking benefits may be beneficial in these cases. However, these conditions may not always be applicable. Marginal changes in emissions averaged over a range of conditions can also be stable, so incorporating multiple scenarios can help to address this concern.

More broadly, there are challenges associated with the choice to include *or* exclude market-mediated effects in policy implementation of an LCFS. Some issues with including market-mediated and other consequential emissions in policy implementation for LCFS are that the estimates rely on modeling assumptions that can be difficult to validate empirically and can be sensitive to a wide range of factors such as commodity prices, consumer behavior, market conditions, the presence of other policies and other actors, and future market trends. As noted in the discussion above of CA-LCFS, consequential impacts (occurring on the margin) can vary on a shorter timescale than do most average impacts, so there is a tradeoff between capturing accurate, up-to-date impacts and providing stability and predictability in how an LCFS is implemented.

On the other side, some issues with excluding market-mediated emissions consequences from policy implementation are that ignoring market-mediated effects does not eliminate these effects, nor the uncertainty surrounding them. Regulating fuels based on only a selected portion of each fuel's emission consequences can create misaligned incentives, favor some fuels over others because of which factors are excluded, or potentially lead to regulations that intend to reduce GHG emissions as measured by metrics such as CIs but end up increasing total GHG emissions, by some estimates (see Figure 3-3) (Holland et al., 2009, 2015; Lark et al. 2022; Plevin et al., 2017; Rajagopal and Plevin, 2013).

Existing policies have so far addressed the challenges of incorporating market-mediated effects, and the drawbacks of excluding them, by mixing ALCA and CLCA approaches. In an LCFS, the mixing of ALCA and CLCA makes the resulting CI estimates methodologically inconsistent in that they do not fully answer a common question with a common frame. The research community broadly, and members of this committee in particular, have different views about the best path forward for weighing these tradeoffs, particularly between ALCA and CLCA, when designing policy.

CONCLUSIONS AND RECOMMENDATION ON THE USE OF DIFFERENT LIFE-CYCLE ASSESSMENT APPROACHES IN LOW-CARBON FUEL STANDARDS

This committee is in agreement that there is an important distinction between using LCA for policy implementation, policy design, and RIA. In all cases, a rigorous accounting of process-level emissions (e.g., fossil fuel combustion emissions at a refinery, N_2O emissions on farms, and CH_4 leakage at compressed natural gas production facilities) is essential. Both ALCA and CLCA approaches rely on accounting of direct emissions in foreground systems in this case, each major stage in the primary fuel supply chain, from raw materials through fuel production and distribution. However, in some cases ALCA estimates and allocates average supply chain emissions and CLCA estimates marginal supply chain emissions, the data inputs and results are likely to differ. For example, electricity used as a transportation fuel in a battery electric vehicle may be assigned the average grid mix in an ALCA and the marginal grid mix in a CLCA.

The committee also agrees that any policy proposal intending to reduce GHG emissions should be able to show through a RIA that the likely consequences of enacting that policy will be to reduce total GHG emissions. CLCA is best suited to answer the fundamental question posed in an RIA: What will the net impact of this policy be on GHG emissions compared with a counterfactual in which no policy is implemented? A rigorous RIA is essential, and it is likely necessary to periodically conduct updated RIA to ensure a policy is continuing to have the intended effects on GHG emissions.

In terms of applying LCA approaches to policy design and policy implementation, opinions differ among committee members regarding the appropriate uses of ALCA and CLCA. Some members hold the opinion that ALCA frameworks are appropriate for use in implementation of an LCFS policy via CI scores for each fuel because the use of straightforward GHG accounting, tied to decisions within the control of fuel producers and primary entities in their supply chain, incentivize fuel producers to reduce their own process emissions and select suppliers with lower GHG footprints. In essence, these members recommend the use of ALCA because it allows for relatively more effective externality management by participants in the policy. However, these members are also of the opinion that CLCA analysis may be used to augment ALCA in some instances, but it should also inform policy design by highlighting risks that can be mitigated through other policy mechanisms (including verification; see Chapter 5), policy "guardrails," and other complementary non-LCFS policies) to manage effects that are not otherwise included in CIs (Khanna et. al., 2017). For impacts that are outside the control of fuel producers and other entities in their primary supply chain, it may be defensible to use "guardrails" and other complementary policy mechanisms to guard against substantial unintended effects. When using multiple interacting policy mechanisms like this, the full committee agrees that such interacting policies should be designed with care to avoid unintended consequences, and the full impacts should be estimated through a RIA.

Other members of the committee are of the opinion that LCFS policy should focus on using CLCA estimates. They argue that it is problematic that ALCA estimates of CIs do not model the policy-relevant factor: the GHG consequences of changing fuel use. These members are concerned about the effects of LCFS policy on competition among fuels: they believe that CIs based on ALCA favor some fuels over others because of modeling boundary choices rather than the GHG emissions consequences of fuel use, creating distorted incentives that, together with other policy factors, can potentially lead to a LCFS policy that increases net GHG emissions (e.g., Holland et al., 2009, 2015; Plevin et al., 2017; Rajagopal and Plevin,

2013). This group is of the opinion that asking how an LCFS affects climate change is a more important aspect of an LCFS than focusing on ALCA-based CIs that may be easier to quantify or verify.

To summarize the key findings in this chapter:

- In existing state LCFS policies, ALCA and CLCA approaches are combined to produce a single CI score. The choice to combine these different approaches was motivated in part, but not exclusively, by a decision to incorporate ILUC GHG emissions into low-carbon fuel policies. Combining results from different LCA modeling approaches in this manner can complicate the interpretation and use of the CI score.
- LCA for RIA is a consequential analysis, estimating how GHG emissions and other factors will change if a policy is enacted. More work is needed to develop national, open-source, transparent CLCA models for use by policymakers and industry experts.
- For policy design, the LCA approach used may depend on a variety of factors including coordination of multiple policy mechanisms, practical issues, limits of regulatory authority, politics, uncertainty, transparency, stakeholder buy-in, avoiding unintended consequences and perverse incentives, and goals and implications beyond GHGs.
- ALCA does not capture market-mediated effects, including the consequences of competition for feedstocks (e.g., electricity, biomass), fuel markets, and land use. ALCA provides insights into choices in the conversion of a feedstock to a fuel pertaining to choice of technology. ALCA may also be used for screening pathways for viability under a particular policy framework.
- Because ALCA and CLCA answer different questions, using both LCA approaches to inform the policy in different ways, together with verification techniques (see Chapter 5), has the potential to strengthen policy design and implementation. As discussed above, there are different views among the committee members on whether or not ALCA and CLCA should be combined in the same analysis.

Conclusion 3-1: The carbon intensities of fuels used in an LCFS are not necessarily equivalent to the full climate consequences of their adoption. Increased use of a fuel with a low carbon intensity, as defined in an LCFS, could potentially decrease or increase carbon emissions relative to the baseline, depending on policy design and other factors. Regulatory impact assessments that use CLCA to project the consequences of policy can help assess the extent to which a given policy design with particular carbon intensity estimates will result in reduced GHG emissions.

Recommendation 3-1: When some emissions consequences of fuel use are excluded from carbon intensity values in an LCFS, the rationale, justification, and implications for these exclusions should be documented.

Conclusion 3-2: More research is needed to evaluate effective methods to collectively leverage the strengths of CLCA, ALCA, and verification methods in achieving LCFS objectives.

Recommendation 3-2: Public policy design based on LCA should ensure through regulatory impact assessment that, at a minimum, the consequential life-cycle impact of the proposed policy is likely to reduce net GHG emissions and increase net benefits to society. Regulatory impact assessments should consider changes in production and use of multiple fuel types (e.g., gasoline, electricity, biofuels, hydrogen).

Recommendation 3-3: LCA practitioners who choose to combine attributional and consequential LCA estimates should transparently document these choices and clearly identify the implications of combining these different types of estimates for the given application, scope and research question.

RECOMMENDATIONS FOR RESEARCH

This chapter concludes with a set of recommendations regarding a research agenda for LCA as it is developed for and applied to fuels. There are many entities that could support and participate in this research. The National Science Foundation supports research to develop and apply LCA methodologies. EPA has developed and applied LCA in transportation fuel policy and is often responsible for developing and implementing policies that use LCA. The U.S. Department of Energy has supported the development and use of LCA for transportation fuels at multiple national laboratories. Finally, because biofuels are a consideration in transportation policy, the U.S. Department of Agriculture, which has a history of supporting and developing biofuel LCAs, is another potential sponsor of LCA research. The Department of Transportation also has a pivotal role in transportation policy and itself implements policies that affect GHG emissions from transportation.

Beyond federal agencies, state environmental and natural resource departments that have or are considering an LCFS may want to support LCA research. Furthermore, non-profit organizations interested in mitigating transportation's effects on global warming may be interested in providing support. The researchers who carry out this work will likely be a comprehensive network of experts from universities, national laboratories, federal and state agencies, non-profit organizations, and industry. It is critical to obtain diverse perspectives and expertise in carrying out the committee's recommended research agenda.

Recommendation 3-4: Research programs should be created to advance key theoretical, computational, and modeling needs in LCA, especially as it pertains to the evaluation of transportation fuels. Research needs include:

- Further development of robust methods to evaluate the GHG emissions from development and adoption of low-carbon transportation fuels, and development or integration of process-based, economic input-output, hybrid, and CLCA methodologies
- Products could include the following:
  o development of national, open-source, transparent CLCA models for use in LCFS development and assessment
  o continued development of national, open-source ALCA models from new or existing models
  o evaluation of different approaches to creating, using, or combining ALCA, CLCA, and verification for evaluation of policy outcomes
  o quantification of variation between marginal and average GHG emissions for various feedstock-to-fuel pathways; and
  o quantification and characterization of the implications of approximations and proxies in LCA, such as comparisons of marginal and average emissions.

Research goals include a comprehensive understanding of the implications of including or excluding various attributional and consequential emissions sources in LCFS CIs on the policy's resulting overall effect on net emissions and social welfare, as well as identification of opportunities to apply ALCA and/or CLCA to design policies that are well-aligned with their intended outcome.

REFERENCES

British Columbia. 2008. Renewable and Low Carbon Fuel Requirements Regulation, BC Reg 394/2008. https://www.canlii.org/en/bc/laws/regu/bc-reg-394-2008/latest/bc-reg-394-2008.html.

CARB (California Air Resources Board). 2015. Detailed Analysis for Indirect Land Use Change. https://ww2.arb.ca.gov/sites/default/files/classic/fuels/lcfs/iluc_assessment/iluc_analysis.pdf.

Corey, R. W. 2022. Petition for Rulemaking to Exclude All Fuels Derived from Biomethane from Dairy and Swine Manure from the Low Carbon Fuel Standard Program. https://ww2.arb.ca.gov/sites/default/files/2022-01/LCFS%20Petition%20Response%202021.pdf.

Ekvall, T. 2019. Attributional and consequential life cycle assessment. In Sustainability Assessment at the 21st Century, M. J. Bastante-Ceca, J. L. Luis Fuentes-Bargues, L. Hufnagel, F-C. Mihai, and C. Iatu, eds. London: IntechOpen. https://www.intechopen.com/chapters/69212.

EPA (Environmental Protection Agency). 2010a. Renewable Fuel Standard Program (RFS2) Regulatory Impact Analysis. https://nepis.epa.gov/Exe/ZyPURL.cgi?Dockey=P1006DXP.txt.

EPA. 2010b. Regulation of Fuels and Fuel Additives: Changes to Renewable Fuel Standard Program; Final Rule. 40 CFR Part 80.

EPA. n.d. Renewable Fuel Annual Standards. https://www.epa.gov/renewable-fuel-standard-program/renewable-fuel-annual-standards.

Farrell, A. E., D. Sperling, A. Brandt, A. Eggert, A. Farrell, B. Haya, J. Hughes, B. M. Jenkins, A. D. Jones, D. M. Kammen, C. R. Knittel, M. W. Melaina, M. O'Hare, R. J. Plevin, and D. Sperling. 2007. A Low-Carbon Fuel Standard for California Part 2: Policy Analysis. UC Berkeley: Transportation Sustainability Research Center. https://escholarship.org/uc/item/1hm6k089.

Government of Canada. 2020. Clean Fuel Regulations. https://pollution-waste.canada.ca/environmental-protection-registry/regulations/view?Id=1170.

Heath, G. A. and M. K. Mann. 2012. Background and Reflections on the Life Cycle Assessment Harmonization Project. *Journal of Industrial Ecology* 16(s1). https://doi.org/10.1111/j.1530-9290.2012.00478.x.

Holland, S. P., J. E. Hughes, and C. R. Knittel. 2009. Greenhouse gas reductions under low carbon fuel standards? *American Economic Journal: Economic Policy* 1(1):106-146. https://www.aeaweb.org/articles?id=10.1257/pol.1.1.106.

Holland, S. P., J. R. Hughes, C. R. Knittel, and N. C. Parker. 2015. Unintended consequences of carbon policies: Transportation fuels, land-use, emissions, and innovation. *The Energy Journal* 36. https://www.iaee.org/en/publications/ejarticle.aspx?id=2626.

Hunt, R. G., W. E. Franklin, and R. G. Hunt. 1996. LCA—how it came about. *The International Journal of Life Cycle Assessment* 1(1):4-7.

Kharecha, P. A., and J. E. Hansen. 2008. Implications of "peak oil" for atmospheric CO_2 and climate. *Global Biogeochemical Cycles* 22. https://doi.org/10.1029/2007GB003142.

Lave, L. B., E. Cobas-Flores, C. T. Hendrickson, and F. McMichael. 1995. Life cycle assessment: Using input-output analysis to estimate economy-wide discharges. *Environmental Science and Technology* 29:420A–426A.

Lark, T. J., N. P. Hendricks, A. Smith, N. Pates, S. A. Spawn-Lee, M. Bougie, E. G. Booth, C. J. Kucharik, and H. K. Gibbs. 2022. Environmental outcomes of the US Renewable Fuel Standard. Proceedings of the National Academy of Sciences 119(9). https://doi.org/10.1073/pnas.2101084119.

Oregon DEQ. 2020. Oregon Clean Fuels Program. https://www.oregon.gov/deq/ghgp/cfp/Pages/default.aspx.

Plevin, R. J., M. A. Delucchi, and M. A. O'Hare. 2017. Fuel carbon intensity standards may not mitigate climate change. *Energy Policy* 105:93-97. https://doi.org/10.1016/j.enpol.2017.02.037.

Rajagopal, D., and R. J. Plevin. 2013. Implications of market-mediated emissions and uncertainty for biofuel policies. *Energy Policy* 56:75-82. https://doi.org/10.1016/j.enpol.2012.09.076.

U.S. Department of Agriculture. n.d. *USDA Global Agricultural Information Network (GAIN) Database*. Greenhouse Gas Reduction (Renewable and Low Carbon Fuel Requirements) Act. https://gain.fas.usda.gov/#/home.

Part II

General Considerations for Life-Cycle Analysis

4

Key Considerations: Direct and Indirect Effects, Uncertainty, Variability, and Scale of Production

This chapter addresses key considerations for applying life-cycle assessment (LCA) methods to transportation fuels, including three topics: (1) the definitions of direct and indirect effects, including supply chain and market-mediated effects triggered by fuel production and use, (2) characterization of uncertainty, and (3) issues with predicting emissions as a function of the scale of fuel production. The committee's conclusions and recommendations on each of the topics are included in their respective sections.

DIRECT AND INDIRECT EFFECTS

Definitions of Direct and Indirect Effects in the Life-Cycle Assessment Context

One of the challenges in discussing the use of LCA in a proposed national low-carbon fuel standard (LCFS) is the inconsistent use of terminology. The guidelines in the International Organization for Standardization (ISO) 14040/14044 (ISO, 2006 2006a,b) provide a common language with which to discuss many aspects of the basic process of conducting an LCA, but they do not define the terms "direct" or "indirect." The terms direct and indirect effects (or emissions) have been frequently used in LCA textbooks, regulatory documents, standards, and studies; however, definitions vary. Hertwich and Wood (2018) note that "different expert communities have developed a bewildering diversity of terms for indirect emissions… which is both a testimony to their importance and an opportunity for a more consistent terminology to ease communication."

Broadly speaking, direct emissions are defined in various references as those emissions released from focal activities, sources or processes, and indirect emissions are those emissions released from non-focal activities, sources, or processes triggered by or induced by the focal activities (Argonne National Laboratory, n.d.-b; Carnegie Mellon University, n.d.; Hertwich and Wood, 2018; Matthews et al., 2014). Here the term "focal" is used to refer to whatever activities are the focus of the study, which differ across study contexts.

Table 4-1 summarizes how various references including regulatory documents, standards, textbooks, and studies define direct and indirect effects. For example, in some definitions the focal activities are those owned or controlled by a focal entity: direct emissions are those emissions released from sources owned or controlled by the focal entity, and indirect emissions are those from sources not owned or controlled by the focal entity but related to its activities (U. S. Environmental Protection Agency [EPA], Greenhouse Gas (GHG) Protocol, Economic Input–Output LCA, Hertwich and Wood [2018]). This definition of activities is often used by companies or government entities who wish to distinguish between emissions they control and emissions triggered by their activities or purchases but released by entities that they do not control. In other definitions, focal activities are those associated with a particular process, a supply chain, or an economic sector.

LCA studies differ substantially in how they choose focal activities and define direct and indirect emissions. For example, an analysis may center on a single facility, and all emissions that occur at the facility itself are considered direct, while all off-site emissions sources (e.g., grid-connected power plants supplying electricity) are indirect. Other analyses may consider all major activities within a fuel's supply chain to be from focal activities (e.g., farming, transportation, biorefining, fuel combustion) and will call any emissions from those activities direct emissions while calling "upstream" supply chain sources, such as fertilizer manufacturing, indirect. Some studies use the term "indirect" to refer to market-mediated effects, such as indirect land use change (ILUC).

TABLE 4-1 Definitions of Direct and Indirect Emissions in the Literature

Source	Definition
Carnegie Mellon University (n.d.): Economic Input Output Life Cycle Assessment	**Economic sector and first-tier suppliers** "in all cases the environmental results are the **total** impacts, directly from the sector of interest and its direct (first-tier) suppliers *and* indirectly from all other sector transactions further up the supply chain."
EPA (2021)	**Entity** "Scope 1 GHG emissions are direct emissions from sources that are owned or controlled by the Agency. Scope 1 includes on-site fossil fuel combustion and fleet fuel consumption. Scope 2 GHG emissions are indirect emissions from sources that are owned or controlled by the Agency. Scope 2 includes emissions that result from the generation of electricity, heat or steam purchased by the Agency from a utility provider. Scope 3 GHG emissions are from sources not owned or directly controlled by EPA but related to Agency activities. Scope 3 emissions include employee travel and commuting. Scope 3 also includes emissions associated with contracted solid waste disposal and wastewater treatment. Some Scope 3 emissions can also result from transportation and distribution (T&D) losses associated with purchased electricity."
EPA (2016)	**Entity** "Indirect emissions are those that result from an organization's activities, but are actually emitted from sources owned by other entities."
GHG Protocol	**Entity** "Direct GHG emissions are emissions from sources that are owned or controlled by the reporting entity. Indirect GHG emissions are emissions that are a consequence of the activities of the reporting entity, but occur at sources owned or controlled by another entity. The GHG Protocol further categorizes these direct and indirect emissions into three broad scopes: Scope 1: All direct GHG emissions. Scope 2: Indirect GHG emissions from consumption of purchased electricity, heat or steam. Scope 3: Other indirect emissions, such as the extraction and production of purchased materials and fuels, transport-related activities in vehicles not owned or controlled by the reporting entity, electricity-related activities (e.g. T&D losses) not covered in Scope 2, outsourced activities, waste disposal, etc."
GREET CCLUB (2016)	**Domestic vs. international land use** "The Carbon Calculator for Land Use Change from Biofuels Production (CCLUB) module was released by Argonne as an Excel spreadsheet that functions both as a standalone model and as a component of GREET. CCLUB estimates the direct (domestic) and indirect (international) emissions that occur as a result of land use changes during the production of ethanol."
Hertwich and Wood (2018)	**Activities** Direct emissions: Emissions directly associated with an activity, a process, or an entity Indirect emissions: Emissions associated with the production of the inputs to an activity or organization

ISO 14040 Series (LCA)	**Not defined**
Lave, Hendrickson and McMichael (1995)	**Economic sector** "The direct economic changes associated with a choice are forecast. For example, switching from steel to aluminum for many automobile components would be represented by an increase in aluminum demand and a decrease in steel demand. An economic input-output model then is used to estimate both direct and indirect changes in output throughout the economy for each sector."
Matthews et al. (2014)	**Activities** "LCA models are able to capture direct and indirect effects of systems. In general, direct effects are those that happen directly as a result of activities in the process in question. Indirect effects are those that happen as a result of the activities, but outside of the process in question. For example, steel making requires iron ore and oxygen directly, but also electricity, environmental consulting, natural gas exploration, production, and pipelines, real estate services, and lawyers. Directly or indirectly, making cars involves the entire economy, and getting specific mass and energy flows for the entire economy is impossible."

A few concrete examples from the literature include (emphasis on terms "direct" and "indirect" added):

(1) In Searchinger et al. (2008): "To produce biofuels, farmers can directly plow up more forest or grass-land, which releases to the atmosphere … carbon … Alternatively, farmers can divert existing crops or croplands into biofuels, which causes … emissions indirectly."

(2) In Bento and Klotz (2014) the delineation between direct and indirect "is based on the application and context being studied and determines the allocation procedures used to assign emissions to a technology, data choice, and the treatment of market-induced, or indirect, adjustments.... For evaluating changes, consequential LCA offers distinct advantages over the attributional approach, particularly in recognizing that market adjustments and indirect effects can be as important as the physical flows captured by attributional LCA, with indirect land use change (LUC) resulting from expanded biofuel production being a key example."

(3) In Pehl et al. (2017): Direct emissions include direct fossil CO_2 (with imperfect carbon capture). Indirect emissions include land use change (LUC), upstream and biogeneic CH_4 operation, construction, bioenergy with carbon capture and storage.

(4) In Hertel et al. (2010) emissions from land use changes are indirect as noted by these authors: "…greenhouse gas (GHG) releases from indirect (or induced) land-use change (ILUC) triggered by crop-based biofuels…" On the other hand, according to these authors emissions from producing feedstock and conversion to biofuel is direct: "Direct releases of GHG also occur during the cultivation and industrial processing of maize ethanol. Estimates of these, not including ILUC."

Figure 4-1 provides examples of how the boundaries between direct and indirect emissions differ across a selection of corn ethanol LCA studies and standards. Note that this figure shows the delineation between what is considered direct versus indirect in each study, not necessarily the system boundary (see Chapter 3).

Despite all of this variation, across all alternative sources that the committee reviewed, some emissions sources are almost universally considered to be direct (e.g., on-site emissions from a biorefinery in a corn ethanol LCA) and some emissions sources are almost universally considered to be indirect (e.g., emissions resulting from induced international LUCs due to market-mediated responses).

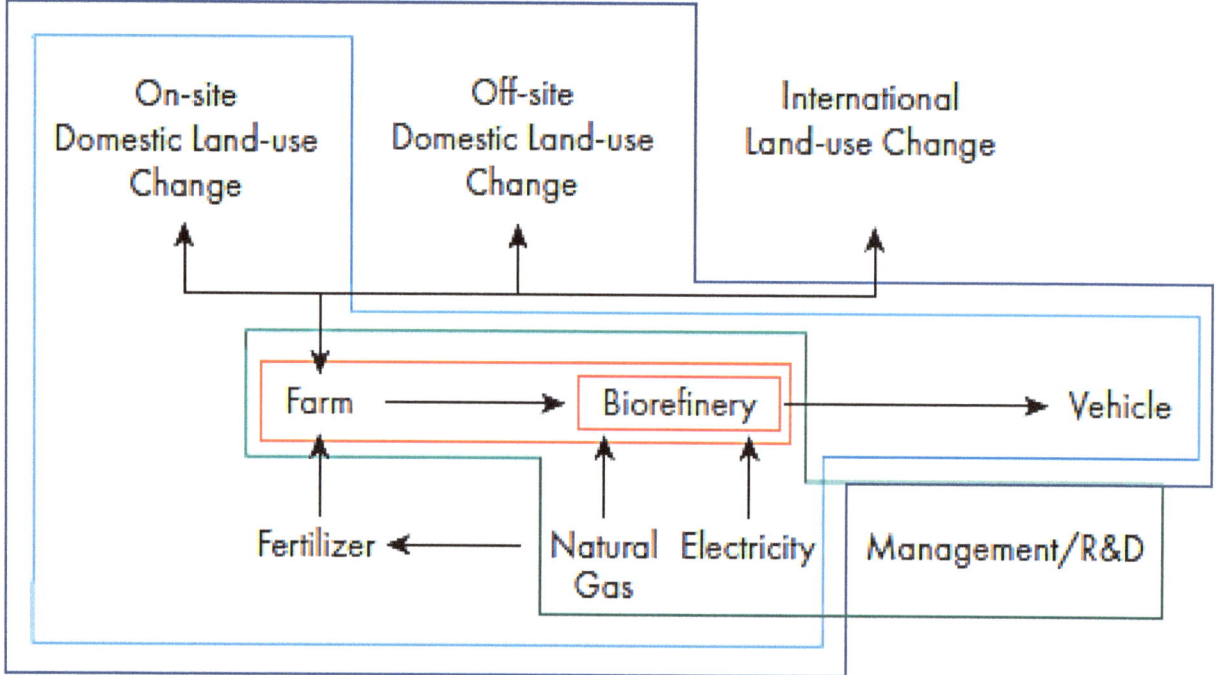

FIGURE 4-1 Illustration of different delineations between direct emissions and indirect emissions for various corn ethanol LCA studies. NOTE: The boxes indicate the activities considered direct in different studies. System boundaries are not shown.

Understanding "Direct" and "Indirect" in Attributional and Consequential Life-Cycle Assessment

The concepts of direct and indirect emissions are distinct from the concepts of attributional and consequential emissions, but in some studies these ideas are conflated. For example, the GHG Protocol states "Indirect GHG emissions are emissions that are a consequence of the activities of the reporting entity." Some LCA studies of biofuels estimate fuel supply chain emissions using attributional LCA (ALCA) and refer to these as "direct emissions" but add consequential estimates of induced LUC as "indirect emissions." Table 4-2 shows the relationship between these concepts. Because some studies use ALCA to model "direct" emissions from focal activities (top left) and use consequential LCA (CLCA) to model "indirect" emissions from non-focal activities (bottom right), the terms can sometimes become conflated in the literature.

TABLE 4-2 Relationship between Direct/Indirect Emissions and Attributional/Consequential LCA

	Attributional	Consequential
Direct	Assigned (e.g., average) emissions from focal processes or activities	Change in emissions from focal sources due to an action or decision
Indirect	Assigned (e.g., average) emissions from other sources induced by the focal processes or activities	Change in emissions from other sources due to an action or decision

Table 4-3 provides some examples of factors that may be considered direct or indirect effects associated with expansions in biofuels and electric vehicles (EVs). These categorizations vary across studies and sources, and this table does not provide a definitive list of which effects should be referred to as direct or indirect: rather, Table 4-3 offers concrete examples of these types of effects. The terms "direct" and "indirect" are used frequently in biofuels LCA studies (see Chapter 9), but they are not used commonly in EV LCA studies (see Chapter 10). This may be, in part, because many biofuels studies use ALCA to model the partial supply chain but include induced LUC as an added consequential element outside the ALCA boundary ever since the issue was raised as a critical factor by Searchinger et al. (2008). In contrast, most modern EV LCA studies model power sector emissions using CLCA, estimating how power grid emissions will change in response to changes in EV charging load (see Chapter 10 for a list of studies) but do not necessarily model other market-mediated effects, such as rebound. Because CLCA is typically not focused on assigning emissions to products or activities, the distinction between direct and indirect has not been useful terminology to adopt in most of the EV LCA literature. However, we provide examples in Table 4-3 to emphasize that both biofuels and EVs (as well as other fuels) have market-mediated effects.

TABLE 4-3 Examples of Factors Often Referred to as "Direct" or "Indirect" in the Transportation Fuels Literature

Direct/Indirect	Biofuel Examples	Electric Vehicle Examples
Potentially referred to as direct effects	Emissions from corn and soybeans in supply chain of an ethanol or biodiesel plant (Plevin et al., 2014a)	N/A
Potentially referred to as indirect effects	Land use emissions and sequestration effects due to changes in demand for a given feedstock: − changes in agricultural biomass, − changes in forgone sequestration, − changes in soil organic carbon, − changes of forest or grassland	Induced emissions due to changes in demand for electricity: − shift in dispatch towards marginal resources to accommodate EVs − adjustments in generation capacity expansion planning (Taheripour et al., 2017) − introduction of Time-of-Use charges and Demand-Side Management Programs (Schmidt et al., 2015). − changes in feedstock sourcing for additional baseload and intermediate electric generating resources (additional fracking, coal mining, liquefied natural gas import/export).
Potentially referred to as indirect effects due to rebound or fuel market price effects	− Higher demand for ag commodities increases or decreases feed/food prices and their consumption (Schmidt et al., 2015) − Yield improvement due to additional investment in agricultural activities induced by biofuel production − Yield reduction due to marginal land being brought into production	− Higher demand for electricity increases or decreases electric rates to consumers, affecting other demand and generation or − Installed capacity costs decrease ($/kW) due to economies of scale resulting in other electricity uses (electric heating, water heating)

continued

TABLE 4-3 continued

Direct/Indirect	Biofuel Examples	Electric Vehicle Examples
	– Reduced gasoline or diesel consumption results in decline in price and an increase in use elsewhere (Martin, 2013)	– Lower vehicle operating costs result in increased travel, reduced range results in reduced travel, or higher upfront costs affect household spending on other items with GHG implications – Reduced gasoline demand results in decline in price and an increase in use elsewhere.

Finally, it should be noted that calling effects "direct" or "indirect" has no bearing on their relative contribution to overall life-cycle impacts. Major sources of emissions can include those within and beyond the focal activities (Matthews et al., 2008; Meinrenken and Lackner, 2015; Meinrenken et al., 2020).

In sum, the terms "direct" and "indirect" are defined and used differently in different regulatory documents, standards, practices, scientific references, and LCA methods, and there are no universally accepted definitions. It is again worth noting that the ISO 14040 series standard for LCA does not use the terms "direct" or "indirect" in relation to effects or emissions.

The distinction between direct and indirect emissions is commonly based on what a given study identifies as focal activities. These are typically based on the source of emissions, the entities that generate the emissions, or, in the case of some approaches that mix ALCA and CLCA, the ALCA system boundary used in the analysis.

Some ALCA studies use the ALCA system boundary to define focal activities but also add other factors beyond the ALCA system boundary, such as CLCA estimates of induced land use change due to market-mediated responses, and label all activities captured by the ALCA system boundary as direct effects. These studies label all activities outside the ALCA system boundary as indirect.

Conclusion 4-1: Dividing emissions into direct and indirect can be used when identifying and classifying sources of emissions, but it can cause confusion, even if carefully defined and transparently presented in an LCA.

Conclusion 4-2: Direct and indirect emissions are concepts distinct from the concepts of attributional and consequential LCA.

Recommendation 4-1: Because the terms "direct" and "indirect" are used differently in different contexts, these terms should be carefully defined and transparently presented when used in LCA studies or policy. Another option is to avoid using the terms "direct" and "indirect" altogether, as they are not considered necessary elements of LCA and may lead to greater confusion.

UNCERTAINTY AND VARIABILITY

LCAs are subject to considerable uncertainty and variability, and LCA methods need to appropriately characterize uncertainty and variability to aid LCA stakeholders' interpretation of LCA results. The appropriate tool for uncertainty and variability analysis depends on the nature of the uncertainty and variability as well as the decisions or policies the LCA is intended to support. Accordingly, this section begins with general background on the types of uncertainty and variability and the methods that are used for uncertainty and variability characterization. It then proceeds to discuss specific uncertainties and variabilities in the context of low-carbon transportation fuels. Finally, it addresses the types of outputs that LCAs need to produce so that policymakers can be appropriately informed as to the range of possible outcomes.

Types of Uncertainty and Methods for Analysis

Uncertainty refers to a lack of knowledge. Variability refers to any difference in outcomes of different trials of a process. The differences may be understood and predictable based on underlying differences in the conditions under which the trials take place. An example of this would be known variation in electric power sources across different regions. In other cases, the variation may be due to inherent randomness of outcomes, generally termed aleatory factors, for example, a roll of the dice. Whether variability contributes to uncertainty depends on context, that is how much of the variation is inherent versus explained by known factors, and whether the explanatory factors are known. Uncertainty can come both from such aleatory factors and also from uncertainty in knowledge of the system in question, generally termed epistemic uncertainty. An example of epistemic uncertainty is the location of Genghis Khan's tomb. There is no variability present as the tomb is located in one specific place, but that location is unknown. Such epistemic uncertainty may be reducible by further research and data collection. Both aleatory and epistemic uncertainty can be present in input parameters and data used in mathematical models.

Different definitions of uncertainty are described in the literature. ISO standard 14040 mentions model imprecision, input uncertainty, and data variability (ISO, 2006a). For LCA, examples of model input uncertainty include data related to process inputs, as well as environmental emissions and technology characteristics, most of which are associated with life-cycle inventory (LCI) data (Lloyd and Ries, 2007). Precision can be understood to refer to variability, such as observations' deviation from their collective mean. Furthermore, ISO 14044 discusses the use of uncertainty analysis to determine the effects of uncertainties in empirical quantities and in modeling assumptions (ISO, 2006b). However, ISO standards do not provide detailed methodological guidance on uncertainty analysis. For example, the weather may influence the yield and ultimately the carbon intensity (CI) of biofuel feedstock cultivation, but the specific weather years in the future is inherently unpredictable. Both aleatory and epistemic uncertainty can be present in input parameters and data used in mathematical models.

Uncertainty is also generally present in the scenarios that analysts should consider. Scenarios consist of coherent sets of model inputs and parameters that reflect conditions of interest such as the times and locations at which activities will take place. They may also include the policy regime or broader market conditions that will apply to the analysis. Examples include common LCA modeling decisions such as choosing a functional unit, allocation methods, end-of-life scenarios, time frame, and geographic scales.

The fundamental assumptions underlying the structure of an analysis represent another form of uncertainty (Lloyd and Ries, 2007; Morgan and Henrion, 1990). In LCA a common example of this is the uncertainty in the structure and mathematical relationships of the models used for deriving emission factors and characterizing life-cycle impacts (Lloyd and Ries, 2007). This uncertainty in the appropriate structure of the analysis is often difficult to characterize or in some cases even to recognize. For example, an LCA may use relatively precise estimates of current process inputs and outputs. However, embedded in the application of such a model for predictions is the assumption that the processes used in the future will resemble the processes used in the past (Morgan and Henrion, 1990). Ascertaining the extent to which this is true often presents a challenge.

Decision variables and value parameters variables are two categories of uncertainty that are defined with respect to the decision maker. Decision variables are those choices or quantities that are under the control of the decision maker. Value parameters represent the preferences of the decision maker or the people on whose behalf the decision is being made. These valuations are sometimes readily expressed and ascertainable by observing market behavior, while in other cases they involve non-market goods (Morgan and Henrion, 1990). A rich literature has been developed on the valuation of non-market goods. Nevertheless, valuation of tradeoffs among far distant futures influenced by planetary processes such as climate change, that the current generation will not even experience, present substantial challenges to any such methodologies.

Stochastic modeling approaches such as Monte Carlo simulation or other Monte Carlo–type propagation techniques have been widely used in LCA to characterize uncertainty in model outputs given uncertainty in model input parameters (Heijungs, 2020). Monte Carlo simulation samples inputs from statistical distributions characterizing their variability and uncertainty, calculates model outputs for those inputs, and repeats the calculations many times (e.g., over 500 runs or more) to estimate the statistical characteristics of the results (e.g., the expected value, standard deviations, and quantiles) (Heijungs, 2020). Most LCA studies apply Monte Carlo simulation for parameter uncertainty (Bamber et al., 2020), although some studies used Monte Carlo to propagate model uncertainty related to emission factors in LCI or characterization factors in life-cycle impact assessment as well as scenario uncertainties due to modeling choices (Meinrenken and Lackner, 2015; Mullins et al., 2011; Usack et al., 2018). A pedigree matrix, in which data inputs are rated on five different factors describing data quality: reliability, completeness, and temporal, geographical and technological representativeness, has been widely used to assess data quality and characterize input uncertainty. Recent developments include the use of the pedigree matrix to provide empirical uncertainty factors for inventory data available in commercial databases such as Ecoinvent (Ciroth et al., 2016). This allows for incorporating parameter uncertainty associated with background LCI data into LCA, but not for model structural or scenario uncertainty.

In addition to Monte Carlo, other forms of probabilistic analyses may be used to propagate input uncertainty through LCA models to estimate output uncertainty. In simple cases, exact analytic approaches may be available. In other cases the output uncertainty can be approximated at least over a small range of output values, using first order uncertainty analysis (Morgan and Henrion, 1990). For uncertainties described by a discrete set of possible outcomes, a model output distribution can be derived by exhaustive enumeration of outcomes. One can discretize continuous distributions (see Clemen and Reilly [2001] for details) and use the exhaustive enumeration of outcomes as an approximation of the output uncertainty distribution. This is often helpful when decision uncertainties are nested within model input parameter or scenario uncertainties.

Input uncertainty is also mentioned in ISO 14040 series regarding the definition of uncertainty analysis as "systematic procedure to quantify the uncertainty introduced in the results of a life cycle inventory analysis due to the cumulative effects of model imprecision, input uncertainty and data variability" (ISO, 2006a). LCI databases (e.g., Ecoinvent) and LCA software such as OpenLCA, SimaPro, GaBi and the Greenhouse gases, Regulated Emissions, and Energy use in Technologies (GREET) model can handle input uncertainty by providing probability distributions of input data that can be used in Monte Carlo simulation (Lloyd and Ries, 2007). However, the potential correlations between inputs are often ignored in LCA, which may lead to under or over-estimation of output variances (although such effect may be minimal in some cases) (Groen and Heijungs, 2017). Global sensitivity analysis is a method in which the impact of multiple inputs over the full range of plausible values is used. Input–output correlations are then often used to identify the most influential inputs in driving the output uncertainty (Cucurachi et al., 2016). Global sensitivity analysis has been widely used in the literature (Gregory et al., 2016; Groen et al., 2014; Guo and Murphy, 2012; Iooss and Lemaître, 2015), and some studies have included correlation analysis in the sensitivity analysis to account for the interdependence of inputs (Bojacá and Schrevens, 2010; Cucurachi et al., 2016; Wei et al. 2015). For parameter uncertainty, different methods have been explored. For example, one study investigated different methods of parameter uncertainty analysis, including Monte Carlo sampling, Latin hypercube sampling, quasi Monte Carlo sampling, fuzzy interval arithmetic, and analytical uncertainty propagation; the authors concluded that more directly usable information can be provided by sampling methods than fuzzy interval arithmetic or analytical uncertainty propagation (Groen et al., 2014).

Model structural uncertainties can be assessed through comparison of the outputs of alternative models with different structural assumptions (Bamber et al., 2020; van Zelm and Huijbregts, 2013). In addition, there are a variety of model averaging approaches available to estimate overall uncertainty given different possible model forms. In some cases models can be weighted according to their fit to data or other indicators of reliability, but it is often difficult to assign such weights objectively. Another issue is that the results of such an analysis are a blend of often mutually contradictory assumptions. Given these issues, in some cases it is more informative to simply compare discrete model runs with different assumptions, rather than parsing the average output of multiple models (Morgan and Henrion, 1990).

Ideally, model structural uncertainty would be assessed through comparisons between models with fundamentally different approaches, such that there would not be common errors made by both approaches. In reality it is often not possible to estimate LCA model outputs through approaches that do not share many of the same assumptions. In such cases one often changes a single assumption at a time in order to clarify the impacts of that one assumption. If synergistic effects of multiple assumptions need to be assessed, then experimental design approaches can be used to efficiently estimate the main effects of differing assumptions and the interactive effects of combinations of assumptions. An example of structural model uncertainty can be found in carbon emissions accounting where land area changes predicted by an economic model are translated into carbon emissions changes. Several major carbon emissions accounting models used in the United States for life-cycle modeling of biofuels include the GREET submodule called Carbon Calculator for Land Use Change from Biofuels Production (CCLUB) developed by Argonne National Laboratory (n.d.), the Agro-ecological Zone Emission Factor (AEZ-EF) model developed by Plevin et al., a data set by the Woods Hole Institute (also an optional parameterization in CCLUB), carbon accounting factors integrated in the Global Biosphere Management (GLOBIOM) model, and others (Dunn et al., 2012; Plevin et al., 2014b, 2022).

> **Recommendation 4-2:** Current and future LCFS policies should strive to reduce model uncertainties and compare results across multiple economic modeling approaches and transparently communicate uncertainties.

Most LCA studies have considered uncertainty associated with different parameters, while evaluation of scenarios (through consideration of alternative, coherent sets of inputs) and structural uncertainties may be less common. A review published in 2020 found that for ALCA, 87 percent of studies reviewed (out of 470 LCA articles) considered parameter uncertainty, while 16 percent and 11 percent of them accounted for scenario uncertainty and model uncertainty, respectively. For consequential LCA, 95 percent of articles (out of 19) included parameter uncertainty and 32 percent of them included scenario uncertainty, while only 3 papers (16 percent) accounted for model structural uncertainty (Bamber et al., 2020). Previous studies have presented different views of the importance of types of uncertainties in LCA. Some studies indicate that parameter uncertainty is more critical than scenario or model uncertainty (Huijbregts et al., 2003; Ziyadi and Al-Qadi, 2019). Other studies made the opposite conclusion that structural or scenario uncertainties dominate (Buchholz et al., 2016) or stated that all of them are important (Huijbregts et al., 2005; Steen, 1997). More recent studies show that the relative contributions of different types of uncertainty are likely to be context-specific.

Essentially no LCA studies can include and explore all sources of uncertainty. The general tendency across a broad set of contexts is to underestimate uncertainty, that is to be overconfident in one's analysis, even despite the analyst's best efforts to accurately represent the degree of uncertainty present (Plous, 1993). This tendency can be documented objectively in some contexts such as the measurement of different physical quantities, including the speed of light and the charge of an electron (Morgan and Henrion, 1990). In policy analysis, evaluation of the accuracy of estimates often requires comparison to a counterfactual scenario that cannot be empirically verified. There is a fairly limited literature attempting to validate the performance of regulatory impacts assessments conducted before the regulation is implemented, and these studies have found that actual impacts often differ substantially from the range of possible impacts that were forecast in advance of the regulation (Gurian et al., 2006; Harrington et al., 2000).

To summarize the key findings in this section:

- Typical uncertainty in LCA includes parameter uncertainty, scenario uncertainty, and model uncertainty. Parameter uncertainty is commonly considered in LCA studies. Scenario and model uncertainty are considered in some studies, but it is rare for LCA studies to explicitly consider all sources of uncertainty.

- Monte Carlo simulation has been widely used in LCA to quantify the statistical characteristics of results under parameter uncertainty, given assumptions about the distribution of input parameters.
- Global sensitivity analysis has been widely used to identify factors driving uncertainty of the results.

Conclusion 4-3: Explicitly considering parametric, scenario, and model uncertainty can help to represent the degree of confidence in model results.

Recommendation 4-3: LCA studies used to inform policy should explicitly consider parameter uncertainty, scenario uncertainty, and model uncertainty.

Recommendation 4-4: When LCA results are used in policy design or policy analysis, the implications of parameter uncertainty, scenario uncertainty, and model uncertainty for policy outcomes should be explicitly considered, including an assessment of the degree of confidence that a proposed policy will result in reduced GHG emissions and increased social welfare.

Uncertainties in Estimating Carbon Intensity

As noted in Chapter 2, the results of LCAs vary substantially based on the methods, data, assumptions, and scope of the analysis. This section describes some of these factors in the context of how they contribute to uncertainties in LCA models of transportation fuels. ALCA uncertainties are addressed first, followed by uncertainties in consequential analyses. This section discusses many sources of uncertainty in ALCAs: system boundaries, representative data, temporal data variation, spatial data variation, nonlinear LCA effects, co-products allocation process, and scale-up uncertainty. Another source of uncertainty, which is estimated in both ALCA and CLCA, is estimating future land use change, which is discussed in a separate section below.

System Boundaries: A fuel LCA usually starts from the feedstock production system, which is considered as the cradle and ends after the fuel is used in a vehicle, which is considered as the grave. For a biofuel LCA, the entire production system must be well defined, for instance, in one study (Sheehan et al., 1998), biodiesel production from soybean oil was defined to include five distinct processes: (1) soybean farming, (2) soybean transportation to the processing facility, (3) oil extraction and purification, (4) conversion of oil into biodiesel (or transesterification), and (5) transportation of biodiesel for distribution. It is nearly impossible to track all the energy used over the life cycle of a product because each input has a life cycle of its own. For instance, machinery used in agriculture has its own life cycle that may involve factory building to computer software. In turn, each of these has its own life cycle and the chain continues through the entire global economy. The boundaries at different levels of production drawn by researchers introduce uncertainty. Economic input–output analysis avoids drawing these boundaries through an integrated approach to understand economy-wide changes in production sectors. However, this approach provides only a highly aggregated depiction of different sectors of the economy that may not accurately characterize changes at the process level. This approach usually relies on the existing input–output tables that do not reflect new technologies under investigation and assumes no substitution in consumption and production.

Representative Data: Much of the data used in LCA comes from several sources such as SimaPro GaBi, GREET, EcoInvent, OpenLCA, Brightway, GHGenius; governmental databases, such as those of U.S. Departments of Agriculture, Energy, Commerce, and Transportation; and global datasets from the Food and Agriculture Organization of the United Nations and the World Bank. The spatial and temporal granularity of these datasets vary considerably, as do the frequency with which they are updated and disseminated. The accuracy and extensiveness of documentation accompanying these datasets also varies. Prior studies have attempted to communicate the representativeness of input data through the use of pedigree matrices, although the scores resulting from such exercises are purely qualitative and cannot be used to capture quantitative uncertainty in modeling outputs.

Temporal Data Variation: Agriculture is weather dependent, and yields vary from year to year. For the same amount of inputs the yield can be different. By picking and choosing data from various years, the result may also be different. Also the current LCA practice may or may or may not take into account long-term changes in agricultural productivity through technology and changes in crop yields due to climate change. These factors suggests that LCA should be conducted frequently to update results in a consistent manner and consider a forward-looking perspective. Lee et al. (2021) document how many LCA inputs for corn ethanol have changed over time and find substantial changes often on timescales of less than a decade. Impacts of other fuel types also change with time (Masnadi and Brandt, 2017). Life-cycle studies can become outdated and inaccurate in domains where technological advancement changes products and processes. It is difficult to identify a basis for judging how quickly such estimates become outdated, but (as noted above) examples of substantial changes in less than a decade are available.

Conclusion 4-4: Up-to-date LCA studies are needed to inform policy.

Recommendation 4-5: Regulatory agencies should formulate a strategy to keep LCAs up to date, which may involve periodic reviews of key inputs to assess whether sufficient changes have taken place to warrant a re-analysis, and agencies should be aware that substantial changes to LCAs on timescales of less than a decade can occur.

Spatial Data Variation: Crops grown around the country from which starch, lipids, or cellulose are derived for biofuel vary significantly in their yield, application of agrochemicals such as lime, consumption of fuel, and water for irrigation. Analysts must find an appropriate degree of granularity for their studies, depending on the research question and scope of their analysis, given that LCA results vary by region and crop. For national assessments, proper aggregations should take into account spatial variation. Transportation of crops varies in distance from zero for on-farm crushing to hundreds of miles in other cases. Different modes of transportation may be available in different regions. Similarly, the transportation of biofuels may vary both in distance and mode.

Conclusion 4-5: LCA studies can produce different estimates depending on regional scope or assumptions.

Recommendation 4-6: LCA studies used to inform transportation fuel policy should be explicit about the feedstock and regions to which the study applies and to the extent possible should explicitly report the sensitivity of the results to variation in these assumptions.

Effects of Scale in LCA: Although LCA defines the environmental impact in terms of functional units, the impact is scale dependent (i.e., per unit impacts may change as a process is scaled up). For example, a process that uses byproducts as a feedstock at small scale may have substantially different impacts if it is scaled up to a point where crops are grown specifically for the process. See "Scale-Up Uncertainty" section below for specific considerations pertaining to that case.

Co-Products Allocation Process: Usually more than one final product is generated during fuel production. ALCA studies must assign emissions to products, and it is unclear how emissions should be divided among the different products. There are primarily two methods to estimate the co-products' share of environmental impact, namely allocation and displacement methods (system expansion) (see also Chapter 6 for additional discussion on this topic). The allocation method allocates the material use, energy use, and emissions between the primary product and co-products based on either mass, energy content, or economic revenue. ISO 14044 (ISO, 2006b) provides the following guidance on allocation procedures:

a. Wherever possible, allocation should be avoided by dividing the unit process to be allocated into two or more sub-processes and collecting the input and output data related to these sub-processes, or expanding the product system to include the additional functions related to the co-products expanding system boundary.

b. Where allocation cannot be avoided, the inputs and outputs of the system should be partitioned between its different products or functions in a way that reflects the underlying physical relationships between them; i.e. they should reflect the way in which the inputs and outputs are changed by quantitative changes in the products or functions delivered by the system.

c. Where physical relationships cannot be established or used as the basis for allocation, the inputs should be allocated between the products and functions in a way that reflects other relationships between them. For example, input and output data might be allocated between co-products in proportion to the economic value of the products.

Despite this guidance, there is no single best method of co-product allocation, and it has been argued that physical relationships should not be prioritized over other allocation methods such as economic value, which is what drives demand, not mass or volume. This inconsistency in how LCA is implemented creates discrepancies among studies and uncertainty in how to estimate the impacts of fuel production (Kim and Dale, 2002). ALCA results can be sensitive to decisions made by modelers about how to assign emissions to co-products, and there is no single "correct" way to do so.

Conclusion 4-6: ALCA studies may produce substantially different results depending on modeling choices about how emissions are assigned to co-products.

Recommendation 4-7: ALCA studies used to inform fuel policy should justify the approach used to handle co-products, and as necessary report sensitivity of results to variation in approaches to assigning emissions to co-products.

Scale-Up Uncertainty: There are various uncertainties about the technologies and methods for scaling up the feedstock production and feedstock conversion from experimental field and laboratory conditions that can affect the CI of the resulting fuel. These uncertainties are described below for the example of biofuels (analysis of the implications of these uncertainties can be found in Shi and Guest, 2020). The committee notes that similar scale-up uncertainties apply to other fuel systems (e.g., EVs, hydrogen fueled vehicles, high octane fuel vehicles).

(a) **Input requirements for feedstock production.** Requirements for fertilizer and other inputs for feedstock production can significantly affect CI of biomass. With feedstocks that are yet to be produced at commercial scale, information about nutrient requirements is typically obtained from experimental field trials that are site-specific, small-scale, and do not provide generalizable information about biophysically or economically optimal input application rates. Machinery requirements for large-scale feedstock production are also uncertain. LCAs are typically static in nature and apply fixed factors for input requirements per unit volume of biofuel at a certain production scale without considering how these might change in a non-linear manner with scale-up and the uncertainties about the process as a transition is made from experimental scale to commercial scale production. This issue can be addressed by careful attention in the goal and scope definition phase.

(b) **Composition of the feedstock.** For example, biomass is primarily composed of lignin, hemicellulose, and cellulose, and this composition can vary across feedstocks and across locations for a particular feedstock. Composition of the biomass can affect the choice of pre-treatment technology, if one is used, and the efficacy of the pre-treatment process, which in turn can affect the methods used for further processing, mass and energy balances throughout the plant, and the overall efficiency of the conversion of biomass to fuel.

(c) **Scale, location, and design of the refinery.** Conducting an LCA of feedstock deconstruction, conversion, and separation processes that are still under development by scaling up process information available at a lab scale requires specifying many decision variables and technology performance parameters. Assumptions include specifying conversion and separation technologies, reaction time, pressure, yield, product formation rate, recovery efficiency, and so forth. Many decisions and performance parameters are uncertain and will vary with the scale of the process, with location specific input availability and costs, and with other factors. Most studies employ a static approach to biorefinery design and simulation and specify a single value for each of these variables/parameters and assume these are invariant to scale, location, and economic conditions. Sensitivity analyses that are limited to changing one variable at a time in discrete steps while leaving all other variables and parameters constant do not capture the potentially complex and non-linear interactions among these variables and parameters, limiting their utility to understand the sensitivity of indicators (e.g., kg CO_2e per MJ of fuel) to individual assumptions and combinations of assumptions for an optimized biorefinery at scale.

While the previous section emphasized ALCA, uncertainties could be compounded or canceled out in CLCA, which involves predicting the induced effects of fuel production based on comparisons between modeled outcomes with and without the fuel production. In considering electricity (Chapter 10), the impact of additional electricity demand on emissions is discussed and shown to depend on the specific generation sources used to meet the additional demand. A somewhat analogous issue exists with biofuels regarding the need to forecast how the additional demand for the fuel will be met in a CLCA, in particular the extent to which the additional demand will be met through the cultivation of additional land (see section on LUC next).

Conclusion 4-7: LCA of commercial-scale production for processes that have not been commercialized involve assumptions that can introduce substantial uncertainty, including effects of interactions among multiple uncertain data or parameters, and so may be particularly sensitive to uncertainty.

Recommendation 4-8: LCA studies used to inform transportation policy regarding processes that do not yet exist at scale should explicitly report sensitivity of findings to uncertainty, in order to produce bounding estimates.

Land Use Change

Over the past 15 years various efforts have been made to assess the magnitudes of GHG emissions induced by changes in land use and land cover due to biofuel production and policy. Regardless of the differences, most such studies follow a common approach consisting of two steps:

(1) Using economic models or other approaches to determine changes in land use and land cover by region for an increase in a given biofuel type (or a set of biofuels).

(2) Using a set of land related emissions factors combined with some supporting assumptions to convert the estimated land use and land cover changes to GHG emissions measured in gCO_2e per MJ of biofuel produced.

Both of these steps are subject to various sources of uncertainties. Some researchers have reported substantial uncertainties in the range of published values for the impact of induced LUC on emission. However, not all of this wide range in the existing estimates of the CI of biofuels associated with induced LUC emissions is due to uncertainties. Some of the variation in CI estimates reflects variation in fuels, feedstocks, and regions considered by different studies. In this section we explore sources of variation and uncertainties in this field of research.

Previous studies have reported a wide range of estimates based in part on variation in the following factors when considering induced LUC:

(1) Variation in the nature of the LUC,
(2) Variation due to the choice of amortization time horizon,
(3) Variation across biofuels produced from various types of feedstocks,
(4) Variation across the results of a model for a given feedstock produced in different regions,
(5) Variations across models for a given biofuel pathway produced in a given region,
(6) Variations across the results of a model for a given biofuel pathway due to model changes in the model parameters and benchmark databases,
(7) Variations across the results of a model for a given biofuel due to changes in the implemented emissions account framework.
(8) Variations in studies may reflect variations in aleatory factors or epistemic uncertainty.

The following sections discuss each of these sources of variation among studies in more detail.

Variation in the Nature of the Land Use Change

There are many possible types of LUC. Crop rotations on existing agricultural lands may be changed. For example, fallow periods may be reduced and winter crops cultivated more frequently, which could affect the soil carbon balance. Pasture land may be converted to cropland, or seldom used land may be used more frequently. Of perhaps most concern is the potential for forest land or peatlands, which generally hold a large accumulated stock of carbon, being converted to cropland. LUC occurs due to many causal factors. These may include economic, biophysical (fire, pests), cultural (communal decision making), technical (slash and burn to boost fertility), demographic (rapid growth of populations and the rural poor), and political (programs to help people who do not own land) factors (Geist and Lambin 2001; Plieninger et al. 2016; Shrestha, 2015; van Vliet et al. 2015; Wubie et al. 2016). Some LUC models attempt to capture economically driven responses such as price–yield elasticity, double cropping, and agricultural expansion to marginal land (Shrestha et al., 2019). A high price–yield elasticity indicates that additional demand for a crop, such as produced by a regulatory incentive for biofuel production, can be met through additional production on the same acreage in response to relatively small price signals. This might be accomplished through more intensive cropping practices such as increasing the frequency of winter crop cultivation on existing agricultural lands (Spatari et al., 2020) or through the use of additional fertilizer inputs, for example. In contrast low yield elasticities would imply that additional land would be brought into production to meet the increased demand. The prospect of changing land use introduces a host of additional uncertainties as to which lands would be brought into production and what the impacts of these land use changes would be. The impacts of bringing abandoned farmland back into production would differ from the impacts of clearing forest land. Furthermore, the impacts of converting land from one use to another may depend on the details of historical uses which impact carbon stocks in the land in question. All of these issues are challenging (see Plevin et al., 2015) and these challenges are amplified for countries where data at best are questionable.

Variation Due to the Choice of Amortization Time Period

The choice of amortization time horizon directly affects the size of ILUC values. While the Intergovernmental Panel on Climate Change (IPCC) approach can be followed to evaluate ILUC values based on the Global Warming Potential (GWP) index over a 100-year time horizon (Schmidt et al., 2015), a second time horizon pertains to the assumed duration of a biofuels policy. Some existing ILUC practices simply amortize induced land use emissions due to a biofuels volume over the number of years the biofuels policy is presumed to be in effect. Some studies have used 20 years for the amortization time horizon following the EU regulatory emissions guidelines. On the other hand, other studies have used 30 years for the time horizon following U.S regulatory emissions guidelines. There are a few exceptions. The Carbon Offsetting and Reduction Scheme for International Aviation (CORSIA, 2021), which represents an international scheme for offsetting and emissions reduction, has applied a 25-year time horizon, a compromise between the U.S. and EU time horizons. Some studies have adopted a 100-year time horizon approach.

The choice of the amortization period over which predicted land use emissions are divided can also introduce significant parameter uncertainty. In the past, amortization periods considered for the EU's Renewable Energy Directive and the California LCFS have differed, 20 years for the former and 30 years for the latter. The Committee on Aviation Environmental Protection for CORSIA agreed to use a 25-year period stating that "this value is a compromise between the European use of 20 years and the U.S. value of 30 years." However, the choice of the amortization periods in ILUC modeling may be a political decision and subject to the time period for policy goals. There is no single correct choice for amortization period. Schmidt et al. (2015) state: "Applying an amortization period, however, introduces arbitrary assumptions, inconsistencies and strange cause-effect relationships (Schmidt et al., 2015)."

One potential alternative is "Baseline Time Accounting" which derives ILUC values independent of amortization periods but takes into account global land use dynamics and the fact that land used for biofuels production can return to food production (Kløverpris and Mueller, 2013; Schmidt et al., 2015). This committee neither endorses nor discourages this alternative. While the approach has received support and criticism, it raises the point that what happens to the land after a policy ends may matter (Kløverpris and Mueller, 2014; Martin, 2013). Baseline Time Accounting also uses assumptions, such as in the determination of counterfactual scenarios. For example, with the electrification of the ground and aviation sectors much smaller land use changes are expected than for biofuels policies, but the land use changes for transmission lines, power plants, and rare earth metal mining may not allow food/feed production after the policy ends. A similar concern exists with the permanence of soil carbon storage when land is converted between row crop and perennial biomass production. This may influence the way we think about amortization of emissions in a one to one comparison.

Variation across Biofuels Produced from Various Types of Feedstocks

The estimated ILUC values can vary across feedstocks significantly. In general, other factors being equal, one may expect variation among ILUC values for biofuels produced from different feedstocks. Differences in yield per unit of land across feedstocks, variations in energy content per ton of alternative feedstocks, differences in fuel production technologies, and differences in properties of animal feed by-products of biofuel pathways are important factors, among others, that explain differences in ILUC values associated with biofuels produced from crops.

Variation across the Results of a Model for a Given Feedstock Produced in Different Regions

A given biofuel could be produced from the same feedstock but cultivated in different countries. Since the economic, agricultural, and land characteristics vary across countries, a model may provide different ILUC values across regions for the same biofuel. Hence, if a model projects two different ILUC values for the same feedstock produced in two different countries, the differences may be due in part to regional variability and not necessarily entirely due to uncertainty.

Variations across Models for a Given Biofuel Pathway Produced in a Given Region

When various models project different ILUC values for a given pathway, that could reflect uncertainty. However, in some cases the sources of differences can be determined and then a check and verification process could help to resolve the sources of differences. Such an evidence-based check and verification does not make the estimated ILUC values deterministic and certain. However, it can narrow down the domain of uncertainty based on the existing evidence.

Variations Due to Model Changes or Changes in the Model Parameters and Benchmark Database

Over the past 15 years various models have been used to assess LUCs for various biofuel pathways. Plevin et al. (2022) noted that "land representation in biofuel LUC models is an important determinant of CI-LUC." Some of these models have been frequently modified over time. Hence, their ILUC projections have changed during the past 15 years. While these types of changes may increase variation in ILUC projections, they could reduce some uncertainties if the model improvements are based on new observations subject to verification.

The existing literature shows that ILUC values are inherently uncertain. It is possible to study the accuracy of the fundamental bases of the theory of induced land use changes due to biofuels and check and verify the accuracy of the implemented assumption that has been used across the board in estimating ILUC values. This effort could help to limit the scope of uncertainty. Evidence on yield changes, changes in land allocation across crops, changes in crop prices in the long run, changes in land cover items over the time, and changes in international trade over time all could be examined to explore the magnitude of uncertainties in ILUC evaluation.

Variations Due to Changes in the Implemented Emissions Account Framework

To calculate ILUC values, after evaluating LUCs for a given biofuel pathway, one needs some emissions parameters and assumptions to convert changes in land areas to GHG emissions. These coefficients measure sinks and sources of GHG gas emissions due to changes in land use and land cover items including above-ground live biomass, below ground live biomass, dead organic matter, soil organic matter, harvest wood during land conversion, non-CO_2 emissions due to land conversion, and forgone carbon sequestration associated with changes in land cover. One needs these data items for alternative land types and climate conditions by region, depending on the spatial resolution of the research at hand. In addition, depending on the implemented modeling approach for land use assessment, a wide range of assumptions are needed to convert the estimated land use changes to land use emissions, measured in terms of ILUC emissions.

Various sources of emissions factors and assumptions have been applied to convert estimated LUCs into ILUC values. Some of the existing sources for emissions factors are: Woods Hole Research Center data, Winrock International data, data sources of the IPCC, data developed by the International Institute for Applied Systems Analysis, data included in the AEZ-EF emissions accounting framework (Plevin et al., 2014a,b) data included in the CCLUB emissions accounting framework – drawing on CENTURY soil organic carbon modeling results (CCLUB Manual, 2017[1]), and many other sources.

The existing literature has extensively highlighted uncertainty in land use modeling. Only a few papers address uncertainty in emissions factors associated with land use changes. Plevin et al. (2016) have developed a sensitivity test on the Global Trade Analysis Project (GTAP)-BIO model parameters and the Agro-ecological Zone Emission Factor (AEZ-EF) model parameters that jointly affect the ILUC values determined by these two models and concluded that "the economic model contributes the majority of the uncertainty" and that the carbon accounting makes only a minor contribution because the carbon account is based on "physical sciences." In contrast, Taheripour and Tyner (2020), and Zhao et al. (2021) have estimated that one parameter of the AEZ-EF emissions model (the share of deforestation that falls on peat land) could largely alter the ILUC values of oilseed-based biofuels. In addition, Chen et al. (2018) and Taheripour and Tyner (2013) have estimated that switching from one set of land use change emissions factors by land type to another set could alter the ILUC values substantially. Another paper that examined changes in emission factors is Leland et al. (2018), who noted that the addition of a new "cropland-pasture" type in AEZ-EF resulted in a 50 percent lower emission factor for the conversion of pasture to cropland.

To summarize, projected ILUC values are subject to various types of aleatory or epistemic uncertainties. Variation across projected ILUC values may reflect uncertainty as well as variation across regions

---

[1] See https://greet.es.anl.gov/publication-cclub-manual-r6-2020.

or scenarios. ILUC values could vary by feedstock, biofuel production technology, and location of feedstock production. Such variation may or may not represent uncertainty, depending on study goals and scope.

A few challenging and important issues that need additional attention regarding the emissions data set in order to reduce uncertainty of ILUC estimates are:

- Matching the land categories defined in the economic models with the land categories included in the emissions data set,
- Assigning emissions coefficients to each type of land category, in particular for alternative types of marginal land (fallow, idle, cropland pasture), low productivity pasture land, managed pasture land, and natural pasture land,
- Changes in soil organic carbon due to LUCs,
- Changes in land use emissions due to changes in land management practices, e.g., changes in tillage practice, cultivation of cover crops, and
- Checking the validity of the assumptions used in the emissions datasets.

In general, the research on ILUC assessment has performed one of three approaches or a combination of them to address uncertainty: scenario-based assessments; detailed sensitivity assessments; and more formal sensitivity tests.

Scenario-Based Assessments: Many studies evaluated ILUC values under alternative scenarios representing a few discrete choices on intensive and extensive margins (yield per harvest, harvest frequency, and productivity of new cropland), land allocation parameters, interactions between food and fuel, demand elasticities (income and price elasticities) for food and fuels, and trade elasticities. These scenario-based uncertainty assessments show that the ILUC values could vary substantially depending on the assumption used in the evaluation.

Detailed Sensitivity Assessments: A few studies developed more detailed sensitivity analyses again on the same factors mentioned above. The findings of these analyses are similar to the findings of the scenario-based assessments mentioned above. Compared with scenario-based approaches, the papers that developed detailed sensitivity assessments examined a larger number of simulations to study the sensitivity of the ILUC value with respect to the changes in various model parameters or a mix of them. In these tests, usually ad hoc distribution functions with arbitrary ranges (or standard deviations) were assigned to each parameter to conduct the sensitivity test. Essentially, these sensitivity tests repeated the scenario-based approach for a larger number of scenarios. These assessments show the same results as the scenario-based practices: (1) the more intensification the lower the ILUC value and (2) the higher the rigidity in market-mediated responses the larger the ILUC values.

More-Formal Sensitivity Tests: Two studies developed more formal sensitivity tests (Plevin et al., 2010, 2015) using two different modeling styles (a simple reduced form model and a comprehensive computable general equilibrium model) and applying a Monte Carlo approach to assess sensitivity of ILUC values with respect to models' assumptions and parameters. Similar to other sensitivity studies in this field, these studies also addressed the importance of yield improvement, the role of land allocation parameters in modeling land use, and the implications of marker mediated responses. These analyses emphasized the role of yield to price response.

Uncertainties in Values

In addition to these uncertainties in empirically determined parameters, there are uncertainties in values. These values may be parameterized relatively explicitly, such as in the discount rate that is chosen, with higher discount rates placing more value on the present, and hence less value on future climate change impacts. Other parameters may reflect values more implicitly such as the selection of an amortization period for emissions from land use disturbance. Longer project periods allow negative values of upfront emissions to be offset by future emissions reductions. A longer project period implies that near term emissions reductions are not highly valued at the margin. Thus, the appropriate project period depends on the role of the particular project in an overall portfolio of efforts to achieve targeted GHG reductions over time.

Summary and Policy Implications

The scope of this report is limited to assessing methods for LCA of low-carbon transportation fuels and does not include recommendations as to appropriate policies regarding transportation fuels. Nevertheless, LCAs are ultimately conducted to inform decisions, including those by governmental policymakers. Accordingly, this report does address some outputs of LCA that may help policymakers manage uncertainty and variability.

Reporting of model outputs should include both the central tendency (mean and median) of estimated values and the credible range of model estimates, such as can be provided by a 90 percent or 95 percent confidence interval. Lower bounds on the CI of fuels (that is, upper bounds on CI) may sometimes be useful for "margin of safety" approaches that penalize options for the amount of uncertainty in their impacts (Springborn et al., 2013). This greater weighting of less favorable outcomes is a feature of risk-averse decision making with the extent of risk aversion constituting a value parameter of the decision maker (see Clemen and Reilly [2001] for a discussion of risk aversion). As risk aversion is a value parameter that is subjective in nature, it falls into the category of uncertainties that are difficult to handle with statistical precision but for which scenario analyses may be informative. Accordingly, LCAs should include multiple uncertainty ranges (90, 95, 99 percent confidence intervals, etc.) for their estimates. However, in an LCFS policy context if a "margin of safety" is applied its consistent use across all LCA pathways (e.g., biofuels, electricity) must be ensured.

Interest rates and planning horizons, which affect the selection of amortization periods and global warming potential time periods of different greenhouse gases, are also complex value parameters. These are generally best handled by reporting values corresponding to multiple scenarios so that decision makers can select the ones most relevant to their policy objectives.

Recommendation 4-9: Modelers should conduct sensitivity analysis to understand implications of variation.

Recommendation 4-10: To effectively inform policy making, LCA studies should document results for a range of input values.

Variability in outcomes can generally be characterized statistically (assuming data are available) and hence handled with the probabilistic methods (see Chapter 3). In some cases, variability will simply be averaged out over many different trials where trials may refer to repetitions at different times or locations. However, if variability is correlated with economic incentives to produce higher or lower CI products, then characterization of variability is paramount. For example, greater use of fertilizer inputs might be associated with greater economic returns for crop production but higher CI of products. Similarly, electricity rate schedules may favor EV charging on fossil-heavy night time baseload electricity grids. In such a situation, variability would not be averaged out as high CI production would be selectively incentivized. In cases where there are relationships between variability in economic returns and CI, LCAs may need to incorporate techno-economic analyses (e.g., Kar et al., 2020) to understand how variability may incentivize production modes with different CIs. Given the potential for variability to influence market behavior and the potential interest by policymakers in risk-averse approaches to manage epistemic uncertainty, explicitly partitioning variability and uncertainty in model outputs is recommended where tractable (for an example analysis, see Gao et al. 2018).

Conclusion 4-8: Variability in methods and circumstances under which fuels are produced may be associated with differential economic returns. When this is the case, a techno-economic analysis may be helpful to understand the conditions under which market actors will produce the fuel.

In conducting LCAs, analysts should be mindful of the implications of uncertainty for policy making, even though their analyses may not recommend any particular policy approach. Analysts may aspire to identify external costs of fuels in a manner that allows policymakers to implement regulatory incentive schemes that internalize these costs for market actors. Given the extent of uncertainty in actual CI, implementing schemes that optimally incentivize market actors may be extremely challenging, particularly as the pace of technological change makes these objectives a moving target. Policymakers may benefit from analyses that explicitly consider technological change and include scenarios that explore alternative future capabilities. In situations of high uncertainty, options have great value, and LCAs that provide policymakers with consideration of alternative future options can inform the development of strategies that are robust with respect to uncertainty.

While options provide value, when the goal of policy is to stimulate private investment, policymakers' options to revise the policy may provoke private-sector investors to require premiums for investment returns to compensate for the risk that regulatory incentives may be withdrawn in the future. To address this, LCAs may be coupled with techno-economic analyses that explicitly consider how maintaining future options to vary regulatory policy may change incentives for market actors. Both the emissions associated with fuels and the economic returns of fuel production may change in uncertain ways on timescales substantially shorter than capital project investment cycles, which often cover decades.

Conclusion 4-9: Research is warranted on how the carbon intensity and economics of fuel production may change over time.

SCALE OF PRODUCTION

Technical Factors

Scale of production can affect the life-cycle implications of a fuel or technology in nonlinear ways. When a fuel is produced at high volumes, it may trigger different effects on the supply chain and different kinds of effects than when produced at low volume. For example, at low volume a process may operate largely using existing feedstock sources, but at high volume a process may induce new feedstock production with qualitatively different emission profiles.

Small changes create marginal effects, but larger changes can create non-marginal consequential effects and may require a different scope of LCA (see Figure 2.3 in Chapter 2). Figure 2.4 shows an illustration of how emissions can be nonlinear with fuel production and the implications for using average ALCA estimates or marginal CLCA estimates. The reasons that emissions can be nonlinear with fuel volume include (but are not limited to) the following examples:

- For biofuels, small increases may come largely (but not exclusively) from domestic production. Large increases may trigger greater international shifts in supply chains, such as shifts from soybean oil to palm oil in Asia, with consequential effects that are highly uncertain and potentially very different.
- For electricity, small increases in demand may primarily trigger increases in dispatchable (fossil) generation, but large changes may trigger construction of new capacity in the power sector and may support increased penetration of intermittent renewables through flexible demand.
- The production of hydrogen may play a complementary role in broader economy-wide decarbonization efforts by providing a variable load that can reduce the cost of integrating a high level of variable renewable energy sources into the grid and reduce the need for battery storage or backup generation. The benefits of sector coupling will depend on the scale of demand for

hydrogen. For example, an analysis of cross sectoral coupling in the Northeast United States (He et al., 2021) found that a peak in the climate benefit of increased system coupling at hydrogen production levels of 5 and 12.5 million tonnes per year at carbon prices of $100 and $1000 per tonne carbon price, respectively.

Large or sudden changes in the scale of utilization of specific inputs to transportation fuel production can have impacts on competing users of these inputs and on the supply chains that produce these inputs. For example, for biofuels, rapid increases in the use of particular agricultural commodities such as grain or vegetable oil can create short-term supply shocks and price increases. Changes of a moderate magnitude may be more readily managed through the inherent flexibility of the supply chain. If there is delayed or slower-than-expected technological progress, over a longer time frame, the total land area devoted to crop production for energy could crowd out food production or land set aside for nature.[2] Conversely, observed yield increases in row crops and increasing feed substitution values from co-products generated during the biofuels production process may at least partially offset demand increases.

Similar concerns have been raised for electricity to charge electric vehicles. For small increases in EV adoption, consequential charging emissions are marginal emissions and can be modeled with a regression or simulation approaches (see Chapter 10), and EVs can be treated as price-takers, with demand too small to influence electricity prices. For large increases in EV adoption, consequential charging emissions may be non-marginal, and EV charging behavior may affect electricity prices, causing changes in emissions from other activities as well. Competing load additions and new uses for electricity (e.g., the current large-scale relocation of crypto currency mining facilities from China) may appropriate some of the renewable resources currently planned to meet load additions to the electricity grid, changing the consequential emissions from EV charging. Conversely, more efficient renewable generating technologies and vehicle-to-grid technologies may offset demand increases, and flexibility of EV charging load could help to support more intermittent renewable capacity, like wind and solar power, which require storage, flexible load, or backup dispatchable generators in order to provide reliable service on demand.

Limitations of Life-Cycle Assessment for Supporting Appropriate Scale of Feedstock Utilization

The debate about fuel life-cycles is broader than methodological issues, often raising questions about other important values. Fuel LCA provides important insight into the climate implications of different fuel production pathways, but this insight is incomplete. The scale of feedstock utilization for transportation fuel has implications that go well beyond LCA methodology and involve many competing priorities and values.

Fuel feedstocks can be used for other purposes, and when a particular feedstock is used for fuel production, there are some impacts on the linked markets that can be modeled through CLCA and others that go beyond life-cycle issues. Increased use of food and feed crops for fuel production can affect crop, food, and cropland prices; the mix and area of crop production can affect air and water quality and biodiversity; while diversion of electricity to transportation will affect other ratepayers, grid stability, the sources of generation, and logistics, such as for emergency evacuations. Fuel LCA for GHG emissions is an incomplete window into these impacts, and a fuel policy based on CIs will affect these concerns.

> **Conclusion 4-10:** The scale of production can affect life-cycle GHG emissions, and current LCA methods often do not explicitly incorporate changes in production scale into their calculations.

> **Conclusion 4-11:** More research is needed to develop LCA methodologies for incorporating scale dependence.

---

[2] In light of this possibility, President Biden issued an executive order on January 27, 2021, with the goal of conserving 30 percent of land and 30 percent of water by 2030: see https://www.whitehouse.gov/briefing-room/presidential-actions/2021/01/27/executive-order-on-tackling-the-climate-crisis-at-home-and-abroad/

Recommendation 4-11: Researchers and regulatory agencies should identify additional information to assess impacts of large changes in fuel systems.

Recommendation 4-12: Because LCA-based carbon intensities in current LCFS policy are often not structured to capture nonlinear and non-life cycle implications of large changes in fuel and fuel pathway production volume, policymakers should consider potential complementary policy mechanisms.

REFERENCES

Argonne National Laboratory. (n.d.). *Carbon Calculator for Land Use Change from Biofuels Production (CCLUB): Users' Manual and Technical Documentation.* file:///C:/Users/Genie/Downloads/CCLUB_Manual_2017_GREET1.pdf.

Argonne National Laboratory. (n.d.-b). *GREET. The Greenhouse Gases, Regulated Emissions, and Energy Use in Technologies Model.* https://greet.es.anl.gov/.

Bamber, N., O. Turner, V. Arulnathan, Y. Li, S. Z. Ershadi, A. Smart, and N. Pelletier. 2020. Comparing sources and analysis of uncertainty in consequential and attributional life cycle assessment: review of current practice and recommendations. *The International Journal of Life Cycle Assessment* 25(1):168-180. https://doi.org/10.1007/s11367-019-01663-1.

Bento, A. M. and R. Klotz. 2014. Climate policy decisions require policy-based lifecycle analysis. *Environmental Science and Technology* 48(10). https://doi.org/10.1021/es405164g.

Bojacá, C. R. and E. Schrevens. 2010. Parameter uncertainty in LCA: Stochastic sampling under correlation. *The International Journal of Life Cycle Assessment* 15(3):238-246. https://doi.org/10.1007/s11367-010-0150-0.

Carnegie Mellon University. (n.d.). *Economic Input-Output Life Cycle Assessment.* http://www.eiolca.net/Method/interp-results/env_airso2.html.

Chen, G., L. D. Knibbs, W. Zhang, S. Li, W. Cao, J. Guo, H. Ren, B. Wang, H. Wang, G. Williams, and N. A. S. Hamm. 2018. Estimating spatiotemporal distribution of PM1 concentrations in China with satellite remote sensing, meteorology, and land use information. *Environmental Pollution* 233:1086-1094.

Chen, G., S. Li, L. D. Knibbs, N. A. Hamm, W. Cao, T. Li, J. Guo, H. Ren, M. J. Abramson, and Y. Guo. 2018. A machine learning method to estimate $PM_{2.5}$ concentrations across China with remote sensing, meteorological and land use information. *Science of the Total Environment* 636:52-60.

Ciroth, A., S. Muller, B. Weidema, and P. Lesage. 2016. Empirically based uncertainty factors for the pedigree matrix in ecoinvent. *The International Journal of Life Cycle Assessment* 21(9):1338-1348. https://doi.org/10.1007/s11367-013-0670-5.

Clemen, R. T. and T. Reilly. 2001. *Making Hard Decisions*, Duxbury, Pacific Grove, CA.

CORSIA. 2021. Carbon Offsetting and Reduction Scheme for International Aviation (CORSIA); https://www.icao.int/environmental-protection/CORSIA/Pages/default.aspx.

Cucurachi, S., E. Borgonovo, and R. Heijungs. 2016. "A Protocol for the Global Sensitivity Analysis of Impact Assessment Models in Life Cycle Assessment." *Risk Analysis* 36(2):357-377. https://doi.org/10.1111/risa.12443.

Dunn, J. B., S. Mueller, M. Wang, and J. Han. 2012. Energy consumption and greenhouse gas emissions from enzyme and yeast manufacture for corn and cellulosic ethanol production. *Biotechnology Letters* 34: 2259–2263.

Dunn, J., A. Qin, S. Mueller, H-Y. Kwon, M. Wander, and M. Wang. 2017. *Carbon Calculator for Land Use Change from Biofuels Production (CCLUB) Manual* (Rev. 4). https://greet.es.anl.gov/publication-cclub-manual-r4.

EPA (U.S. Environmental Protection Agency). 2016. *Greenhouse Gas Inventory Guidance: Indirect Emissions from Purchased Electricity.* https://www.epa.gov/sites/default/files/2016-03/documents/electricityemissions_3_2016.pdf.

EPA. 2021. *Greenhouse Gases at EPA.* https://www.epa.gov/greeningepa/greenhouse-gases-epa.

Gao, S., P. L. Gurian, P. R. Adler, S. Spatari, R. Gurung, S. Kar, S. M. Ogle, W. J. Parton, and S. J. Grosso. 2018. Framework for improved confidence in modeled nitrous oxide estimates for biofuel regulatory standards. *Mitigation and Adaptation Strategies for Global Change* 23:1281-1301. https://doi.org/10.1007/s11027-018-9784-1.

Geist. H. J. and Lambin. E. F. 2001. "What Drives Tropical Deforestation? A Meta-Analysis of Proximate and Underlying Causes of Deforestation Based on Subnational Case Study Evidence." Louvain-la-Neuve (Belgium): LUCC International Project Office, LUCC Report Series No. 4.

Gregory, J. R., A. Noshadravan, E. A. Olivetti and R. E. Kirchain. 2016. A methodology for robust comparative life cycle assessments incorporating uncertainty. *Environmental Science & Technology* 50(12):6397-6405.

Groen, E. A. and R. Heijungs. 2017. Ignoring correlation in uncertainty and sensitivity analysis in life cycle assessment: What is the risk? *Environmental Impact Assessment Review* 62:98-109. https://doi.org/10.1016/j.eiar.2016.10.006.

Groen, E. A., R. Heijungs, E. A. M. Bokkers, and I. J. M. de Boer. 2014. Methods for uncertainty propagation in life cycle assessment. *Environmental Modelling & Software* 62:316-325. https://doi.org/10.1016/j.envsoft.2014.10.006.

Guo, M., and R. J. Murphy. 2012. LCA data quality: Sensitivity and uncertainty analysis. Science of the Total Environment, 435-436, 230-243.

Gurian, P. L., R. Bucciarelli-Tieger, M. Chew, A. Martinez, and A. Woocay. 2006. Validating pre-regulatory cost estimates for the revised arsenic MCL. Proceedings of the AWWA Annual Conference and Exposition, San Antonio, TX.

Harrington, W., R. D. Morgenstern and P. Nelson. 2000. On the accuracy of regulatory cost estimates. *Journal of Policy Analysis and Management* 19(2):297-322.

He, G., D. S. Mallapragada, A. Bose, C. F. Heuberger-Austin, and E. Gençera. 2021. Sector coupling via hydrogen to lower the cost of energy system decarbonization. *Energy & Environmental Science* 14:4635. https://pubs.rsc.org/en/content/articlelanding/2021/ee/d1ee00627d.

Heijungs, R. 2020. On the number of Monte Carlo runs in comparative probabilistic LCA. *The International Journal of Life Cycle Assessment* 25(2):394-402. https://doi.org/10.1007/s11367-019-01698-4.

Hertel, T., A. A. Golub, A. D. Jones, M. Hare, R. Plevin, and D. M. Kammen. 2010. Effects of US maize ethanol on global land use and greenhouse gas emissions: Estimating market-mediated responses. *Bioscience* 60:223-231. https://doi.org/10.1525/bio.2010.60.3.8.

Hertwich, E. G. and R. Wood. 2008. The growing importance of scope 3 greenhouse gas emissions from industry. *Environmental Research Letters* 13:104013. https://iopscience.iop.org/article/10.1088/1748-9326/aae19a.

Huijbregts, M. A. J., W. Gilijamse, A. M. J. Ragas, and L. Reijnders. 2003. Evaluating uncertainty in environmental life-cycle assessment. A case study comparing two insulation options for a Dutch one-family ewelling. *Environmental Science & Technology* 37(11):2600-2608. https://doi.org/10.1021/es020971.

Huijbregts, M. A. J., L. K. J. Geelen, E. G. Hertwich, T. E. McKone, and D. Van De Meent. 2005. A comparison between the multimedia fate and exposure models caltox and uniform system for evaluation of substances adapted for life-cycle assessment based on the population intake fraction of toxic pollutants. *Environmental Toxicology and Chemistry: An International Journal* 24(2):486-493.

Iooss, B., and P. Lemaître. 2015. A review on global sensitivity analysis methods. In *Uncertainty Management in Simulation-Optimization of Complex Systems* 101-122. New York: Springer.

ISO (International Organization for Standardization). 2006a. *ISO 14040-2006: Environmental Management - Life Cycle Assessment - Principles and Framework.* https://www.iso.org/standard/37456.html.

ISO. 2006b. *ISO 14044:2006 Environmental Management — Life Cycle Assessment — Requirements and Guidelines.* https://www.iso.org/standard/38498.html#:~:text=ISO%2014044%3A2006%20specifies%20requirements,and%20critical%20review%20of%20the.

Lee, U., Y. Kwon, M. Wu, M. Wang. 2021. Retrospective analysis of the U.S. corn ethanol industry for 2005–2019: implications for greenhouse gas emission reductions. *Biofuels Bioproducts & Biorefining.* https://doi.org/10.1002/bbb.2225.

Martin, J. 2013. Letter to the editor: Regarding your article 'Baseline time accounting: Considering global land use dynamics when estimating the climate impact of indirect land use change caused by biofuels. *International Journal of Life Cycle Assessment* 18(2):319–330. https://doi.org/10.1007/s11367-012-0488-6.

Kar, S., B. Riazi, P. L. Gurian, S. Spatari, P. R. Adler, and W. J. Parton. 2020. An optimization framework to identify key management strategies for improving biorefinery performance: A case study of winter barley production. *Biofuels, Bioproducts and Biorefining,* 14(6):1296–1312. https://doi.org/10.1002/bbb.2141.

Kim, S., and B. E. Dale. 2002. Allocation procedure in ethanol production system from corn grain. *International Journal of Life Cycle Assessment* 7:237–243.

Kløverpris, J. H., and S. Mueller. 2013. Baseline time accounting: Considering global land use dynamics when estimating the climate impact of indirect land use change caused by biofuels. *International Journal of Life Cycle Assessment* 18:319–330. https://doi.org/10.1007/s11367-012-0488-6t.

Kløverpris, J. H., and S. Mueller. 2014. Baseline time accounting–Reply to the letter to the editor of Martin. *International Journal of Life Cycle Assessment* 18(7):1279. *International Journal of Life Cycle Assessment* 19:257–259.

Lave, L. B., C. T. Hendrickson and F. C. McMichael. 1995. Using input-output analysis to estimate economy-wide discharges. *Environmental Science and Technology* 29:420A–426A. https://pubs.acs.org/doi/pdf/10.1021/es00009a748.

Kwon, H., M. Wu and M. Wang. 2021. Retrospective analysis of the U.S. corn ethanol industry for 2005–2019: Implications for greenhouse gas emission reductions. *Biofuels, Bioproducts and Biorefining* 15:1318–1331. https://doi.org/10.1002/bbb.2225.

Lloyd, S. M., and R. Ries. 2007. Characterizing, propagating, and analyzing uncertainty in life-cycle assessment: A survey of quantitative approaches. *Journal of Industrial Ecology* 11(1):161-179. https://doi.org/10.1162/jiec.2007.1136.

Malins, C., R. Plevin, and R. Edwards. 2020. How robust are reductions in modeled estimates from GTAP-BIO of the indirect land use change induced by conventional biofuels. *Journal of Cleaner Production* 258.

Masandi, M. S. and A. R. Brandt. 2017. "Climate impacts of oil extraction increase significantly with oilfield age." *Nature Climate Change* 7:551–556.

Matthews, H. S., C. T. Hendrickson, and D. Matthews. 2014. *Life Cycle Assessment: Quantitative Approaches for Decisions that Matter.* https://www.lcatextbook.com/.

Matthews, H. S., C. T. Hendrickson, and C. L. Weber. 2008. The importance of carbon footprint estimation boundaries. *Environmental Science & Technology* 16(42):5839–5842.

Meinrenken, C. J., Rauschkolb, N., Abrol, S., Chakrabarty, T., Decalf, V.C., Hidey, C., McKeown, K., Mehmani, A., V. Modi, and P. J. Culligan. 2020. MFRED, 10 second interval real and reactive power for groups of 390 US apartments of varying size and vintage. *Scientific Data* 7:375. https://doi.org/10.1038/s41597-020-00721-w.

Meinrenken, C. J. and K. S. Lackner. 2015. Fleet view of electrified transportation reveals smaller potential to reduce GHG emissions. *Applied Energy* 138:393-403. https://doi.org/10.1038/s41598-020-62030-x.

Morgan, M. G. and M. Henrion. 1990. *Uncertainty: A Guide to Dealing with Uncertainty in Quantitative Risk and Policy Analysis.* Cambridge, England: Cambridge University Press.

Mullins, K. A., W. M. Griffin, and H. S. Matthews. 2011. Policy implications of uncertainty in modeled life-cycle greenhouse gas emissions of biofuels. *Environmental Science & Technology* 45(1):132–138. https://doi.org/10.1021/es1024993.

Pehl, M., A. Arvesen, F. Humpenöder, A. Popp, E. G. Hertwich, and G. Luderer. 2017. Understanding future emissions from low-carbon power systems by integration of life cycle assessment and integrated energy modelling. *Nature Energy* 2:939-945. https://doi.org/10.1038/s41560-017-0032-9.

Plevin, R. J., H. K. Gibbs, J. Duffy, S. Yui, and S. Yeh. 2014a. *A Model of Greenhouse Gas Emissions from Land-Use Change for Use with AEZ-based Economic Models.* https://ww2.arb.ca.gov/sites/default/files/classic/fuels/lcfs/lcfs_meetings/aezef-report.pdf.

Plevin, R. J., H. K. Gibbs, J. Duffy, S. Yui, and S. Yeh. 2014b. *Agro-Ecological Zone Emission Factor (AEZ-EF) Model* (v47). https://www.gtap.agecon.purdue.edu/resources/res_display.asp?RecordID=4346.

Plevin, R. J., J. Beckman, A. A. Golub, J. Witcover, and M. O'Hare. 2015. Carbon accounting and economic model uncertainty of emissions from biofuels-induced land use change. *Environmental Science & Technology* 49(5):2656-2664.

Plevin, R. J., A. D. Jones, M. S. Torn and H. K. Gibbs. 2010. Greenhouse gas emissions from biofuels' indirect land use change are uncertain but may be much greater than previously estimated. *Environmental Science & Technology* 44(21):8015–8021. https://doi.org/10.1021/es101946t. The Psychology of Judgment and Decision Making (McGraw-Hill Series in Social Psychology) 1st Edition. New York: McGraw-Hill.

Plieninger, T., H. Draux, N. Fagerholm, C. Bieling, M. Bürgi, T. Kizos, and T. Kuemmerle. 2016. The driving forces of landscape change in Europe: A systematic review of the evidence. *Land Use Policy* 57: 204-214. https://doi.org/10.1016/j.landusepol.2016.04.040.

Plous, S. 1993. The Psychology of Judgment and Decision Making (McGraw-Hill Series in Social Psychology) 1st Edition. New York: McGraw-Hill.

Schmidt, J. H., B. P. Weidema and M. Brandão. 2015. A framework for modelling indirect land use changes in life cycle assessment. *Journal of Cleaner Production* 9:230-238. https://doi.org/10.1016/j.jclepro.2015.03.013.

Searchinger T., R. Heimlich, R. A. Houghton, F. Dong, A. Elobeid, J. Fabiosa, S. Tokgoz, D. Hayes, and T. H. Yu. 2008. Use of U.S. croplands for biofuels increases greenhouse gases through emissions from land-use change. *Science* 319(5867):1238-1240. https://doi.org/10.1126/science.1151861.

Sheehan, J., V. Camobreco, J. Duffield, M. Graboski, and H. Shapouri. 1998. *Life Cycle Inventory of Biodiesel and Petroleum Diesel for Use in an Urban Bus.* No. NREL/SR-580-24089. National Renewable Energy Laboratory (NREL), Golden, CO.

Shi, R., and J. S. Guest. 2020. BioSTEAM-LCA: An integrated modeling framework for agile life cycle assessment of biorefineries under uncertainty. *ACS Sustainable Chemistry & Engineering* 8(51):18903-18914.

Shrestha, D. S. 2015. *Biofuel and Indirect Land Use Change.* Moscow, Idaho: University of Idaho.

Shrestha, D. S., B. D. Staab, and J. A. Duffield. 2019. Biofuel impact on food prices index and land use change. *Biomass and Bioenergy* 124:43-53.

Spatari, S., V. Larnaudie, I. Mannoh, M. C. Wheeler, N. A. Macken, C.A. Mullen, and A. A. Boateng. 2020. Environmental, exergetic and economic tradeoffs of catalytic-and fast pyrolysis-to-renewable diesel. *Renewable Energy* 162:371-380.

Springborn, M., B.-L. Yeo, J. Lee, and J. Six. 2013. Crediting uncertain ecosystem services in a market. *Journal of Environmental Economics and Management* 66(3):554–572. https://doi.org/10.1016/j.jeem.2013.07.005.

Steen, B. 1997. On uncertainty and sensitivity of LCA-based priority setting. *Journal of Cleaner Production* 5(4):255-262.

Taheripour, F. T., X. Zhao, and W. E. Tyner. 2017. The impact of considering land intensification and updated data on biofuels land use change and emissions estimates. *Biotechnology for Biofuels and Bioproducts* 10:191. https://doi.org/10.1186/s13068-017-0877-y.

Taheripour, F. T. and W. E. Tyner. 2013. Induced land use emissions due to first and second generation biofuels and uncertainty in land use emission factors. *Economics Research International.* https://doi.org/10.1155/2013/315787.

Taheripour, F. T. and W. E. Tyner. 2020. US biofuel production and policy: implications for land use changes in Malaysia and Indonesia. *Biotechnology for Biofuels* 13(1):1-17.

Taheripour, F. T., T.W. Hertel, and W. E. Tyner. 2011. Implications of biofuels mandates for the global livestock industry: a computable general equilibrium analysis. *Agricultural Economics* 42(3):325-342.

Usack, J. G., L. G. Van Doren, R. Posmanik, R. A. Labatut, J. W. Tester and L. T. Angenent. 2018. An evaluation of anaerobic co-digestion implementation on New York State dairy farms using an environmental and economic life-cycle framework. *Applied Energy* 211:28-40.

van Zelm, R. and M. A. J. Huijbregts. 2013. Quantifying the trade-off between parameter and model structure uncertainty in life cycle impact assessment. *Environmental Science & Technology* 47(16): 9274-9280. https://doi.org/10.1021/es305107s.

Wei, W., P. Larrey-Lassalle, T. Faure, N. Dumoulin, P. Roux, and J. D. Mathias. 2015. How to conduct a proper sensitivity analysis in life cycle assessment: Taking into account correlations within LCI data and interactions within the LCA calculation model. *Environmental Science & Technology* 49(1):377-385. https://doi.org/10.1021/es502128k.

Wubie, M. A., M. Assen and M. D. Nicolau. 2016. Patterns, causes and consequences of land use/cover dynamics in the Gumara watershed of Lake Tana basin, Northwestern Ethiopia. *Environmental Systems Research* 5(8):1-12.

Zhao, X., F. Teheripour, R. Malina, M. D. Staples, and W. E. Tyner. 2021. Estimating induced land use change emissions for sustainable aviation biofuel pathways. *Science of the Total Environment* 779. https://pubmed.ncbi.nlm.nih.gov/33744564.

Ziyadi, M., and I. L. Al-Qadi. 2019. Model uncertainty analysis using data analytics for life-cycle assessment (LCA) applications. *The International Journal of Life Cycle Assessment* 24(5):945-959. https://doi.org/10.1007/s11367-018-1528-7.

5
Verification

THE IMPORTANCE OF VERIFICATION

Life-cycle analysis (LCA), whether attributional or consequential, will consider emissions across various activities that occur in a variety of different sectors and geographic locations. Some of the activities may already be actively monitored and regulated, as is the case with many power plants across the United States. Many greenhouse gas (GHG)-emitting activities are not regularly monitored, meaning that LCAs must rely on data from theoretical calculations, experimental measurements, or a small number of field measurements to approximate the magnitude of their emissions. Confirming LCA results through direct measurement of all activities for an entire fuel pathway is outside the realm of feasibility, let alone possibility. However, knowledge gathered to date using LCA can guide the targeted allocation of resources toward verification of emissions sources, key data, and other effects that have the greatest impact on a given fuel's net climate implications. A verification strategy can also play a valuable role in monitoring changes over time and prompting action to help limit any negative unintended consequences of a low-carbon fuel standard (LCFS).

Verification could be developed and used in an LCFS for two purposes. The first purpose is to verify that at least some of the conditions defined in a policy are met and that critical systems (e.g., food production, ecosystem services, and provision of affordable energy) that may be affected by the policy—in particular through market-mediated mechanisms—are not exhibiting undesirable effects that exceed certain thresholds. If the threshold is exceeded, corrective action may be warranted as defined in the policy to limit adverse policy effects. The U.S. Renewable Fuel Standard (RFS) contains a very rudimentary example of this type of verification. It requires that the U.S. Environmental Protection Agency (EPA) verify that domestic agricultural expansion has not exceeded 402 million acres, which was identified as the baseline amount of agricultural land in 2007, when the policy was implemented. It should be noted that EPA's definition of agricultural land includes cropland, pastureland, and Conservation Reserve Program land (40 CFR Part 80 2010, 14669-15320), and changing among these different land uses can have large effects on net GHG emissions. Furthermore, this threshold is a net value and does not preclude land moving in and out of agricultural production, which can contribute to GHG emissions. This committee is not commenting on the efficacy of this particular use of verification, but rather acknowledging its existence. It should also be noted that agricultural expansion is affected by many factors beyond fuel policy. The RFS also provided permissions to alter and adjust the mandated level of biofuels consumption, if needed. Aside from land use, LCFS policy effects on other types of critical systems might be verified through collecting data on electricity generation by type, energy costs (electricity, natural gas), or food prices. Related to this committee's charge, these verification steps could be used to evaluate whether models that investigate market-mediated effects associated with the use of transportation fuels and the GHG implications of these effects are reflecting trends (e.g., in land use, energy prices) observed in the real world. If models produce results that verification strategies do not support, the models or the parameters they contain may need to be adjusted. Importantly, it is not always possible to establish a causal relationship between policy or fuel use and measured outcomes, such as land use change (LUC), without modeling assumptions, so verification can provide useful information but may not definitively establish consequences of fuel use or policy intervention.

A second objective of using verification in an LCFS could be to confirm that individual supply chain actors are meeting certain standard requirements or adopting practices that reduce GHG emissions beyond baseline values for fuel pathways. For example, in the RFS, energy consumption at ethanol plants is assigned a default value. Verification that an individual ethanol plant is consuming less or more energy than this default value, and therefore possibly lowering or raising corn ethanol life-cycle GHG emissions below the default value, can serve to incentivize companies to reduce emissions or flag whether action is needed to reduce plant emissions if a less GHG-intensive fuel earns more credits (as in an LCFS). This approach can also provide agencies implementing an LCFS with data about the state of an industry. Furthermore, it can reduce uncertainty in parameters used in baseline LCAs the agencies would develop — though it is important to be cautious of reporting bias (e.g., entities with lower GHG intensity than the default have incentives to report, while entities with equal or higher GHG intensity do not). Such outcomes support mandatory reporting where possible. The last motivation is especially important for parameters that greatly influence LCA results (e.g., methane leakage rates from natural gas infrastructure). Information obtained for this purpose could be used to improve both attributional and consequential LCA models.

The distinction and interaction between verification and certification merits further explanation. "To verify" is generally defined as to substantiate or prove the truth of a claim. The California Air Resources Board (CARB) states: "The LCFS verification program, under CARB oversight, provides confidence and reliability in reported data for stakeholders, market participants, and the public." In this chapter, we refer to verification as a process through which auditors collect data that substantiate a claim on the part of an actor in the fuels supply chain that the requirements of a certification standard are met. Certification involves attesting (thorough verification) that a product, service, organization, or person has met an official standard established in an LCFS.

This chapter provides examples of verification's use in contemporary low-carbon fuel policies. The chapter includes recommendations for how verification might be used in the future to reduce potential undesirable effects of LCFS and improve low-carbon fuel LCAs that inform policy.

CURRENT USE OF VERIFICATION

The next three subsections will address how existing LCFS incorporate verification at the national level to account for market-mediated effects and at the individual fuel supply chain level.

National-Level, Market-Mediated Effects

The predominant market-mediated effect that has been subject to verification is LUC. While LUC in the context of biofuel production has been known and studied for some time, increased attention given to LUC in the context of biofuels production stems from the 2007 Energy Independence and Security Act (EISA), which limited the types of feedstocks that can be used to make renewable fuel that is eligible for renewable identification numbers (RINs) under the RFS (see Chapter 2). Specifically excluded for biofuels production under the EISA definition was virgin agricultural land or land cultivated after December 19, 2007 in the United States. These land types were defined to include pastureland and land from the Conservation Reserve Program while native grasslands and forests would be excluded.[1] In the final rule documenting changes to the RFS in 2010 (40 CFR Part 80 pages 14669–15320), EPA established a national-level baseline for agricultural land in the United States at 402 million acres based on three data sources: the Farm Service Agency Crop History, the U.S. Department of Agriculture (USDA) Census of Agriculture, and the USDA Cropland Data Layer. It established an aggregate compliance approach in which, if the amount of agricultural land in the United States that is eligible for growing what the RFS defines as renewable biomass stays below 402 million acres, aggregate compliance has been achieved and RIN-eligible renewable fuels are being produced with what the policy defines as renewable biomass. However, if the amount of agricultural land exceeds 402 million acres, then renewable fuel producers would need to undertake verification processes to demonstrate that renewable fuel is made from renewable biomass. Notably, this is a threshold amount and does not preclude the possibility of land entering and leaving agriculture, which can contribute to GHG emissions. While this threshold has not yet been exceeded, in 2010 a lower threshold (397 million acres) was exceeded, which triggered a requirement that EPA should revisit the ability of the aggregate compliance approach. There is no public record of EPA's response. Beyond EPA's quantification of LUC,

---

[1] See https://www.epa.gov/laws-regulations/summary-energy-independence-and-security-act.

a series of studies seeking to quantify LUC produced widely differing results (Copenhaver et al., 2021; Dunn et al., 2015; Lark et al., 2015, 2022; World Wildlife Fund, 2021; Wright and Wimberly, 2013).

In 2018, EPA reported the environmental and resource conservation impacts of biofuels to-date and into the future (EPA, 2018). Part of this assessment includes evaluating LUC. In this context, EPA uses the term "land use change (LUC)" to include different types of land uses and how land cover changes to meet these uses. In its 2018 triennial report to Congress, EPA described its use of five data sources to characterize LUC, displayed in Table 5-1. These sources include USDA's Major Uses of Land in the United States (Bigelow and Borchers, 2017), USDA's 2012 Census of Agriculture (USDA, 2014), USDA's 2012 National Resources Inventory (USDA, 2015), the U.S. Geological Survey's (USGS) U.S. Conterminous Wall-to-Wall Anthropogenic Land Use Trends 1974-2012 (Falcone, 2015), and two studies from academic institutions (Lark et al., 2015; Wright et al. 2017).

TABLE 5-1 Studies Used in EPA's 2018 Triennial Report to Congress

Study	Comparable term(s)	Definition	Years Reported	Changes in Million Acres (%)
USDA Major Uses of Land in the United States (2017)	Cropland used for crops	Three of the cropland acreage components – cropland harvested, crop failure, and cultivated summer fallow – are collectively termed cropland used for crops, or the land used as an input in crop production.	2007 – 2012	+5 (1.5%)
USDA Census of Agriculture (2014)	Harvested cropland + failed/abandoned + summer fallow	Harvested cropland – includes land from which crops were harvested and hay was cut, land used to grow short-rotation woody crops, Christmas trees, and land in orchards, groves, vineyards, berries, nurseries, and greenhouses. No separate definition for failed/abandoned, or summer fallow cropland.	2005 – 2012	+7.8 (2.4%)[a]
USDA National Resources Inventory (2015)	Cultivated cropland	Cultivated cropland comprises land in row crops or close-grown crops and also other cultivated cropland, for example, hayland or pastureland that is in a rotation with row or close-grown crops.	2007 – 2012	+4.3 (1.4%)
USGS U.S. Conterminous Wall-to-Wall Anthropogenic Land Use Trends	Production, Crops	Areas used for the production of crops, such as corn, soybeans, wheat, vegetables, or cotton, as well as perennial woody crops such as orchards and vineyards. Includes cultivated crops, row crops, small grains, and fallow fields.	2002 – 2012	+3.9 (1.2%)
Lark et al. 2015	Net cropland	Net cropland increases (gross expansion – gross abandonment) of lands in the lower 48 states that have no evidence of cultivation since 1992.	2008 – 2012	+3 (1%)[b]
Wright et al. 2017	Net cropland	Net cropland increases (gross expansion – gross abandonment) of lands in the lower 48 states that have no evidence of cultivation since 1992.	2008 – 2012	+4.2 (N/A)[c]

[a] Harvested cropland, failed/abandoned cropland, and summer fallow changed by +5.4, +4.0, and -1.5 million acres, respectively between 2007 and 2012 according to the Census of Agriculture.

[b] Estimates from Lark or Wright are likely to be lower because they focus on a subset of lands that had no evidence of cultivation for 20 years or more, rather than all land. The committee include these in the table for convenience and completeness.

[c] The committee could not calculate the percentage increase from Wright et al. (2017) because the 2008 baseline acreage within 100 miles of a biorefinery was not reported.

EPA noted several challenges in using the data sources in Table 5-1 to assess aggregate compliance. First, many of these data sources are not collected annually. Additionally, over time, the methods used to compile these data in the different studies enumerated in Table 5-1 change. This is true of the multiple data sources used in the USDA Major Uses of Land in the United States. These methodological changes themselves—not actual changes in land use—can drive LUC analysis results. The USDA National Resources Inventory and USGS U.S. Conterminous Wall-to-Wall Anthropogenic Land Use Trends, however, retroactively adjust results to account for changes in methodology over time. Approaches to modifying the USDA Cropland Data Layer to reduce error, particularly in areas such as grassland–cropland land types that may alternate between cultivation and idle or fallow states annually, can help reduce error estimates but it is unclear to what extent. Comparing estimates of agricultural land among data sources shows better agreement at the national level (e.g., for the Census of Agriculture and the USDA National Resources Inventory) but with less agreement at the state level (Copenhaver et al., 2021). In fact, agricultural expansion may occur in small parcels that would be hard to detect with data sources such as the USDA Cropland Data Layer, which has a 30 m resolution (Dunn et al., 2017). Many non-agricultural land types, such as wetlands, hold high ecosystem value even at small sizes (López-Tapia et al., 2021; Van Meter and Basu, 2015). Moreover, the USDA Cropland Data Layer relies on the National Land Cover Database for non-agricultural land type data, but the National Land Cover Database currently suffers from low accuracy when applied to LUC analyses. One way remote sensing datasets report accuracy is with user and producer accuracies. User accuracy reflects the reliability of a remote sensing dataset's classification from the perspective of the dataset user. It corresponds to error of commission and accounts for the occasional inclusion of land types that are not actually in a given category (e.g., inclusion of alfalfa pixels in the grassland category). Producer accuracy corresponds with the error of omission and accounts for a given category of land cover omitting pixels that it should contain (e.g., exclusion of land planted in soybeans from cropland). The National Land Cover Database reports producer and user accuracy for various land categories. For example, using the National Land Cover Database to estimate grassland loss between 2011 and 2016 had a producer accuracy of 80 percent and a user accuracy of 34 percent (Wickham et al., 2018). In the case of agricultural gain, these accuracies are 23 percent and 33 percent, respectively (Wickham et al., 2018). These very low accuracies for the very types of change the EPA is required to assess under the requirements of the RFS limit the ability of the National Land Cover Database (and accordingly, the USDA Cropland Data Layer) to serve as a primary resource for use in land cover monitoring (Wickham et al., 2018). The major challenge that arises in the use of remote sensing data to evaluate LUC is that distinguishing between land that is eligible (e.g., pasturelands) versus ineligible (e.g., natural grasslands) for producing renewable fuel feedstocks is an error-prone process. Different types of grasslands pose a particular challenge given the relatively low accuracy and differences in reported area in various data sources. For example, the USDA Cropland Data Layer has a 30 percent commission error for grasslands in the southern plains (Wang et al., 2022). Furthermore, the USDA Cropland Data Layer (2017)[2] estimate of grasslands (including natural and other grasslands) is 250 million acres greater than that in the 2015 National Resources Inventory.

To summarize, the remote sensing–derived data sources available at the time of RFS implementation, and still today, have some technical limitations on their use for particular applications as a source of information on LUC, particularly at the high resolution necessary to understand effects on small but important ecosystems. However, they have been deemed sufficiently useful by EPA (2018) for the agency to have concluded in its Second Triennial Report that there has been a change in cropland area since the passage of EISA. Signals such as this can be useful in determining the direction of change, if not the absolute amount, of LUC.

The committee notes again that estimates of changes in agricultural land area in the United States depend on the data source. These have differences in definitions and methodology, data collection frequency, spatial resolution, and other factors. Accordingly, current data allow for some insight into LUC, but challenges stemming from data frequency and methodological differences exist.

---

[2] https://www.nass.usda.gov/Research_and_Science/Cropland/Release/

Conclusion 5-1: In verification to evaluate land use change at a national level, specifying the approach used to evaluate the extent, location, and type of agricultural expansion and the degree of uncertainty aids in transparency and clarity.

Conclusion 5-2: Insight into the degree of agricultural expansion domestically into ecologically important, but potentially small, land parcels requires more frequent data with higher spatial resolution and ideally high producer and user accuracy.

Recommendation 5-1: Estimates of historical land use change—which may be used to inform economic models that evaluate market-mediated land use change—based on survey or remote sensing data should rely on more than one data source and should include estimates of uncertainty. Higher resolution, higher accuracy, and more frequently collected data sources should be made accessible to the public.

It should be noted that even if it were possible to evaluate expansion of agricultural area without error, it is not possible to verify in most cases, using the strict definition used here, that agricultural expansion was driven by a policy that encourages the use of biofuels (which is the case for other market-mediated effects as well). When changes in agricultural patterns (either through expanded agricultural area or changes in crop rotation patterns) occur near biofuels facilities, it may be possible to attribute these changes partially to the demand from the biorefinery, especially when these insights are accompanied with survey data. However, there are many other factors that drive LUC including urbanization and industrialization.

It is important to verify and determine the source(s) of discrepancy in data on changes in area of U.S. cropland. Beyond LUC, other national-level market-mediated effects may be evaluated under LCFS policies. Expanded use of electric vehicles could result in incremental load additions that may put increased pressure on the use of marginal generating resources. One study has shown that marginal resources often result in higher carbon intensities (Ryan et al., 2016). Conversely, smart charging technologies can provide load shaping and reduce ramp rates on the power infrastructure (van Triel and Lipman, 2020). Others have shown potential stress to power grids from electric vehicles (EVs), or result in reassignment of resources to EV charging from other loads and adjustments in electricity prices[3] (Brown, 2020; Garcia and Freire, 2016; Graff Zivin et al., 2014; Vivanco et al., 2014). Electricity price adjustments from load shaping programs, renewable portfolio standards, and other measures aimed at addressing market-mediated effects associated with EVs must be carefully monitored to ensure transparency for rate payers.

Conclusion 5-3: In verification to evaluate electricity load shifts from national electric vehicle policies, specifying the approach used to evaluate the extent, location, and type of load expansion to be verified and the degree of uncertainty will aid transparency and clarity.

Conclusion 5-4: While smart charging has potential to provide information about the carbon intensity of retail electric vehicle load, the assignment of specific generators to specific loads relies on assumptions from either an attributional frame (e.g.: under what conditions renewable generation should be assigned to electric vehicle load or to another load) or from a consequential perspective (e.g.: what emissions would look like in a counterfactual scenario without electric vehicle load).

International Land Use Change

There has long been concern that demand for biofuels in the United States stemming from renewable fuel policies would lead to international agricultural expansion into forests, grasslands, and wetlands

---

[3] Ryan et al. (2016) found an up to 68 percent difference between marginal and average emissions factors for an individual charging station. For the U.S. average they found that marginal emissions factors are 21 percent higher than average emissions factors. These findings were generally corroborated by others but the inverse can also occur.

(e.g., Searchinger et al., 2008 and other citations in Chapter 9). In recent years, many satellite-based tools have been used to monitor LUC globally and particularly in regions sensitive to biofuel-induced LUC. In the case of global LUC, optical observations from sensors on satellites including the Advanced Very High Resolution Radiometer, the Moderate Resolution Imaging Spectroradiometer, Sentinel data, and the Landsat Enhanced Thematic Mapper Plus among others have been used to create global land use land cover maps that allow for detection of change (Song et al., 2018). As Karra and Kontgis (2021) describe, these types of maps enable non-geospatial experts to track changes in land use and land cover that might require policy or other interventions to slow it. Further, to enable use of satellite data to monitor land use and land cover change at a global scale (and even a national one) automated methods of converting the pixels in spatial observation data to land categories needs to be automated to produce a map with high accuracy and resolution on a policy-relevant time scale.

While monitoring land use and land cover for the entire globe is now a possibility with this technology (e.g., Potapov et al., 2022), it may be desirable within the context of an LCFS to focus attention on regions that have historically been prone to deforestation or regions with high carbon stock lands, at least in part as a result of demand for biofuel feedstocks, or that economic modeling point to as potential hotspots of carbon stores or LUC. Driven by recognition that preserving the Amazon rainforest is critical to slow global warming and for maintaining the many other ecosystem services it provides, there are long-standing efforts to build a near real-time ability to monitor deforestation in the Amazon using satellite imagery (Finer et al., 2018). Similarly, the GHG implications of palm oil plantation expansion into peatland in Indonesia have been a long-term concern around the use of palm as a biofuel feedstock (Fargione et al., 2008). Efforts to use satellite data to track the history of palm oil plantations in this region are also a reality (Danylo et al., 2020; Irvin et al., 2020).

Clearly, it is possible to monitor international land use using satellite data. There are several questions that remain regarding how to use this information in an LCFS because of challenges in establishing causality of international LUC, as with any market-mediated effect, or to tie it to any one policy. This concern is important in understanding whether an LCFS is having its intended effect and not inducing unintended effects. It may be possible within an LCFS to set a threshold for international LUC of concern (e.g., occurring in a biofuel feedstock producing region) because the GHG emissions that would occur if that threshold were exceeded would jeopardize the climate benefits of an LCFS. The use of satellite data to guide design and implementation of a U.S. LCFS—together with or in place of international GHG emissions from LUC assigned to a biofuel based on economic modeling—is relatively unexplored and deserves attention from the research and policy communities, especially as it might be used to complement modeling efforts that aim to predict the location of future LUCs.

Conclusion 5-5: Since satellite data allow for monitoring of international land use change, it would be possible to use satellite data to monitor international land use change, support calculations of land use change impacts, and support results from economic models used to estimate international land use change GHG emissions.

Recommendation 5-2: The research and policy communities should develop frameworks and methodologies for use of satellite data to characterize national and international land use change that may be in part attributable to an LCFS. Examples of framing questions include:

- Should an LCFS include measures to mitigate undesirable international land use change, or is it sufficient to monitor international land use change that may be due to the LCFS and these GHG emissions to the associated fuel?
- What are the guardrails (e.g., amount and type of land converted to agriculture in a certain region) that a monitoring approach would put in place and, if approached or exceeded, what action would be undertaken as a result?
- How can satellite data and economic modeling be most effectively used synergistically to limit GHG emissions from international land use change?

- What public data sources will be used to track land use change?
- How should uncertainty in land use change estimates be reported?

Fuel Supply-Chain Emissions Verification

As described previously, an objective of using verification in an LCFS could be to provide agencies implementing an LCFS with data about the state of an industry that informs parameters used in transportation fuel LCAs. Furthermore, it can reduce uncertainty in parameters used in baseline LCAs the agencies would develop, particularly for the parameters that greatly influence LCA results (e.g., soil carbon changes, methane leakage rates from natural gas infrastructure). Information obtained for this purpose could be used to improve LCA models.

For individual fuel supply chains, certification protocols lay out the criteria datasets that are required to achieve certification under various certification programs that may be part of policies. Auditors collect the required data under the protocol and verify the conditions within the fuel pathway that determine the fuels' carbon intensity (CI). Certification schemes are akin to certification protocols, although the term "scheme" is used more often in a European context. The committee therefore uses the term "protocol."

A non-comprehensive example list of currently operating certification protocols in the sustainability field include International Sustainability and Carbon Certification (ISCC), Bonsucro, Red Cert, and Roundtable on Sustainable Biofuels. These systems certify compliance with defined specifications for feedstock production but also conversion into other products including biofuels. Separately, other protocols focus on parts of the supply chain such as the ANSI/LEO-4000 Comprehensive Sustainable Agriculture Standard. A non-comprehensive example list of certification bodies includes SCS Global Services, SGS, Peterson Control Union, and Technischer Überwachungsverein.

Among the policies, the European Union's (EU) Renewable Energy Directive (RED), the Carbon Offsetting and Reduction Scheme for International Aviation (CORSIA), and the California Low Carbon Fuel Standard (CA-LCFS) currently have the most robust approaches to verification, although none of these policies are without their shortcomings. RED and CORSIA utilize a system of recognized certification protocols and auditors who are trained by the protocol developers to verify the intended application of the protocols. The EU and CORSIA regularly review and accredit ("re-recognize") certification protocols. All EU-recognized certification protocols are listed publicly at the European Commission's website (European Commission, n.d.). Likewise, all CORSIA-recognized certification protocols are listed on their website.[4] Certification protocols, in turn, then list all of the certification bodies that have been trained to follow and implement an individual protocol. In the case of the ISCC protocol, for example, their collaborating certification bodies are listed at their website (ISCC, n.d.-a).

CARB, which administers the CA-LCFS, developed its own set of certification protocols, using established sustainability auditing firms to verify compliance with them (CARB, 2020). CARB publishes the list of verification bodies and verifiers accredited to perform LCFS verification services (CARB, 2022). CARB states that its verification is based on the International Organization for Standardization (ISO) 14064 standard (ISO, 2020). Beginning in 2019, verifiers started applying for CARB accreditation and taking required training and exams.

On-farm practices are a commonly-verified aspects of low-carbon transportation fuel policies (e.g., RED and CORSIA). For example, the EU's RED provides for recognized EU certification protocols that have established specific guidelines on how to assess soil carbon levels for inclusion in biofuels life- cycle modeling, although this committee notes caveats on the functionality of this approach as described elsewhere in this report. Within the protocols' framework, improvements in soil carbon on individual farms can generate incremental credits to meet, for example, the qualifying 50 percent GHG reduction in CI threshold set under the EU RED for corn ethanol blending into the transportation fuel supply (ISCC, 2021). In the RED, soil carbon credits are specifically required to be assessed on a field-by-field basis.

---

[4] See https://www.icao.int/environmental-protection/CORSIA/Documents/ICAO%20document%2004%20-%20Approved%20SCSs.pdf.

In a letter to CARB by 20 stakeholder groups, the Midwestern Clean Fuels Policy Initiative requested that CARB allow farmers to generate individual farm-level GHG life-cycle emissions values rather than the current fixed average agriculture phase value. This would allow farmers to generate incremental credits for enhanced agricultural practices (Great Plains Institute, 2020). CARB, to date, has not acted on this request. Inclusion of soil carbon credits under the proposed Midwest Low Carbon Fuel Standard remains a topic of discussion mostly around the topic of permanency and additionality.

On-farm audits can also verify whether an individual property has increased the amount of agricultural land or changed the type of crop being grown. For example, it is relatively straightforward for a farmer or auditor to report changes in land use on private farmland. These changes could include idling a field or expanding agriculture into forested land on farm property. In the EU RED, under various certification protocols accredited by this policy, on-farm LUC is assessed via self-declaration forms as well as on-farm audits. In this policy, the number of required on-farm audits by protocols is a function of a region's risk for land use conversion. Audit frequency is higher in areas that are near rainforests or sensitive grasslands. The number of farm audits in low-risk regions for agricultural expansion is the square root of all growers delivering to a first gathering point (e.g., a grain elevator or biofuels plant) whereas 100 percent of plantations supplying palm oil producers in Southeast Asia must be audited. In policy frameworks where agricultural feedstock producers can generate reductions in GHG emissions from the feedstock production stage and thereby extra revenue, conservation management practices are encouraged (Piñeiro et al., 2020).

In contrast to the extent of the previously-described verification approaches related to biofuels, only the LCFS addressed verification approaches relative to EV life-cycle GHG emissions. Under the LCFS credit, generators may submit applications to verify a lower CI rating for unique pathways that would result in greater potential GHG reductions (California Low Carbon Fuel Standard Regulation, CCR 17 §). Within that framework, a relatively new provision under the CA-LCFS allows for incremental credits for electric vehicles.[5] Utilities, automakers, and others can use meters or vehicle data to claim additional credits from residential EV charging that uses lower carbon electricity. CARB states that, "Any equipment that is capable of measuring electricity used for residential EV charging, and for tracking and recording the amount of electricity dispensed to that vehicle over a specific time period, may be registered in the LCFS Reporting Tool (LRT) as a piece of Fueling Supply Equipment (FSE)."[6] Incremental credits are calculated based on the difference between the low-carbon electricity used to fuel vehicles and the California average grid electricity (CARB, 2020b, n.d.). Utilities are required to reinvest a share of the credits on customers, EV drivers, and to support transport electrification broadly. Further, utilities must dedicate a minimum portion of credit revenues to EV rebates and equity projects. In essence, verified retail metering and the credits generated are used for reinvestment in transportation electrification. This effort can generate clear benefits to offset the costs of verification (see Kelly and Pavlenko, 2020).

In summary, protocols exist that could inform LCA for an LCFS for individual fuel producers including estimated on-farm soil carbon level changes and risk-based LUC verification. This does not preclude new protocols being created independently of existing ones. In addition, as practiced under the CA-LCFS for residential EV charging, incremental credits can generate extra revenue to offset verification costs and encourage additional renewable infrastructure.

Conclusion 5-6: Certification and verification approaches have been implemented in contemporary LCFSs to inform values for many parameters that influence emissions.

CHALLENGES IN IMPLEMENTING VERIFICATION APPROACHES

Challenges have arisen in the implementation of the verification strategies described previously, including managing competing certification protocols, potentially high costs, trade effects, inadvertent pre-

---

[5] See https://ww2.arb.ca.gov/sites/default/files/classic/fuels/lcfs/guidance/lcfsguidance_19-01.pdf.
[6] See https://ww2.arb.ca.gov/sites/default/files/classic/fuels/lcfs/guidance/lcfsguidance_19-03.pdf.

ferred treatment for certain technologies, baseline selection, and timing of verification and associated corrective action.

Competing Certification and Verification Systems and Their Level of Quality

As described in the section on verification in LCFS, there are multiple certification protocols for biomass and biofuels alone. One challenge in implementing verification protocols within an LCFS is deciding whether to adopt one or more existing protocols or to devise a new one.

In the U.S. RFS, EPA has pathways that are generally applicable to renewable fuels that any producer using a certain feedstock-conversion-fuel pathway can use.[7] Alternatively, fuel producers can apply for individual pathway assessments to demonstrate that they are able to achieve life-cycle GHG emissions, as estimated by EPA, for a pathway that differ from the generic pathways. EPA has established the Quality Assurance Plan, which is a voluntary program that is meant to verify if conditions for a preferred pathway are met. When they are, the fuel is eligible to receive RINs under the RFS (see Chapter 3). Quality assurance plans are customized to individual producers aiming to demonstrate their fuel meets the qualification for RINs in certain categories such as advanced (50 percent GHG reduction threshold) or cellulosic (60 percent GHG reduction threshold) as opposed to renewable fuel (20 percent GHG reduction threshold). Through the quality assurance plan process, verification data are collected. EPA has minimum requirements for quality assurance plans but it does not specificy their exact content or form.

While devising a new protocol may allow for incorporation of criteria that are important to the regulatory body implementing a policy, new protocols may create barriers for new and existing fuels and companies to participate in a policy. One approach (Moosmann et al., 2020) to managing the existence of multiple protocols is for different policies to accept certifications undertaken under a separate policy. This can reduce costs and administrative burden on producers and enables coverage across long supply chains, but it comes with the drawbacks of possible inconsistencies in accounting and accepted levels of deviation.

In Brazil's RenovaBio program, life-cycle GHG emissions are calculated for fuel produced from each participating plant. The RenovaBio policy does not allow third-party protocols. Instead, they developed their own protocol but utilize existing auditing groups (e.g., SCS, SGS, Peterson Control Union) (USDA, 2021a). Each grower must provide their farm-level data to the government. In Colombia's case, the sustainability protocol was developed by the government but third-party auditors are enlisted to ensure sustainability and compliance with ISO standards (USDA Foreign Agricultural Service, 2018). In contrast, the Japanese allow the use of the already established ISCC Plus protocol (a European Protocol derivative) for ethyl tert-butyl ether sustainability certification (ISCC, n.d.-b; USDA Foreign Agricultural Service, 2020). With the ISCC Plus certification requirement, biofuel deliveries into Japan are subject to farm-level audits.

CARB was interested in pursuing verification strategies but desired to develop its own certification standard. This effort is on hold in part because of the labor involved in establishing a certification protocol that is unique to CARB.

It should be noted that certification protocols may be carried out by private companies within rules set by public, policy-setting entities (e.g., co-regulation). In a regulatory context, the term co-regulation (German Federal Ministry for Economy Cooperation and Development, 2013) indicates that regulators have defined sustainability criteria for certain economic sectors or activities and recognize verification processes carried out by private sector auditors that ensure compliance with those criteria. In co-regulation, these private actors carry out this control and reduce the regulatory burden of regulators while increasing engagement within the private and nongovernmental sectors in achieving the policy's objectives. Alternatively, policymakers may prefer these functions to be carried out by the public sector.

Co-regulation via third party verification protocols needs to be adequately organized. Verification systems should be independent, third party systems with multi-stakeholder governance (including large nongovernmental organization shares). They should also incorporate internal integrity auditing systems. In

---

[7] See https://www.epa.gov/renewable-fuel-standard-program/approved-pathways-renewable-fuel.

these procedures, the certification protocol periodically audits their own recognized auditors to ensure that the protocols are followed correctly. For example, the American National Standards Institute supports verification protocols in assessing the "competence of verification bodies in accordance with international standards for accreditation" (ANSI, 2015). Furthermore, there is a need for ongoing auditing of certification protocols through mechanisms like ISO 17065 or ISO 17021 and the principles of the ISEAL Alliance Code of Good Practice for Assuring Conformance with Social and Environmental Standards (ISEAL, n.d.; ISO 2020a,b).

Finally, co-regulation protocols need to be rated in benchmarking efforts that provide comprehensive, verified, and transparent information on voluntary sustainability standards.[27]

Administration of Certification

Some data used to satisfy one certification protocol could be applied to being certified for another.

Trade

Verification can be used to allow fuels produced in one country to qualify for inclusion in LCFS in other countries. In the case of the U.S. ethanol plants described previously, verification enabled trade of biofuels between countries. Conversely, verification systems can also act as non-tariff trade barriers particularly when a multitude of systems competes for legitimacy. Narenda points out: "thus, rather than filling institutional voids, the multiplicity of competing standards creates additional non-tariff barriers for emerging economy firms." (Montiel et al., 2018).

Brazil's RenovaBio program is an example of a policy that has raised concerns for preferential treatment of domestic fuels. This concern has arisen even though the country is dependent upon international supply chains. Fuel standard programs that rely on international supply chains may have to be mindful whether a policy acts as an implicit trade barrier. However, RenovaBio is a relatively young policy, and future adjustments may potentially level the playing field.

Conclusion 5-7: Certification through protocols and methods that are consistent or compatible across regions and countries may mitigate global trade barriers.

Inadvertent Favoring of Individual Fuels

Observation of CA-LCFS, which allows for individual, company-specific fuel pathways to become eligible based on their CI, has highlighted how an LCFS policy might inadvertently or intentionally favor one fuel pathway over another. Such a concern has been raised about the LCFS because it awards CI credits for activities that could reduce transportation GHG emissions but are not directly tied to the process of selling low-carbon fuels themselves. The CA-LCFS, however, currently only applies this approach to EV-related pathways. It awards rebates for installing charging stations funded by selling credits generated by supplying electricity to EVs. Within the LCFS, there is no comparable incentive for infrastructure related to biofuels like e85 pump installation (Bushnell et al., 2021) or other fuels. If using verification to award credits for non-fuel-sales related activities is inconsistent across different fuel types, the policy may not be technology-neutral.[8]

A second and more serious issue of asymmetry arises because residential retail metering for EVs is instituted as an incentive rather than a requirement. In the absence of full retail metering as a requirement, CARB relies on EV usage surveys, which have been shown to be inaccurate (Davis, 2019). Accordingly, it is possible that vehicle miles traveled in EVs is lower than the policy is counting as it calculates its GHG-

---

[8] Energy Institute WP 318R: "EnPolicymakers have begun to treat the LCFS as a means for directing resources to preferred technology solutions, such as ZEVS in California, setting the policy on a path different from the science-based, technology-neutral fuel standard it was originally positioned to be."

reducing benefits. Differences in driving patterns among EV owners and gasoline vehicle owners is an area of active research.

Recommendation 5-3: If applied, verification requirements should be used consistently and comparably across pathways to encourage technology development and deployment.

Benchmark Selection

Policies can define default baseline values for parameters that are to be certified. For example, a policy may set a default amount of diesel consumed in the harvesting of corn. The certification process will then establish whether a farm consumes less or more than this default amount. This ability to verify lower or higher emissions can result in economic gain or loss for a supply chain actor, which will motivate them to pursue certification or to produce a fuel that complies with the policy. Accordingly, default values must be chosen judiciously such that they reflect common practice rather than adoption of new technologies or practices that lower or raise emissions. If a default value reflects what can be achieved with emerging technologies and practices that lower emissions, a supply chain actor may lose the economic incentive to undertake certification, let alone underestimating actual emissions. One example of this situation is the previously-described incremental credits awarded under the CA-LCFS for retail EV charging. Emissions from residential charging are assessed relative to those from the average grid electricity (rather than the marginal mix). The average grid electricity CI can sometimes (but not always) be significantly lower than the marginal mix for EV charging (Mueller and Unnasch, 2021; Tamayao et al., 2015). As a result, utilities and automakers may be less encouraged to generate the extra, incremental credits. Therefore, calculating incremental credits based on the difference between renewable electricity that is presumably used to charge vehicles at motorists' homes and the CA average marginal mix (rather than the average mix) could result in a much stronger incentive for incremental credits. This approach could further incentivize verified retail metering.

As illustrated in Figure 5-1, producers with emission rates above or near the default have incentives to use the default benchmark emission rates, whereas producers with emission rates sufficiently below the default have incentives to collect data and document emission rates.

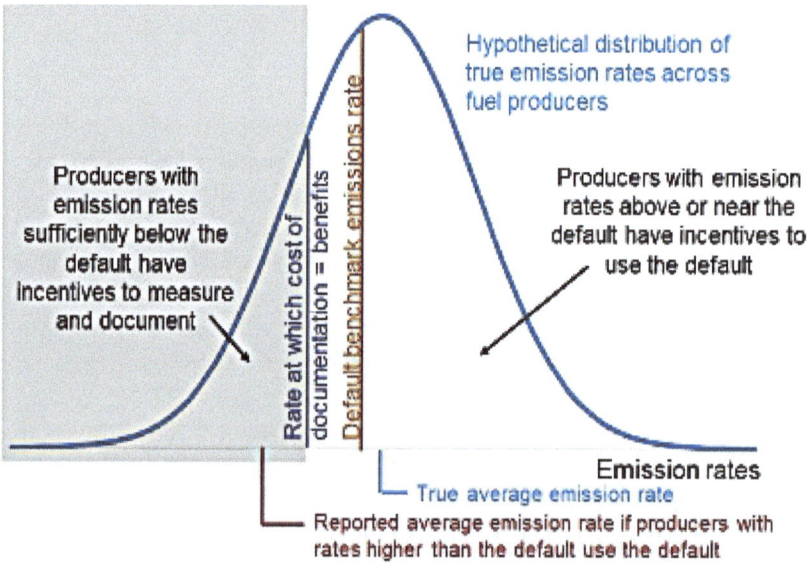

FIGURE 5-1 Hypothetical illustration that when producers with emission rates higher than the default use the default and producers with emission rates lower than the default measure and document, reported average emission rates can be substantially lower than true average emission rates.

Recommendation 5-4: Baselines, if used, should consider (1) the state of technology, (2) inputs from multiple stakeholders, (3) implications for cost of implementation, and (4) incentives that the baselines create for innovation to reduce emissions and for data collection to demonstrate emissions reductions.

Timing of Verification and Associated Corrective Action

The frequency of verification of national-level effects and potential corrective action requires careful consideration. One viewpoint is that if verification and corresponding corrective action is too frequent, unclear policy signals will be sent to the actors who produce fuels, and efforts to achieve policy objectives could be destabilized. Another viewpoint is that frequent and ongoing verification and corresponding corrective action provides a clear policy signal for compliance, promoting and stabilizing policy objectives. On the other hand, if verification occurs too infrequently, undesirable effects may be ongoing and irreversible. The influence of frequency of verification and corrective action on verification program outcomes requires research.

FUTURE TECHNOLOGY FOR VERIFICATION

Verification of national-level, market-mediated effects of LCFS policies and parameters used in LCA values requires reliable and consistent data gathering from in-person audits, through apps, database systems, sensors, and a combination thereof. Sensor data can come from satellites, airplanes, and unmanned aerial vehicles, for example, to measure land use and land management practices as well as emissions from oil and natural gas extraction. As these data become increasingly available with high-resolution and frequency, automated methods and artificial intelligence algorithms are increasingly needed to process and publish them efficiently, in a simple form to be used. Satellite data should be converted to simple tables to be compared with other available data sources for verification. Internet of Things sensors are ubiquitous and can be used to measure, assess, and verify a host of processing variables (e.g., at biorefineries). Finally, "smart metering" can be used to verify natural gas use and control electric smart charging, vehicle-to-grid, and vehicle-to-everything charging. Sensor technologies are continuously improving and the cost for all three groups has declined significantly (Microsoft, 2019; Spaceref, 2021).

While technologies that support verification of sustainability requirements set by policies keep continuously improving, policymakers need to be mindful of their limitations as well, including their accuracy and resolution.

To date, technologies in the sensing area have been substantially improving as described in the following state-of-the-art examples.

Technologies to Support Verification of Land Use Changes

Increasingly, high-resolution, high-frequency imagery that could be used to monitor national- or global-scale LUCs is becoming available. Critically, automated techniques that use artificial intelligence-based techniques are needed to process this imagery rapidly. Research is underway with this purpose (Lopez-Tapia et al., 2021). These techniques rely on ground-truth data in multiple years. These data, which are fairly limited, are needed to build training and testing datasets for artificial intelligence techniques.

At the farm level, some remote sensing tools have been specifically developed to support LUC audits. The Global Risk Assessment Services tool uses satellite and remote sensing data on a global level to assess, monitor, and manage the risk of biofuel producers to cause deforestation and LUC (GRAS, n.d.). A mapping app complements the Global Risk Assessment Services to enable users to collect farmer land use and equipment activity data, fields polygons, and pictures. Remote sensing tools have also been developed to specifically track reclaimed lands that qualify as a special land use category (unused, underused) for biofuels production under several policies (RED, CORSIA) (Mueller et al., 2021).

On-farm emissions associated with management practices may also be important to verify. Farm equipment manufacturers and technology providers have developed a host of data gathering tools to optimize farm operations that can be used to inform sustainability practices. All of them include apps, desktop applications, and data from equipment sensors. Examples include (no endorsement by this committee is implied):

- Original equipment manufacturers make decisions to include sensors on their farm equipment that can track the type of data that are needed in biofuel LCA. These manufacturers are helping generate high-quality data by their selection of accurate sensors. One example of this advancement is John Deere's Operations Center, which is an online farm management system that integrates field equipment data.
- Several organizations have close interactions with growers and can therefore directly verify LCA-relevant conservation management data.
    - The Bayer Carbon View Program provides nutrient and conservation management insights and monetary incentives via the Climate Field View platform. The strength of the platform for LCA verification is the direct large-scale interaction with growers.
    - The Farmer Business Network's Gradable Carbon platform aims to integrate practices with voluntary carbon management programs. Notably, this system has a parallel tie-in with soil carbon sampling.
    - Precision Conservation Management is supported by Midwest grower organizations who work directly with farmers to input management data into the Precision Conservation Management Farmer Portal. The portal generates a "sustainability-focused Resource Analysis and Assessment Plan (RAAP)."[9] This program works directly with growers to show the value of conservation management practices. The strength of the program is the direct, personal interaction with growers.
- Other organizations collect equipment and logistics data that is relevant for LCA.
    - Bushel's FarmLog app collects data on standard activities like planting, fertilizing, spraying, irrigating, and harvesting operations. Of particular interest for LCA parameter verification is that the company is working with delivery points to collect weight-scale data from, for example, grain elevators.[10]

These technologies support easier collection of individual LCA-relevant data points. Additionally, database systems and blockchain technologies originally developed for supply chain applications can ensure that those individual data points gathered can be more easily inserted, linked, and verified (e.g., across feedstock growing, processing, transportation, and blending). USDA has repeatedly researched and showcased the use of blockchain technology (USDA, 2021b; USDA Agricultural Marketing Service, 2020). GEVO, a U.S.-based ethanol and aviation fuel producer formed a joint venture to track land use and production practices using blockchain (Verity Tracking, 2022). Alternatively, or in complement, database systems like Trace-Your-Claim have been integrated successfully into certification protocols (Global Risk Assessment Services, 2018).

Farm apps have also been examined in peer-reviewed publications, and some gaps have been identified. Eichler Inwood and Dale (2019) state: "Nearly all of the apps we found can be characterized as "single solution" approaches that provide limited data to improve one specific aspect of efficiency—and often sustainability—but they are not effectively designed to integrate sustainability concerns from multiple dimensions or themes of indicators for sustainable agricultural landscapes." Ideally, the data collected to support an LCFS would be available to the public for analysis.

---

[9] See https://www.precisionconservation.org/.
[10] See https://farmlogs.com/essentials.

Technologies to Support Data Verification of EV Pathways

The International Renewable Energy Agency (IRENA) states that "smart charging means adapting the charging cycle of EVs to both the conditions of the power system and the needs of vehicle users" (IRENA, 2019, p. 3). The different charging technology types and potential for detailed charging data collection have been widely documented in the peer-reviewed literature (see Barone et al., 2020; Haotian Liu et al., 2020). Many smart chargers enable participation in utility demand response programs. Moreover, smart chargers are also helpful to track the EV vehicle class and types as well as verified electricity use for more accurate tracking of energy efficiency ratio groups by LCFS regulators. Both user-managed charging and supplier-managed charging technologies can be used for smart grid management.

Some selected technologies have been reviewed to showcase their integration with LCA:

- Chargepoint offers residential and fleet smart chargers for EVs. The chargers come with an app that allows tracking time-of-use charging and "real-time access to estimated environmental impact data" (Chargepoint, 2022).
- Enel X JuiceBox features load-sharing, which enables the owner to use one dedicated circuit for multiple units. This is useful for two-EV parties sharing a dwelling without the capacity for separate dedicated circuits.
- Nuvve offers a Heavy-Duty Charging Station designed specifically for vehicle-to-grid applications with smart charging capability for heavy-duty fleet vehicles such as electric school buses. The charging stations are controlled through Nuvve's fleet management app that allows bidirectional vehicle-to-grid and vehicle-to-building services when connected to a vehicle-to-grid compatible vehicle.
- Siemens Vericharge offers modular charging stations at the residential and commercial level that can be integrated with building management systems. This technology enables tracking of commercial charging where electric feeders serve multiple charging stations.

Accuracies of verification technologies for EVs are increasing over time as a host of promising new technologies has emerged over the last decade.

Recommendation 5-5: Combinations of newly developed sensor (including satellite) and supply chain technologies (e.g., database systems, blockchain) could be considered to improve land use change assessments. Policies need to be consistent with verification technology and set realistic expectations for verified LCA values. Data should be made publicly available for external verification. The GHG footprint of verification technologies should be included in the LCA as well.

Emerging Satellite Technologies to Support Verification of Crude Oil, Natural Gas Pathways, and Existing Emissions Models

New satellite sensors can directly measure GHG emissions from oil and natural gas fields. Some selected technologies in this area include the following:

- The Carbon Mapper Consortium is the first-ever public–private–nonprofit satellite collaborative that is measuring methane and CO_2 from oil and gas and other assets (waste, coal mines, dairies, etc.). The consortium includes Carbon Mapper and RMI (nongovernmental organizations), NASA Jet Propulsion Lab and CARB (government), Planet (private), and philanthropies (including High Tide Foundation and Bloomberg). Carbon Mapper is now conducting extensive flyovers with methane sensors in the United States and Canada (and planning international flights for 2022), in advance of its first two satellites launching in 2023. A total of 25 satellites are ultimately planned in the buildout (Carbon Mapper, n.d.).

- European Space Agency Sentinel-5P Satellite. This satellite sensor collects methane and aerosol data that can be used to verify models or emissions from large sources (The European Space Agency, n.d.).

In sum, new satellite technologies are emerging that enable direct GHG measurements of some emissions sources. In addition, verification of some CI LCA emissions components can be one tool to enable implementation of an LCFS to move beyond use of fixed default values. Verification programs have incentivized fuel providers to adopt emissions-reducing technology; there is a large body of verification efforts to draw best practices from and to apply in an LCFS.

Conclusion 5-8: There are a number of issues relating to the choice of certification protocols that use verification, including the cost to fuel providers, the benefits of reciprocity among protocols, and whether protocols act as trade barriers. These should be weighed against the net costs or benefits that verification provides to society including the carbon footprint of the certification process itself.

Conclusion 5-9: Certification protocols that use verification strategies can complement initial fuel pathway modeling with LCA and associated models (e.g., economic models used to estimate land use changes) to lessen the impacts of uncertainty in LCA results and to inform policymakers of the effects of an LCFS as they unfold. This insight can aid in policy adjustments if undesirable effects arise over the course of the policy.

Recommendation 5-6: An LCFS should consider inclusion of a certification protocol with verification. The protocol and its implementation should be overseen by an agency or group of agencies with the complementary expertise sets needed for success. These expertise sets include insights into multiple energy systems and new technologies, economics, environmental effects of fuels and their production routes, agriculture, fossil fuel production, and electricity generation.

Recommendation 5-7: Certification protocols should be revisited periodically to adapt to the emergence of new verification technology, national and global trends in the energy, transportation, and agriculture sectors, and to update baselines as needed based on evolving common practice.

Recommendation 5-8: Economic modeling and verification processes are complementary to each other and should both be used. Verification processes to assess international- and national-level land use change should use state-of-the art remote sensing technologies, when appropriate, which are evolving toward increased frequency and spatial resolution.

REFERENCES

ANSI (American National Standards Institute). 2015. *ANSI Launches Pilot Program: Accreditation to International Sustainability and Carbon Certification (ISCC) Certification System for Sustainability and Greenhouse Gas Emissions.* https://www.ansi.org/news/standards-news/all-news/2015/02/ansi-launches-pilot-program-accreditation-to-international-sustainability-and-carbon-certification-i-05.

Barone, G., G. Brusco, D. Menniti, A. Pinnarelli, G. Polizzi, N. Sorrentino, P. Vizza, and A. Burgio, A. 2020. How smart metering and smart charging may help a local energy community in collective self-consumption in presence of electric vehicles. *Energies* 13:4163. https://doi.org/10.3390/en13164163.

Bigelow, D., and A. Borchers. 2017. *Major Uses of Land in the United States, 2012.* Economic Information Bulletin Number 178, U.S. Department of Agriculture. https://www.ers.usda.gov/webdocs/pub

lications/84880/eib-178.pdf?v=0#:~:text=In%202012%2C%20the%20major%20land,)%3B%20miscellaneous%20uses%20(such%20as.

Brown, A. 2020. *Electric Cars Will Challenge State Power Grids.* https://www.pewtrusts.org/en/research-and-analysis/blogs/stateline/2020/01/09/electric-cars-will-challenge-state-power-grids.

Bushnell, J., E. Muehlegger, E. Rapson, and J. Witcover. 2021. *The End of Neutrality? LCFS, Technology Neutrality, and Stimulating the Electric Vehicle Market.* Energy Institute WP 318R. https://haas.berkeley.edu/wp-content/uploads/WP318.pdf.

Carbon Mapper. n.d. Carbon Mapper: Accelerating local climate action, globally. https://carbonmapper.org/.

Chargepoint. 2022. Meet ChargePoint Home Flex. https://www.chargepoint.com/drivers/home.

Copenhaver, K., Y. Hamada, S. Mueller, and J. B. Dunn. 2021. Examining the Characteristics of the Cropland Data Layer in the Context of Estimating Land Cover Change. *ISPRS International Journal of Geo-Information* 10(5):281.

Danylo, O., J. Pirker, G. Lemoine, G. Ceccherini, L. See, I. McCallum, Hadi, F. Kraxner, F. Achard, and S. Fritz. 2021. "A map of the extent and year of detection of oil palm plantations in Indonesia, Malaysia and Thailand." *Scientific Data* 8(96).

Davis, L. W. 2019. How much are electric vehicles driven? *Applied Economics Letters* 26(18):1497-1502. https://doi.org/10.1080/13504851.2019.1582847.

Dunn, J. B., Z. Qin, S. Mueller, H. Y. Kwon, M. M. Wander, and M. Wang. 2017. *Carbon Calculator for Land Use Change from Biofuels Production (CCLUB) Users' Manual and Technical Documentation* (No. ANL-/ESD/12-5 Rev. 4). Argonne National Laboratory. file:///C:/Users/Genie/Downloads/CCLUB_Manual_2017_GREET1.pdf.

Dunn, J. B., S. Mueller, and L. Eaton. 2015. *Comments on Cropland Expansion Outpaces Agricultural and Biofuel Policies in the United States.* April 29. https://greet.es.anl.gov/files/comments-cropland-expansion.

Eichler Inwood, S. E., and V. H. Dale. 2019. State of apps targeting management for sustainability of agricultural landscapes. A review. *Agronomy for Sustainable Development* 39:8. https://doi.org/10.1007/s13593-018-0549-8.

EPA (US Environmental Protection Agency). 2018. Biofuels and the Environment: Second Triennial Report to Congress (Final Report, 2018). U.S. Environmental Protection Agency, Washington, DC, EPA/600/R-18/195, 2018.

European Commission - Joint Research Centre - Institute for Environment and Sustainability. 2010. International Reference Life Cycle Data System (ILCD) Handbook - General guide for Life Cycle Assessment - Detailed guidance. First edition March 2010. EUR 24708.

EN. Luxembourg. Publications Office of the European Union. https://eplca.jrc.ec.europa.eu/uploads/ILCD-Handbook-General-guide-for-LCA-DETAILED-GUIDANCE-12March2010-ISBN-fin-v1.0-EN.pdf.

European Commission. n.d. https://energy.ec.europa.eu/index_en.

Falcone, G. 2015. Proposal of a consistent framework to integrate geothermal potential classification with energy extraction. *Geothermal Energy Science* 3(1):7-11.

Fargione, J., J. Hill, D. Tilman, S. Polasky, and P. Hawthorne. 2008. Land clearing and the biofuel carbon debt. *Science* 319(5867):1235-1238.

Finer, M., S. Novoa, M. J. Weisse, R. Petersen, J. Mascaro, T. Souto, F. Stearns, and R. G. Martinez. 2018. "Combating deforestation: From satellite to intervention." *Science* 360(6395):1303-1305. https://doi.org/10.1126/science.aat1203.

Garcia, R., and F. Freire. 2016. Marginal life-cycle greenhouse gas emissions of electricity generation in Portugal and implications for electric vehicles. *Resources* 5:41. https://doi.org/10.3390/resources5040041.

German Federal Ministry for Economy Cooperation and Development. 2013. Recognition of private certification schemes for public regulation Lessons learned from the Renewable Energy Directive.

http://bioresproject.eu/wp-content/uploads/2015/08/3-Recognition-of-private-certification-schemes-for-public-regulation_R....pdf.

Graff Zivin, J., M. J. Kotchen, and E. T. Mansur. 2014. Spatial and temporal heterogeneity of marginal emissions: Implications for electric cars and other electricity-shifting policies. *Journal of Economic Behavior & Organization* 107(Part A):248-268.

GRAS (Global Risk Assessment Services). n.d. *Sustainable, Transparent & Deforestation-Free Supply Chains*. https://www.gras-system.org/.

Great Plains Institute. 2020. A Clean Fuels Policy for the Midwest. https://www.betterenergy.org/wp-content/uploads/2020/01/Clean-Fuels-Policy-for-the-Midwest.pdf.

IRENA (International Renewable Energy Agency). 2019. *Electric-Vehicle Smart Charging Innovation Landscape Brief*. https://irena.org/-/media/Files/IRENA/Agency/Publication/2019/Sep/IRENA_EV_smart_charging_2019.pdf?la=en&hash=E77FAB7422226D29931E8469698C709EFC13EDB2.

Irvin, J., H. Sheng, N. Ramachandran, S. Johnson-Yu, S. Zhou, K. Story, R. Rustowicz, C. Elsworth, K. Austin, and A. Y. Ng. 2020. Forestnet: Classifying drivers of deforestation in Indonesia using deep learning on satellite imagery. arXiv preprint arXiv:2011.05479.

ISEAL. n.d. ISEAL Codes of Good Practice. https://www.isealalliance.org/defining-credible-practice/iseal-codes-good-practice.

ISCC (International Sustainability and Carbon Certification). System Update. 2021. Gains from esca are accounted only for the fields with no tillage (in case farmer have some fields with conventional tillage and others with no tillage), and only for the years the bioenergy crop is planted on those fields. https://www.iscc-system.org/update/10-march-2021.

ISCC. n.d.-a. "Recognised CBs." https://www.iscc-system.org/process/certification-bodies-cbs/recognized-cbs/.

ISCC. n.d.-b. "Japanese Government recognises ISCC." https://www.iscc-system.org/japanese-government-recognises-iscc/.

Karra, K., and C. Kontgis. 2021. Global land use/land cover with Sentinel-2 and deep learning? In IGARSS 2021–2021 IEEE International Geoscience and Remote Sensing Symposium. https://ieeexplore.ieee.org/document/9553499.

Kelly, C., and N. Pavlenko. 2020. *Assessing the Potential for Low-Carbon Fuel Standards as a Mode of Electric Vehicle Support*. ICCT Working Paper 2020-29. https://theicct.org/wp-content/uploads/2021/06/LCFS-and-EVs-dec2020.pdf.

Lark, T. J., J. M. Salmon, and H. K. Gibbs. 2015. Cropland expansion outpaces agricultural and biofuel policies in the United States. *Environmental Research Letters* 10(4):044003.

Liu, H., X. Zhu, Z. Xiao, Y. Wu, Y. Li, D. Li, K. Denga, and S. Liu. 2020. The application of smart meter in the management of electric vehicle charging facilities. *Procedia Computer Science* 175:774-777.

López-Tapia, S., P. Ruiz, M. Smith, J. Matthews, B. Zercher, L. Sydorenko, N. Varia, Y. Jin, M. Wang, J. B. Dunn, and A. K. Katsaggelos. 2021. Machine learning with high-resolution aerial imagery and data fusion to improve and automate the detection of wetlands. *International Journal of Applied Earth Observation and Geoinformation* 105:102581. https://doi.org/10.1016/j.jag.2021.102581.

Microsoft. 2019. *Manufacturing Trends Report*. https://info.microsoft.com/rs/157-GQE-382/images/EN-US-CNTNT-Report-2019-Manufacturing-Trends.pdf.

Montiel, I., P. Christmann, and R. Zink. 2018 (revised: 2021). The Effect of Sustainability Standard Uncertainty on Certification Decisions of Firms in Emerging Economies. Baruch College Zicklin School of Business Research Paper No. 2018-02-03. *Journal of Business Ethics* 154(3):667-681.

Moosmann, D., S. Majer, S. Ugarte, L. Ladu, S. Wurster, and D. Thrän. 2020. "Strengths and gaps of the EU frameworks for the sustainability assessment of bio-based products and bioenergy." *Energy, Sustainabilty and Soceity* 10(22).

Mueller, S., and S. Unnasch. 2021. *High Octane Low Carbon Fuels: The Bridge to Improve Both Gasoline and Electric Vehicles*. University of Illinois at Chicago. https://erc.uic.edu/wp-content/uploads/sites/633/2021/03/UIC-Marginal-EV-HOF-Analysis-DRAFT-3_22_2021_UPDATE.pdf.

Mueller, S., R. Pearson, and J. Pristolas. 2021. *U.S. Reclaimed Coal Lands: An Analysis of Low Risk for Indirect Land Use Change under the Carbon Offsetting and Reduction Scheme for International Aviation (CORSIA)*. https://erc.uic.edu/wp-content/uploads/sites/633/2021/08/ISCC_Corsia_US_Coal-Lands_UIC_GeoMARC_Final.pdf.

Piñeiro, V., J. Arias, J. Dürr, P. Elverdin, A. M. Ibáñez, A. Kinengyere, C. M. Opazo, N. Owoo, J. R. Page, S. D. Prager, and M. Torero. 2020. A scoping review on incentives for adoption of sustainable agricultural practices and their outcomes. *Nature Sustainability* 3:809–820. https://doi.org/10.1038/s41893-020-00617-y.

Potapov, P., Li, X., Hernandez-Serna, A., Tyukavina, A., Hansen, M.C., Kommareddy, A., Pickens, A., Turubanova, S., Tang, H., Silva, C.E. and Armston, J., 2021. Mapping global forest canopy height through integration of GEDI and Landsat data. Remote Sensing of Environment, 253, p.112165.

Potapov, P., S. Turubanova, M. C. Hansen, A. Tyukavina, V. Zalles, A. Khan, X.-P. Song, A. Pickens, Q. Shen, and J. Cortez. 2022. Global maps of cropland extent and change show accelerated cropland expansion in the twenty-first century. *Nature Food* 3:19–28. https://doi.org/10.1038/s43016-021-00429-z.

Precision Conservation Management. 2022. *Economically Viable Sustainable Farming: PCM - An Innovative Farm Service Program*. https://www.precisionconservation.org/.

Ryan, N. A., J. X. Johnson, and G. A. Keoleian. 2016. Comparative assessment of models and methods to calculate grid electricity emissions. *Environmental Science & Technology* 50:8937–8953. https://doi.org/10.1021/acs.est.5b05216.

Searchinger, T., R. Heimlich, R. A. Houghton, F. Dong, A. Elobeid, J. Fabiosa, S. Tokgoz, D. Hayes, and T. H. Yu. 2008. Use of US croplands for biofuels increases greenhouse gases through emissions from land-use change. *Science* 319(5867):238-1240.

Song, X. P., M. C. Hansen, S. V. Stehman, P. V. Potapov, A. Tyukavina, E. Vermote, and J. R. Townshend. 2018. Global land change from 1982 to 2016. *Nature* 560(7720):639-643.

Spaceref. 2021. *Global Small Satellites Market Is Projected to Reach at a Market Value of US$360.5 Billion by 2030*: Visiongain Research Inc. http://www.spaceref.com/news/viewpr.html?pid=58466.

Tamayao, M.-A.M., J. J. Michalek, C. Hendrickson, and I. M. L. Azevedo. 2015. Regional variability and uncertainty of electric vehicle life cycle CO2 emissions across the United States. *Environmental Science & Technology* 49(14):8844-8855. https://doi.org/10.1021/acs.est.5b00815.

The European Space Agency. n.d. Sentinel-2. https://sentinel.esa.int/web/sentinel/missions/sentinel-2.

USDA Foreign Agricultural Service. 2018. Colombia Annual Biofuels 2018. https://apps.fas.usda.gov/newgainapi/api/report/downloadreportbyfilename?filename=Biofuels%20Annual_Bogota_Colombia_7-6-2018.pdf.

USDA Foreign Agricultural Service. 2020. Biofuels Annual. https://apps.fas.usda.gov/newgainapi/api/Report/DownloadReportByFileName?fileName=Biofuels%20Annual_Tokyo_Japan_10-28-2020.

USDA. 2021a. US Department of Agriculture, Foreign Agricultural Service Voluntary Report. February 25, 2021 Report Number: BR2021-0008 Report Name: Implementation of RenovaBio - Brazil's National Biofuels Policy. Prepared By: Sergio Barros. https://usdabrazil.org.br/wp-content/uploads/2021/05/Implementation-of-RenovaBio-Brazils-National-Biofuels-Policy_Sao-Paulo-ATO_Brazil_02-25-2021.pdf.

USDA (US Department of Agriculture). 2014. *2012 Census of Agriculture*. https://www.nass.usda.gov/AgCensus/.

USDA. 2015. *2012 National Resources Inventory*. https://www.nrcs.usda.gov/Internet/FSE_DOCUMENTS/nrceprd396218.pdf.

Van Meter, K. J., and N. B. Basu. 2015. Catchment Legacies and Time Lags: A Parsimonious Watershed Model to Predict the Effects of Legacy Storage on Nitrogen Export. *Ecological Applications* 25(2):451–65. https://doi.org/10.1371/journal.pone.0125971.

van Triel, F., and T. E. Lipman. 2020. Modeling the future California electricity grid and renewable energy integration with electric vehicles. *Energies* 13:5277. https://doi.org/10.3390/en13205277.

Vivanco, D. F., J. Freire-González, R. Kemp, and E. van der Voet. 2014. The remarkable environmental rebound effect of electric cars: A microeconomic approach. *Environmental Science & Technology* 48:12063–12072. dx.doi.org/10.1021/es5038063.

Wang, M., M. Wander, S. Mueller, N. Martin, and J. B. Dunn. 2022. Evaluation of survey and remote sensing data products used to estimate land use change in the United States: Evolving issues and emerging opportunities. *Environmental Science & Policy* 129:68-78.

Wickham, J., S. V. Stehman, and C. G. Homer. 2018. Spatial patterns of the United States National Land Cover Dataset (NLCD) land-cover change thematic accuracy (2001–2011). *International Journal of Remote Sensing* 39(6):1729-1743.

World Wildlife Fund. 2021. *Plowprint Report*. https://www.worldwildlife.org/projects/plowprint-report.

Wright, C., and M. Wimberly. 2013. Recent land use change in the Western Corn Belt threatens grasslands and wetlands. *Proceedings of the National Academy of Sciences*. https://doi.org/10.1073/pnas.1215404110Corpus ID: 12215032.

Wright, C. K., B. Larson, T. J. Lark, and H. K. Gibbs. 2017. Recent grassland losses are concentrated around U.S. ethanol refineries. *Environmental Research Letters* 12(4):044001.

6
Specific Methodological Issues Relevant to a Low-Carbon Fuel Standard

This chapter discusses several methodological issues that are relevant to assessing life-cycle greenhouse gas (GHG) emissions for transportation fuels: (1) allocation applied to multi-output systems that produce fuels or provide feedstocks for fuel production; (2) negative emissions and the implications of negative carbon intensity (CI) values for fuels; (3) biogenic carbon, including accounting methods for carbon in short- and long-rotation biomass and soil; (4) indicators and temporal aspects of GHGs and climate change; and (5) vehicle–fuel combinations and efficiencies.

ALLOCATION TO AND FROM OTHER PRODUCTS

Allocation is a fundamental aspect of attributional life-cycle analysis (ALCA). In contrast to consequential LCA (CLCA), which estimates differences in net emissions across counterfactual scenarios without necessarily attributing emissions to products, ALCA requires that decisions are made about how to allocate emissions to co-products and byproducts.

The process to quantify the emissions of a specific type of fuel or energy source needs to include all the inputs of materials and energy required to produce that fuel, or, alternatively, the emissions that result from producing and using that fuel. Many industrial processes, including fuel manufacturing, produce not just a single product but a variety of outputs. Allocation is the process of dividing the total process emissions or otherwise attributing portions of the total emissions to individual output streams. Allocation by a variety of methods is common in ALCA. The concept of multiple outputs—each carrying responsibility for a portion of total emissions— is also critical for system expansion and CLCA methods as practitioners seek to assign emissions to causational factors that can be addressed to reduce net emissions in the context of climate change.

Output streams can be classified as products, co-products, byproducts, residues, and wastes. These classes of outputs exist on a continuum. It can present a challenge to objectively classify output streams along this continuum, particularly when the economic value of a waste stream changes or is relatively close to the value of byproducts and co-products.

Economic incentives to optimize net profits from manufacturing often translate to minimizing waste. However, the efficiency of waste minimization is subservient to the overall profit incentive. For instance, when fossil energy prices are relatively cheap, net profit may incentivize the use of more cheap energy to convert feedstock into greater amounts of higher value products. Existing regulations that address the environmental or public safety consequences of waste may shift the economic incentives by adding cost to waste disposal. Public policy may also seek to create markets for the utilization of wastes. Such successful programs may shift the economic value along the waste-to-product continuum. However, the classification of the waste material rarely changes. The original motive for manufacturing operations remains of interest for the maintenance of ongoing policies and the development of new policies.

These concepts can be illustrated by examining some common fuel types and their associated output streams. Petroleum transportation fuels are co-produced with a range of outputs including butane, gasoline, naphtha, kerosene, diesel fuel, heavy gas oil, and residual fuel oil. All of these may come from a single unit of refined crude oil. Each of these may be sold as a marketable product or further refined into products with greater market demand. These examples do not include additional waste products that also result from refining the same unit of crude oil. In order to quantify the life-cycle emissions for gasoline, diesel, or kerosene jet fuel, the total process emissions for the refining process needs to be determined, as well as the transportation and other effects of crude oil extraction. Then, those total emissions need to be allocated to each of the process outputs (Bredeson et al., 2010).

There are multiple ways to perform these allocations, and the choice of method can have a significant impact on the results (Wang et al., 2004, 2011). Therefore, it is important to pair allocation methods with the policy objective. It is common practice not to allocate process emissions to waste products. Doing so would reduce the emission burden of marketable products. By allocating emissions only to marketable products, the opportunity is preserved to influence emissions through market strategies. When the policy goal is to reduce emissions, the policy has more economic leverage to affect that outcome when emissions impacts are carried by the most valuable products in high demand.

For waste, residues, and byproducts of fossil fuel origin, the emissions of combusting those products should be added to the net emissions that are allocated to the marketable products. Failing to do so would under-represent the actual emissions of producing petroleum fuels. Special attention is also needed to avoid double counting when byproducts and residues are consumed internally. Emissions from combusting biomass or biomass-derived products are discussed in the Biogenic Emissions section.

Allocation is often quantified by ratios between products based on stoichiometry, mass, volume, energy content, economic value, or a combination of these and other factors. Significant literature exists about the pros and cons of allocation choice (Cherubini et al., 2018; Schrijvers et al., 2016; Tillman, 2000; Wang et al., 2011). Mass ratio has often been used as a simple, consistent method but can also distort results for some applications and neglect to account for the relative value of products, which can drive demand. Energy ratio has also been applied commonly for energy policy such as the Renewable Energy Directive (RED) of the European Union (EU). Economic ratios reflect the impetus for different products but can be challenging to apply because they may change over time. They also require the consideration of how to address market changes as a result of the policy itself.

Allocation by economic ratio could be more philosophically consistent with the aims of market-based approaches to reducing emissions, such as a low-carbon fuel standard (LCFS). In recent history, in some countries, gasoline demand was thought to be a critical motivation for the refinement of crude oil; in other countries, other distillates, liquefied petroleum gas, or petrochemical feedstocks were thought to be the motivation. In reality, factors such as system requirement, the net profit margin from every stream of refined output, emission caps, seasonable shift, and other strategic and tactical drivers determine the business decision of refinery operators. A fuel policy that mandates increased volume of ethanol may erode the economic incentive to produce gasoline but does nothing to address the other output streams from refining a unit of crude oil. If demand for distillate fuels including diesel, heating oil, and aviation fuel remains steady with few low-carbon alternatives, then distillate demand may take over as the leading profit center for petroleum refining. Refiners could continue processing the same fossil throughput with higher profit margin on distillate fuels and lower profit or even loss on gasoline. Co-products or byproducts such as residual gas oil may have weaker demand or market value. In lieu of their production, users of those products could switch to other alternatives with less influence on the overall refining business. Increasingly, olefin production for chemicals feedstocks drives product slate shifts in refineries. As gasoline and distillate demand decreases, there could be increased focus on petrochemical production.

Allocation by energy content alone deprives the market from acknowledging that some fuels have greater versatility. Liquid fuels, for instance, have more power in economic decisions, because they are more versatile to a wider market base. Liquid fuels are easier to handle than solid fuels. Liquid fuels can be pumped and stored or shipped in containers of various sizes or shapes, which makes them less costly to move. Liquid fuels are also more energy dense than gaseous fuels. However, fuel prices are not determined on parity with energy density, but by relative supply and demand.

Corn ethanol is the most common biofuel used in U.S. transportation today. The outputs of ethanol production include distillers grains, technical corn oil, biogenic carbon dioxide (CO_2), and corn stover that remains on the field. Other biofuel feedstocks and conversion processes produce a variety of animal feed and energy coproducts including protein meal, glycerin, propane, and naphtha.

For biofuels, emissions can be allocated for the cultivation of feedstock as well as the processing and transportation of materials. Most current ethanol facilities in the United States receive whole kernels of corn as their input of raw materials. This corn can come directly from nearby farms, or it can come by truck, rail, or even barge from distant storage facilities. Unlike ethanol facilities that mill and process grain, biodiesel and renewable diesel processors generally take in vegetable oil, not oilseeds. Some biodiesel processors are co-located with oilseed crushers, but they retain the capacity to operate completely independently depending on economic conditions. Allocation can still be carried out, even though ethanol and biodiesel and renewable diesel facilities receive different kinds of feedstock.

Common agricultural commodities such as corn, soybeans, wheat, sorghum, and canola are generally stored, transported, and traded as whole grain. Their natural seed structure is relatively stable with a protective coating that allows grains to be transported and stored for long periods with low rates of degradation. The same is not true for biofuel commodities such as sugar cane and palm fresh fruit bunches, which need to be processed quickly after harvest. When an agricultural commodity is processed into usable components, the emissions of planting, cultivating, harvesting, and transportation can be allocated to those various components. Such processing is common for soybeans for instance.

Approximately half of the soybeans produced in the United States are exported to other countries as whole beans (Denicoff et al., 2014). The majority of soybeans that are used in the United States are first processed to separate the constituents. Soybeans can be consumed whole as livestock feed. However, the ratio of nutrients is not perfectly matched to an optimized diet. More flexibility for livestock growers and more value for processors can be achieved when grains are processed to separate various components.

While the EU has applied allocation according to energy content to all fuels participating in RED, such allocation has a distorting effect when the supply chain affects food production and not merely energy production. The nutritional (and thereby economic) value of food and feed co-products is affected by more than their simple calorific content. Not all calories are equal in nutritional value. Protein calories tend to be the most expensive calories for food and feed; carbohydrate and fat calories are considerably less expensive; and energy from fat is generally the cheapest of all edible energy because fat is more energy dense. Because of this, energy allocation puts more burden on diesel fuels derived from fats and vegetable oils than it does on gasoline replacements derived from carbohydrate sources. In contrast with energy allocation, mass allocation assigns lower burdens to fat-based fuels because fat contains more energy per unit of mass.

Allocation by economic ratios is more complicated to conduct, but it has the potential to capture nuances that transcend food and energy markets. Proper economic analysis should be able to recognize that protein is valued more highly than agricultural outputs of fats and carbohydrates. The relative value of protein, fats, and carbohydrates is a consequence of supply and demand. Human nutrition requires calories from each macronutrient group. Plant physiology supplies a different ratio of these macronutrients than required for human nor animal nutrition. Plants tend to optimize the storage of fat and carbohydrate energy in seeds.

Allocation is a critical approach for attributing emissions for product systems with multiple products that have comparable values (whether in terms of mass, energy, or market value); however, there are complexities that arise when attributing emissions to product system outputs that only comprise a small share of the overall a product system. For example, inedible tallow from beef production has a high market value and has a well-developed market for secondary uses, but it accounts for less than 2 percent of the overall value for beef production based on a study in 2015 (ICF International, 2015). Given its high market value and multiple uses, tallow and similar materials occupy a niche somewhere between a co-product and a waste, depending on their relative value to all other meat products.

The material definition for outputs for a given product system can have substantial implications for the assessed emissions of those materials, with high relevance for the assessment of different biofuel feedstocks. The standard of the International Organization for Standardization (ISO) for LCA broadly defines co-products as "any two or more products coming from a unit process or product system," and wastes as "substances or objects which the holder intends or is required to dispose of," but does not provide any clarity on materials whose significance in the product system is more ambiguous (ISO, 2006). In the absence of a formal definition, common terminology for these kinds of products in the literature includes the terms *byproducts, residues, and wastes*. For these materials, system expansion can sometimes be preferable to allocation (Weidema, 1999); in other cases, they are treated similarly to wastes.

Existing biofuel policies have not developed a consistent approach for material categorization for biofuel feedstocks. The GREET model,[1] developed by Argonne National Laboratory, distinguishes between co-products and "waste byproducts," but does not have a middle category. For feedstocks in the latter category, upstream emissions are only attributed to those feedstocks from the point of diversion; in the case of tallow, GREET models it as a waste byproduct of cattle production and subsequently, a co-product of the rendering process (Wang et al., 2020). The U.S. Renewable Fuel Standard (RFS) and the California Low-Carbon Fuel Standard (CA-LCFS) draw on this approach. In its LCAs for the RFS, the U.S. Environmental Agency (EPA) allocates upstream emissions to co-products, but not for either byproducts or wastes. EPA uses some flexibility in classifying feedstocks as byproducts or wastes, taking into account their market value, existing uses, and whether those feedstocks would be incinerated or landfilled at end of life (ICF International, 2015).

The CA-LCFS does not have a formal framework to determine the categories for different biofuel feedstocks; they are determined on a case-by-case basis. As in GREET, the CA-LCFS does not allocate emissions toward byproducts or wastes. Secondary products, byproducts and wastes do not bear indirect land use change (ILUC) emissions, but co-products do. The CA-LCFS also includes another category called secondary or incremental products, which are products produced from existing processes after modifications to the process—such as corn oil produced from existing corn ethanol production. In the case of corn oil a displacement approach is used, in which the emissions impact of a reduction in wet distillers grains with solubles credits from the corn oil produced is added to the corn oil feedstock (CARB, 2018).

The EU RED also assesses emissions differently based on material categories, distinguishing between products/co-products, wastes/residues from processing, and residues from agriculture, aquaculture, fisheries, and forestry. The RED LCA approach considers wastes and residues to have zero GHG emissions up to the process of collection of those materials. RED defines co-products as the materials that are "the primary aim of the production process" and utilizes the energy allocation method in most cases to attribute emissions.[2] RED specifies that allocation shall not be used for co-products that are not the intended output for the product system such as crop residues, or for processing residue, which "is a substance that is not the end product(s) that a production process directly seeks to produce. It is not a primary aim of the production process and the process has not been deliberately modified to produce it."[3]

The U.K.'s Renewable Transport Fuel Obligation identifies outputs that have a high economic value relative to primary products and have uses beyond energy recovery as products, rather than wastes or residues (U.K. Department for Transport, 2021). In the Renewable Transport Fuel Obligation, wastes and residues are considered to have zero life-cycle GHG emissions up to the process of collection of those materials. For materials not explicitly included in RED, the Renewable Transport Fuel Obligation guidance suggests that materials trading for 15 percent or higher of the price of the primary product is an indicator of economic significance, though other factors may be taken into account, such as the quantity of the material and its other uses (U.K. Department for Transport, 2018).

Expanding on the RED approach, one potential method to classify feedstocks would be to assess the extent to which they motivate production for an overall product system. For example, a material with high per-kilogram market value that comprises a low overall share of value for a product system is likelier to be a byproduct or secondary product than a co-product or primary product. An economic cut-off may be straightforward to use, but as with other allocation methods it may be arbitrary, and a material's value relative to other outputs may change over time (U.K. Department for Transport, 2015). ICF International (2015) proposes an approach in which a product is understood to be a valuable material whose supply is elastic with demand, while a byproduct (or secondary product) has some economic value, but its supply is not elastic to demand. Wastes and residues have lower value and inelastic supply, and are primarily disposed of or combusted for energy recovery.

---

[1] Formally, the greenhouse gases, regulated emissions, and energy use in technologies model.
[2] See Directive (EU) 2018/2001 on the promotion of the use of energy from renewable sources. See https://eur-lex.europa.eu/legal-content/EN/TXT/PDF/?uri=CELEX:32018L2001.
[3] See Directive (EU) 2018/2001; 2010/C 160/02.

An illustrative example evaluates the impacts of different product definitions on life-cycle emissions for tallow-derived sustainable aviation fuel and diesel (Seber et al., 2014). The well-to-wheel emissions of the fuel using a co-product allocation is more than twice as large as the same fuel treating the tallow as a byproduct.

The choice of allocation method can affect the relative CI of fuels under an LCFS. Allocation will also affect the CI assigned to co-products that fall outside of fuel markets and low-carbon fuel policy. For instance, in the example of tallow-derived biofuel, treating tallow as a co-product rather than a waste reflects that these fuels have become responsible for more of the emissions of cattle production. Any increase in emissions allocated to fuel accompanies a corresponding decrease in the emissions assigned to meat production, which lies outside the system boundaries of RFS, CA-LCFS, and RED.

For some byproducts, wastes, and residues with existing uses outside of alternative fuel production, allocation may not be a suitable method for attributing upstream emissions. They may be either low-value or account for a small share of the total output of a product system; nevertheless, their use for fuel production may divert them from an existing use and necessitate material substitution (ICF International, 2015). For example, EPA estimates displacement emissions for sorghum oil (a byproduct of sorghum grain) diverted from animal feed, as a component of its LCA of sorghum-derived biodiesel for the RFS (83 FR37735).[4] In that analysis, the displacement emissions for sorghum oil are estimated on the nutritional deficit from its diversion from animal feed and an assumption that it would be replaced by corn. This substitution approach is one type of system expansion application (Heijungs et al., 2021).

Allocation is often used in ALCA. The decision to define a fuel feedstock as a co-product, byproduct, or waste can have implications for its LCA, and therefore on the understanding of its climate implications and its role in a future fuels policy. In practice, upstream emissions for most co-products are subject to allocation whereas upstream emissions attributable to waste-based feedstocks are only included from the point of collection onward. For byproducts and wastes with existing uses prior to fuel conversion, a system expansion-substitution approach is an option.

NEGATIVE EMISSIONS

Fuels assigned net negative CI values raise important questions that warrant special scrutiny to distinguish between actual CO_2 removal and storage and fuels pathways that include credits for avoided emissions. To date, the life-cycle CI of most fuels is a positive number. However, there are already several fuel pathways assigned negative CI scores under the CA-LCFS, and there are many other proposed future fuel pathways that may achieve negative emissions. In addition to fuels with a negative CI score on an overall basis, the use of CO_2 capture, utilization, and storage in fuel supply chains can add a carbon removal term to existing pathways if they are shown to be effective at a large scale.

The most significant fuels with large negative CI scores under existing policy are based on biomethane from manure. (A few pathways for hydrogen from landfill gas also have received much smaller negative CI scores). The use of manure-based biomethane for transportation (either as compressed natural gas, liquefied natural gas [LNG], electricity or hydrogen) has been assigned CI scores under the CA-LCFS as low as -630 g CO_2e/MJ for dairy biomethane-based electricity, -530 g CO_2e/MJ for dairy manure biomethane based compressed natural gas and -360 g CO_2e/MJ for swine manure-based biomethane LNG (CARB, 2021).

These large negative scores arise from the assumption that confined animal feeding operations will store manure in lagoons, thus emitting methane to the atmosphere, and LCFS support for the use of manure-derived biomethane will result in the construction of enclosed anaerobic digesters that would otherwise not have been constructed. If, for example, more stringent policies were to require all confined animal feeding operations to manage manure in enclosed anaerobic digesters, this assumption about the counterfactual would no longer hold true. Similarly, by capturing the reduction in emissions for the LCFS, no reduction in emissions can be assigned to the animal food product pathways. Policy or practice changes that reduce

---

[4] Renewable Fuel Standard Program: Grain Sorghum Oil Pathway, Final Rule.

assumed manure methane emissions outside their use for transportation fuel would require an update to the life cycle, which could reduce or eliminate the negative emissions presently associated with the fuel pathway. This phenomenon is not unique to biomethane, and other LCA pathways will also depend on the external policy and technology landscape. However, the impact on manure methane pathways is potentially among the largest of pathways because the GHG potential of the avoided methane emissions is so large. This dependence of the large negative CI scores on methane pollution policy is especially important for planning over a time frame in which such policies may change.

> **Conclusion 6-1:** The carbon intensity of fuels derived from methane that would otherwise be released (e.g., methane from manure or landfill) is strongly influenced by assumptions in the LCA of the alternative fate of methane pollution and is subject to dramatic change if relevant regulations or practices change.

In addition to the fuels with net negative CIs because of methane destruction, the California Air Resources Board (CARB) has also developed protocols that allow crediting for carbon capture and sequestration either as part of the production of other fuels sold in California, or from direct air capture provided the project satisfies the permanence certification requirements (CARB, 2018). Pathway applications submitted by at least one ethanol producer showed a substantial reduction in the CI score (> 40 percent), but not a negative CI score (Red Trail Energy, 2019). In addition to existing pathways for carbon sequestration, several stakeholders in the biofuels and agricultural sector are advocating for crediting of carbon removal through soil carbon sequestration associated with biofuel feedstock production (see comments from Gevo and Indigo Ag in CARB [2020]) and building of pipelines to move CO_2, which can facilitate CO_2 sequestration and/or utilization including for enhanced oil recovery. The integrity of any carbon removal elements in the biofuel life cycle depends on the permanence of the carbon storage, which is especially uncertain for soil carbon and which could be lost following changes in tillage. Furthermore, the use of CO_2 for enhanced oil recovery affects both the CI of the ethanol (one CO_2 source) and the petroleum products.

Many transportation fuel producers in the biofuel and petroleum industries are emphasizing the importance of carbon removal in their long-term decarbonization plans. It is important to note that the negative elements of fuel LCAs arising from avoided methane emissions, avoided CO_2 emissions, or carbon capture and sequestration in geologic reservoirs or soil are subject to different regulatory contexts that are evolving over time.

> **Recommendation 6-1:** LCA for LCFS policies should provide as much transparency as possible on the different carbon removal elements of fuel life cycles allowed under the policy, as well as insight into how these may change over time, to inform policymakers and stakeholders. Specifically, LCA pathway analyses used to determine carbon intensity scores should separately indicate the contributions from negative elements (if any) and the counterfactual scenarios, such as avoided CO_2 emissions, avoided methane emissions, carbon capture and sequestration in geologic reservoirs or soil, and use in enhanced oil recovery.

BIOGENIC EMISSIONS

Biomass removes atmospheric carbon (biogenic carbon) through the photosynthesis process, and part or all of the biogenic carbon is released during biomass conversion, transportation, decay, and biofuel combustion; see Figure 6-1. Fossil-based carbon may also be released in the same system, such as GHG emissions from burning fossil fuels to supply heat for biomass drying and conversion. Estimating the carbon uptake and biomass growth is the first step to track the biogenic carbon associated with a biomass feedstock. Carbon stock changes in ecosystems are also important given their role in determining GHG fluxes associated with land use change (LUC).

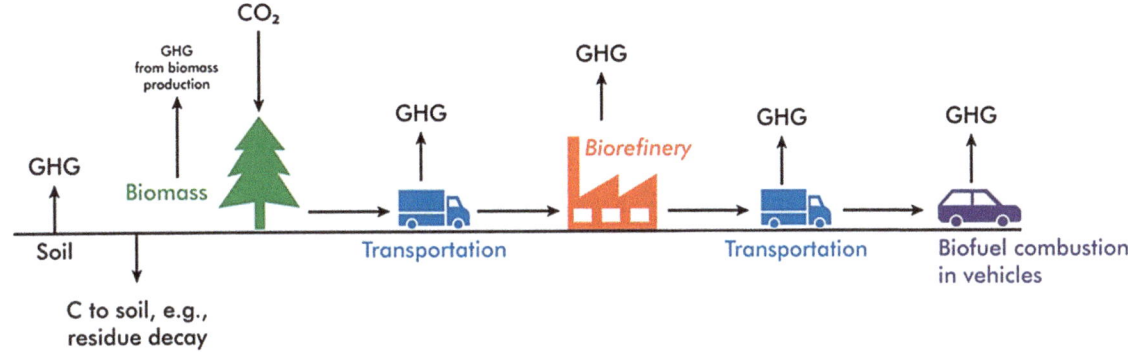

FIGURE 6-1 GHG emission flows across the life cycle of biofuel.

Different methods of evaluating carbon flows for short-rotation crops and long-rotation woody biomass are assessed and discussed in this section. Changes in soil carbon caused by LUC can generate significant GHG emissions; the soil carbon section discusses different models and methods to estimate soil carbon change and the challenges in modeling and data collection. GHG emissions associated with growing and harvesting are discussed in Chapter 9. Emissions are generated during the process of combustion of biofuels in a vehicle and during the process of conversion of feedstock (e.g., corn) to fuel in a refinery and its transportation to the consumer. These emissions can be referred to as biogenic carbon, contemporary carbon, or organic carbon. In contrast, the carbon in petroleum, natural gas, and coal is referred to as fossil carbon.

Biogenic CO_2 emissions are the release of the carbon embodied in the feedstock that was removed from the atmosphere through the process of photosynthesis and stored in the biomass. These emissions are returned to the atmosphere during the process of conversion of the feedstock to biofuel in a refinery and through combustion of biofuel in a vehicle. These biogenic emissions are sometimes excluded from the estimation of the GHG intensity of corn ethanol (Wang et al., 2012). This approach is controversial, and some studies accounted for these biogenic emissions (during conversion and combustion) to determine the extent to which they are offset through sequestration during biomass regrowth. The feedstock for biofuels from annually harvested agricultural crops has sometimes been treated as carbon neutral in that the annual biogenic carbon uptake by the feedstock from the atmosphere is considered to offset the emissions released during conversion and combustion of the biofuel. It is important to note that, based on this reasoning, only the biogenic carbon emissions embodied in the feedstock and released during its production and consumption as fuel are considered carbon neutral; it is not intended to imply carbon neutrality of the life-cycle emissions (De Kleine et al., 2016). This approach of treating all biogenic carbon as carbon neutral ignores more potent GHG emissions, particularly methane (CH_4), that may be generated in the process of conversion and release of biological carbon. Thus, with different global warming potentials for different forms of carbon, the fuels with non-CO_2 GHG emissions cannot be considered carbon neutral. This is one of the core reasons in the argument against assuming carbon neutrality of biogenic carbon, and there are other reasons such as the impacts of land use and temporal aspects associated with long-rotation feedstocks (Fargione et al., 2008; Lan et al., 2021; Searchinger et al., 2008; Wiloso et al., 2016).

Biogenic Emissions from Annually Harvested Feedstocks

The treatment of biogenic CO_2 emissions from agricultural feedstocks, specifically food crops, converted to biofuels as carbon neutral has been questioned by DeCicco (2016), Searchinger (2010), Searchinger et al. (2009), and others (Fargione et al., 2008; Lan et al., 2021; Searchinger et al., 2008; Wiloso et al., 2016). When food crops, such as corn and soybeans, are used to produce biofuels, two things happen in order to meet the demand for biofuel. First, there is some diversion of the crop from food/feed use to produce biofuels. Second, this raises crop prices (given a fixed demand for food/feed) and creates incentives to bring additional land to be planted under this crop to produce feedstock for the biofuel. Thus the total feedstock used for biofuel production is partly coming from the existing production and partly from new (additional) production of feedstock.

Searchinger (2009) suggests that "biomass should receive credit to the extent that its use results in additional carbon from enhanced plant growth or from the use of residues or biowastes." In other words, the carbon uptake by the crop that would have been produced anyway for use as food/feed should not be used as an offset for the carbon emissions during production and combustion of the biofuel. The additional carbon may be generated from the increase of biomass uptake due to changes in land management or from the utilization of biomass that would otherwise emit GHG emissions through rapid decomposition (Searchinger, 2009). DeCicco et al. (2016) use an analysis of direct carbon exchanges by comparing only the additional biogenic carbon emissions uptake during the production of the feedstock directly converted to biofuel and the biogenic emissions released during the production and combustion of the biofuel. DeCicco et al. (2016) mentioned their work as a narrow analysis that examines carbon neutrality by evaluating the extent to which feedstock CO_2 uptake offsets biogenic CO_2 emissions from fuel combustion.

De Kleine et al. (2017) have argued that for crops being grown and harvested annually, all carbon taken up by the crop, whether the crop is used for food, feed, or fuel will be returned to the atmosphere within a year or in a short period. This contemporary carbon, when released by a biorefinery or vehicle, should not be considered as net addition of carbon to the atmosphere. They explained that biofuel production leads to an exactly equal increase in net ecosystem production; therefore a 100 percent biogenic carbon offset should be applied. Net ecosystem production was used by DeCicco et al. (2016) to estimate the additional carbon, which is a portion of carbon taken up by biomass that becomes material available for local sequestration or other disposition. In response to this argument, DeCicco (2017) pointed out the need for careful analysis to circumscribe the temporal and spatial scope for any net ecosystem production increase instead of assuming that net ecosystem production increases occur somewhere and justify the full biogenic carbon offsets. Furthermore, the use of biomass for one purpose affects the production of biomass for other purposes (plant growth), so there can be an opportunity cost from reducing biomass productivity. The assessment of carbon emissions of non-food crops used as feedstocks for biofuels may be considered to depend on the alternative use of the land on which they are grown. Future research should clarify how changes in carbon stock (including soil carbon change) are being considered in assessing changes in net ecosystem production (DeCicco and Schlesinger, 2018; Field et al., 2020, 2021; Haberl et al., 2012; Kalt et al., 2019; Khanna et al., 2020; Searchinger, 2012; Searchinger et al., 2018).

Different frameworks exist, and one of them was developed by EPA for biogenic CO_2 emissions from stationary sources. This framework "assesses the extent to which the production, processing, and use of biofuels results in a net atmospheric contribution of biogenic carbon emissions". This assessment framework does not "include a full LCA, which would take into account all upstream and downstream GHG emissions and sequestration related to feedstock production and use, including from all fossil fuel inputs used, for example, to power machinery used to harvest and transport biogenic feedstocks." The framework mentions the key decisions for biogenic CO_2 flux assessment, including the choice of baseline, temporal and geospatial scales, and feedstock categories. In this framework, the baseline approach is used to evaluate the landscape emissions associated with the feedstock growth, potential leakage, biogenic emissions that would have occurred on the feedstock landscape without the use of biogenic feedstock, and changes in land use or land management. The EPA framework presents two baseline approaches. The first is the Reference Point Baseline approach that measures the net change in carbon between two points in time. The second approach is the Anticipated Baseline approach, where the carbon stocks in a baseline scenario that "establishes historic or simulates future anticipated biogenic feedstock use and related environmental and socio-economic conditions and impacts along a specified time scale" is compared with an "alternative scenario of changed (e.g., increased or decreased) biogenic feedstock demand." The EPA framework suggests no preferences between the two approaches. The Reference Baseline approach has several limitations. First, it attributes all of the changes in carbon stocks in the after–biofuel production scenario to the level of biofuel production relative to the start year; this disregards other factors that could have changed over time, such as, weather, market conditions and technology that affect carbon stocks. Second, the time period chosen as the reference point can influence the magnitude of biogenic emissions that are considered additional. The limitation of the Anticipated Baseline approach is the uncertainty associated with assumptions for future scenarios. These future anticipated baselines need to be developed using some modeling or analytical approaches, such as dynamic modeling or extrapolation of historical trends, which bring in different kinds of uncertainty (EPA, 2014).

Biogenic Emissions from Long Rotation Feedstocks

Demand for long-rotation feedstock such as forest feedstocks for bioenergy could be met by some combination of increasing intensity of forest harvest; diverting biomass from other uses such as pulp, timber, and mill residues; and by planting more land under forests. Increasing the intensity of forest harvesting will reduce the stock of carbon on existing forestland and create a carbon debt. Diversion of biomass from other uses, such as pulp and timber that can be stored for forest carbon, can also create a carbon debt (Fargione et al., 2008).

Researchers use a variety of methods to estimate carbon debt in long-rotation feedstocks. Two commonly used approaches are stand-level and landscape-level assessments, and a third less-common approach is the dynamic landscape-level assessment, which are all described below. These approaches are points on the LCA spatial spectrum, as defined in the goal and scope definition phase, that can range from a forest stand to global forested land. A forest stand is "the fraction of a landscape belonging to one age class" (Berndes et al., 2013; Peñaloza et al., 2019) while a forest landscape is an area with different age classes. Stand-level assessment models the forest system as a single stand of trees or an increasing stand of trees. Therefore stand-level carbon accounting can show the carbon dynamics of sequential events such as site preparation, plantation, thinning, tree growth, and final harvest. The starting point of analysis strongly affects the results of stand-level assessment (Cowie et al., 2021). Landscape-level assessment considers a fixed landscape of several stands that are being managed jointly to meet the demand for forest biomass in a continuous manner, or a region in which land can move in and out of forest production. With a single stand scale of accounting, increase in the harvest of trees for bioenergy will result in a carbon debt and then a dividend. This is because at a stand level, trees can take many years to grow back after harvest; there is a time interval between carbon release and re-absorption of that carbon from the atmosphere, which can temporarily increase GHG emissions in the atmosphere. As a result forest bioenergy may not necessarily be carbon neutral or it may only be carbon neutral over longer time frames (Cherubini et al., 2011; McKechnie et al., 2011). The magnitude of this debt may depend on the counterfactual level of carbon stock that is considered to prevail in the absence of the demand for bioenergy. This could be one where the forest is being cut for timber only, or it could be a forest that is never harvested; the carbon debt created by cutting a natural forest is larger than with cutting a plantation forest.

One alternative to the stand-level view of forest management is the landscape view of the forest that recognizes a forest that is being managed in a way to generate biomass continuously to feed an industrial operation. In this case if the demand for bioenergy leads to more intensive harvesting then the carbon stock in the landscape could decrease, but the carbon debt created would be spread out across the landscape and be recovered over a shorter time frame (as in Dwivedi et al., 2016). In this case the landscape is managed in an integrated way like a plantation that is part of a bioenergy supply chain and the whole fuel shed is considered together when examining the carbon balance (Jonker et al., 2014).

A third approach is one that allows the size of the landscape to change dynamically in the scenario with forest bioenergy use compared to the baseline with no harvesting of the forest for bioenergy. The dynamic landscape view incorporates market-driven effects that arise when the demand for bioenergy affects the returns to land and creates incentives for LUC as well as for changing forest management practices. In this view, biogenic carbon accounting involves accounting not only for changes in carbon stock in standing biomass but also for changes in carbon stock due to changes in land use as land moves in and out of forestry. It also accounts for changes in carbon stock as forest biomass is diverted from wood products, which provide long-term storage for carbon, to bioenergy.

The differences between stand-level and landscape-level approaches have been widely discussed in literature (Cintas et al., 2016; Cowie et al., 2021; Peñaloza et al., 2019). Jonker et al. demonstrated how stand-level and landscape-level approaches yield different carbon debts. Using the stand-level approach, the study estimated the carbon debt based on the forest carbon stock changes of stands at the same age. Using the landscape-level approach, they estimated the carbon debt as the average carbon debt of unevenaged trees (in this study, 0–25 years for low management intensity and 0–20 years for high management intensity) (Jonker et al., 2014). Stand-level and landscape-level approaches answer different questions (Cowie et al., 2021). Stand-level assessment provides in-depth understandings of growth patterns and interactions between different carbon pools in the forests (Cowie et al., 2021). Landscape-level assessment better represents the dynamics of the forest system managed at a landscape level in which fluctuations observed at the stand level are evened out (Cowie et al., 2013). It can consider changes in forest management in response to retrospective or anticipated bioenergy demand, as well as natural disturbance such as fire. Previous studies suggest stand-level assessment may be useful when one specific forest stand can be traced for a forest product (Peñaloza et al., 2019) or when the purpose is to inform forest management; while landscape-level assessment is appropriate for assessing the large-scale impacts of bioenergy policy (Cowie et al., 2021).

Forest Bioenergy

In the case of forest bioenergy it is important to integrate both LCA of supply chain emissions during the production of the feedstock as well as biogenic carbon due to changes in forest carbon stocks (which could be biogenic carbon emissions or sinks, depending on the net forest carbon stock changes) to provide a complete assessment. This accounting is needed because life-cycle related carbon emissions from bioenergy production can be expected to be accompanied by changes in forest management practices that will affect biogenic carbon stocks, for example by affecting the age at which trees are harvested, the intensity with which forests are managed, the species that are planted, the collection of residues, and a change in land use between forestland and cropland. Life-cycle accounting of the CI of using pulpwood for bioenergy needs to consider the forgone sequestration of biogenic carbon in forest products due to the diversion of biomass from forest products to bioenergy. Collection of logging residues for bioenergy will involve emissions in the collection and transportation of biomass but lead to avoided biogenic carbon emissions from gradual decay of forest biomass. This practice will lead to biogenic emissions during combustion of the residues but can displace fossil emissions from generating an equivalent amount of energy. Since changes in land use and in the production of forest products in bioenergy scenarios are likely to be induced by market price changes caused by the increased demand for bioenergy, including their carbon implications requires a dynamic landscape-based accounting approach.

In some methods of accounting, it is necessary to cumulate the changes in biogenic carbon over a specified time horizon and add them to the life-cycle emissions over the same horizon. This approach requires specifying a time frame over which to cumulate the negative and positive changes in biogenic carbon over time to estimate the total carbon impact of increased demand for bioenergy (McKechnie et al., 2011). Typically, for a forestland, carbon stocks could decrease in the near term after harvest and then grow in the future with forest regrowth. The estimated overall impact of bioenergy demand will depend on the time horizon used to account for the total effect. A short time horizon for cumulating changes will ignore longer term impacts and could under- or over-estimate the carbon impacts of forest bioenergy. Studies differ in the time horizon used for cumulating the impact of a demand for forest bioenergy on forest carbon stocks. Jonker et al. (2014) use a 75-year horizon as the time frame while McKechnie et al. (2011) use a 100-year horizon; these studies provide no scientific criteria to choose the justification for the choice of time horizon. Thus, using different time frames may be appropriate to explore the sensitivity of results. The magnitude of the carbon savings with using forest biomass for bioenergy relative to the baseline increases as the length of the time horizon increases.

Sometimes woody biomass is assumed to be carbon neutral if it is harvested from a forest with stable or increasing carbon stock, which is questionable if the life-cycle GHG emissions and all biogenic carbon flows are not considered and compared with a realistic counterfactual scenario (Cowie et al., 2021). This carbon neutral assumption is often used as a basis for ignoring and not reporting biogenic carbon. However, whether and how long the biogenic carbon can be fully refilled by growth or re-growth of forest depends on many factors such as the temporal and geospatial scales of the carbon analysis and the impacts of biofuel production on forest management, as discussed previously. Optimistic assumptions (Giuntoli et al., 2020) of forest management practices in counterfactual scenarios assumed in the literature may overstate the carbon savings associated with forest biomass. Stand-level analysis is more often used in LCAs for forest bioenergy, given its capacity to investigate sequential forest management activities, growth patterns, and different carbon pools. Landscape-level analysis allows for exploring forest dynamics, especially the changes in forest management and carbon stocks in response to biofuel demand (Cowie et al., 2021). The landscape-level analysis may be more challenging given the additional data needs of forest carbon stocks. Historical data available from the U.S. Forest Service could be used as a reference to understand the potential impacts of using woody biomass for biofuel production.

In sum, different biogenic carbon accounting methods exist but there is no method widely agreed upon (Brandão et al., 2013). This committee does not endorse any biogenic accounting method discussed in this section. Biogenic carbon has been included in some studies with a simultaneous consideration of carbon uptake and release, while some other studies exclude biogenic carbon.

For long-rotation woody biomass, stand-level and landscape-level approaches have been used; both methods consider forest carbon stock changes but differ in the temporal and spatial scale of accounting for these changes.

Conclusion 6-2: Different biogenic carbon accounting methods exist and the choice of method affects the carbon intensity of fuels.

Recommendation 6-2: All biogenic carbon emissions and carbon sequestration generated during the life cycle of a low-carbon fuel should be accounted for in LCA estimates.

Recommendation 6-3: Research should be conducted to improve the methods for accounting and reporting biogenic carbon emissions.

Soil Carbon

LUC, land management, and land management change (e.g., reducing tillage frequency, applying manure as a soil amendment) can alter soil carbon (NASEM, 2019). These changes and corresponding GHG emissions (or carbon sequestration) can be accounted for in LCA. For example, if a biofuel is to be made from an energy grass feedstock that is planted on land that was previously planted in corn, the changes in soil carbon per unit land area could be multiplied by the amount of land previously in corn that is now planted in energy grasses, but there would also be changes in land use elsewhere due to market-mediated effects. More broadly, the results of LUC models that predict the amounts, types, and locations of land that would be used for feedstock production can be used in conjunction with soil organic carbon (SOC) modeling results to estimate soil carbon stock changes that would accompany widespread LUC.

There are some efforts (Ledo et al., 2019) to gather and distribute the data for soil carbon changes (Xu et al., 2019). It is common to estimate soil carbon changes. One way of doing this is to use the Intergovernmental Panel on Climate Change (IPCC) emission factors. The simplest approach the IPCC offers is use of Tier 1 emission factors, which allows for SOC change estimation even in the absence of site-specific data. In the 2019 Refinement to the 2006 IPCC Guidelines for Natural Greenhouse Gas Inventories,[5] IPCC puts forward reference SOC stocks for six different soil types in 10 climate regions. These values are accompanied by an uncertainty value that in some cases is the 95 percent confidence interval but may also be 90 percent of the mean value when limited data are available. If more data are available (e.g., country-

---

[5] See http://www.ipcc-nggip.iges.or.jp/.

specific data on tillage practices or reference soil carbon stock) IPCC defines two additional levels of SOC calculations, Tier 2 and Tier 3.

Another approach is to use a soil carbon model such as DAYCENT (Colorado State University, n.d.) or CENTURY (Parton, 1996) although many other models exist (e.g., MEMS [Zhang et al., 2021]) or to build a model for specific use in a study. There are many parameters involved in SOC modeling. Documenting them transparently can improve the LCA community's ability to interpret the defensibility of the SOC modeling approach. For example, the choice of soil depth can influence results. Historically (see EPA, 2010), a soil depth of 30 cm was often used, but SOC changes further below the surface (e.g., 100 cm) can be notable, particularly for deep-rooted energy crops (Qin et al., 2014, see Figure 6-2).

Modeling results that choose a final year soon after a disruption to historical land use may report SOC changes that are very large and not reflective of long-term trends. On the other hand, if land use, cover, or management practices change frequently, the soil carbon model will not reach equilibrium. In that case, choosing the starting and final SOC values can be challenging. One notable challenge in using soil carbon models is determining the appropriate land use history. This can be particularly challenging for land that may be idled in some years and planted in crops in others. Some studies show that SOC modeling results can be insensitive to assumptions about land use history (Emery et al., 2017). However, this is not true in every instance, and transparency in the choice of land use history, and other SOC modeling parameters, is important. Some have argued that using SOC modeling approaches that assume too detailed of an approach to land use history could introduce a false sense of confidence in results and that a more general approach is more appropriate. Further study supported by SOC data is needed to support a harmonized approach to addressing land use history modeling, particularly for marginal lands or cropland-pasture that varies in how it is used.

Recently, the permanence of any soil carbon change has come into question for at least two main reasons. Even if land management practices do not change, climate change can "undo" SOC stock gains. Permanence issues have been raised in multiple literature reports (Dynarski et al., 2020; Leifeld and Keel, 2022). Accordingly, permanence needs to be raised as a methodological issue that arises in biofuel LCA. First, economic or political conditions may change, which may affect land use. For example, soil carbon stored due to conversion of row cropland to perennial species, such as with the Conservation Reserve Program of the USDA Farm Service Agency, may be lost when land is returned to production. Second, given the influence of climate on SOC (Hicks Pries et al., 2017), future SOC modeling efforts may need to address the changing climate.

Given the uncertainty arising in SOC modeling, reporting of uncertainty assists the community in interpreting SOC modeling results that are used in LCA.

Overall, more data are needed to inform both high-level approaches such as those in the IPCC Tier 1 methodology, to inform the parameters used in SOC modeling, and to provide ground truth to validate SOC modeling results. It should be noted that IPCC Tier 1 methodology may have limited value for assessing SOC effects of dedicated perennial bioenergy crops, which could be treated as generic "managed grassland" or "managed forest," which would not account for the high productivity of these purposefully-selected crops. These data could be collected from emerging, remote-sensing based methods that may be more cost effective and rapid than conventional sampling methods.

In sum, changes in SOC can be a significant contributor to the life-cycle GHG emissions of a biofuel. SOC modeling results are sensitive to parameters including soil depth and land use history. The permanence of modeled SOC changes is uncertain given climate change and other factors that could include market drivers or policy changes that would influence land use. Researchers have varying views on the level of detail that is appropriate in modeling SOC, in particular for land use history. Despite efforts to collect additional data, many data and knowledge gaps remain regarding SOC changes for land with varying land use histories used to grow different biofuel feedstocks. These data gaps impede calibration and validation of soil carbon models.

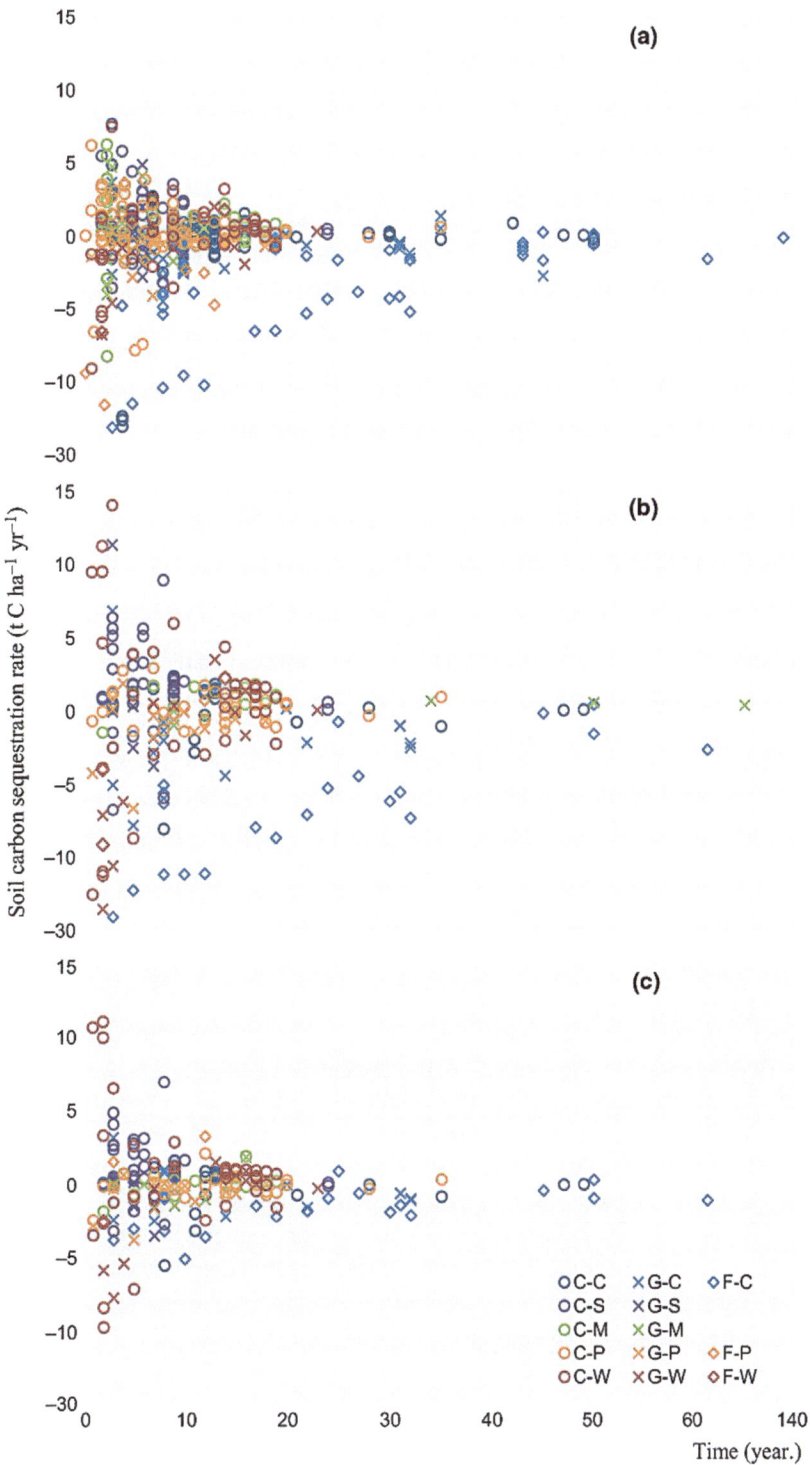

FIGURE 6-2 Soil carbon sequestration rate changes with time. The rates were estimated for three soil depths: (a) 0–30 cm, (b) 0–100 cm, and (c) 30–100 cm, corresponding to Fig. 7 (Qin et al., 2014, p. 75). Land use types changed from cropland (C), grassland (G), and forest (F) to corn (C), switchgrass (S), Miscanthus (M), poplar (P), and willow (W). These estimates are for the specific conditions and assumptions described in the paper. SOURCE: Qin et al. (2014, p. 75). Reprinted with permission © 2014 John Wiley & Sons Ltd.

Conclusion 6-3: Given the importance of soil organic carbon changes in influencing life-cycle GHG emissions of biofuels, investments are needed to enhance data availability and modeling capability to estimate soil organic carbon change. Capabilities to evaluate permanence of soil organic carbon changes should also be developed.

Recommendation 6-4: Research should be conducted to collect existing soil organic carbon data from public and private partners in an open source database, standardize methods of data reporting, and identify highest priority areas for soil organic carbon monitoring. These efforts could align with the recommendations made in the 2019 National Academies report on negative emissions technologies to study soil carbon dynamics at depth, to develop a national on-farm monitoring system, to develop a model-data platform for soil organic carbon modeling, and to develop an agricultural systems field experiment network. These efforts should also be extended internationally.

Recommendation 6-5: Research should be conducted to explore remote-sensing and in situ sensor-based methods of measuring soil carbon that can generate more data quickly.

INDICATORS, OTHER CLIMATE FORCERS, AND TIMING OF EMISSIONS

Fuels lead to the emission and uptake of CO_2 (biofuel) and other GHGs in every life-cycle stage. These emissions and uptakes must then be aggregated into a common unit so that the climate change impact of different fuels can be compared. To aggregate different GHG emissions into a common unit, namely CO_2 equivalent (CO_2e), metrics expressing the relative contribution of GHGs to climate change are needed. Since its publication in the First Assessment Report of the IPCC in 1990 (IPCC, 1990) and its adoption for the application of the Kyoto Protocol, global warming potential (GWP)[6] calculated for a 100-year time horizon (GWP_{100}) established itself as the most commonly used metric in LCA (Levasseur et al., 2016a).

GWP is the cumulative radiative forcing caused by a unit-mass of GHG released to the atmosphere, integrated over a given time horizon, relative to that of a unit-mass of CO_2. However, as stated by the authors of the First Assessment Report: "It must be stressed that there is no universally accepted methodology for combining all the relevant factors into a single global warming potential for greenhouse gas emissions. A simple approach (GWP) has been adopted here to illustrate the difficulties inherent in the concept" (IPCC, 1990). Despite this warning, the use of GWP for a 100-year time horizon in LCA has rarely been challenged. In its Fifth Assessment Report, the IPCC reminds that "the most appropriate metric will depend on which aspects of climate change are most important to a particular application, and different climate policy goals may lead to different conclusions about what is the most suitable metric with which to implement that policy" (Myhre et al., 2013).

Since the IPCC First Assessment Report, the science regarding climate metrics has evolved, and many other metrics have been proposed to compare the climate impact of different GHGs. For instance, in its Fifth Assessment Report (Myhre et al., 2013), IPCC proposes and discusses the use of global temperature change potential (GTP) for 20-, 50-, and 100-year time horizons in addition to GWP for 20- and 100-year time horizons. GTP is defined as the instantaneous global temperature change caused by a unit-mass of GHG released to the atmosphere a given number of years following the emission corresponding to the time horizon selected, relative to that of a unit-mass of CO_2 (Shine et al., 2005). Additional metrics, such as global precipitation potential (Shine et al., 2015), are presented in the very recent IPCC Sixth Assessment Report (Forster et al., 2021).

Metrics vary according to the element of the cause–effect chain that it quantifies (e.g., radiative forcing for GWP, temperature change for GTP or precipitation change for global precipitation potential), the time horizon selected, and its cumulative or instantaneous nature. An instantaneous metric quantifies the change at a particular time after the emission, expressing the effect of the emission persisting after a

---

[6] A measure of how much energy the emissions of 1 ton of a gas will absorb over a given period, relative to the emissions of 1 ton of CO_2.

given time horizon. A cumulative metric integrates the impact over the selected time horizon following the emission, that is, it expresses the total effect from the time of the emission up to the given time horizon. Any metric can be presented as an absolute value, or as a relative value, dividing the absolute value by the equivalent value for CO_2 (Forster et al., 2021). Instantaneous metrics could be deemed more appropriate if the goal is to not exceed a fixed target at a specific time, while cumulative metrics could better suit the need to reduce the overall damage when the impact depends on how long the change occurs for (Forster et al., 2021). The selection of a time horizon can also be driven by the types of impact to be captured by the metric. On the one hand, metrics addressing shorter-term climate change are more appropriate to capture impacts affected by the rate of change such as the adaptation of species to changing habitat, heat stress, or extreme weather events. On the other hand, metrics addressing long-term climate change are more suitable for impacts associated to sea level rise or coral bleaching, among others (Levasseur et al., 2016a).

In its Sixth Assessment Report, IPCC also discusses new emission metric approaches, such as GWP*[7] (Cain et al., 2019; Smith et al., 2021) and combined global temperature change potential (Collins et al., 2020), that have been developed to better account for the different physical behaviors of short- and long-lived GHGs. They found that these new approaches can improve the evaluation of the contribution to global warming of different GHGs within a cumulative emission framework, as pulse-based emission metrics (e.g., GWP, GTP) do not represent accurately the effect of sustained short-lived GHG emissions (Forster et al., 2021). The combined global temperature change potential values are published in the IPCC Sixth Assessment Report and are to be applied to a change in emission rate rather than a change in emission amount, as it has been designed for cumulative emission frameworks such as national-level inventories, which is different from the LCA framework.

From 2014 to 2016, the Life Cycle Initiative, hosted by the United Nations Environment Programme (UNEP), led an international initiative aiming at developing consensus-based metrics for use in LCA for climate change. A task force composed of researchers from both the climate metric and LCA fields performed a critical analysis of the most recent scientific findings and developed recommendations (Cherubini et al., 2016; Levasseur et al., 2016b). This was followed by a consensus-finding workshop (Jolliet et al., 2017; Levasseur et al., 2016a). Indicators were evaluated according to their environmental relevance, that is their capacity to cover the broad spectrum of relevant long- and short-term impacts, as well as reliability. The recommendation from this international consensus-finding workshop is to use two different indicators in parallel: GWP for 100 years (GWP_{100}) for shorter-term impacts, and GTP for 100 years for long-term impacts, including climate–carbon feedbacks for both.

For very short-term impacts, another recommendation from the international consensus-finding workshop is to perform a sensitivity analysis using GWP for 20 years, including emissions from near-term climate forcers. Near-term climate forcers (CO_2, NO_x, SO_x, volatile organic compounds, black carbon, and organic carbon) affect the climate through different physical and chemical mechanisms (e.g., changes in methane lifetime or cloud cover). These pollutants have lifetimes in the atmosphere of days to weeks, which is too short to allow them becoming well mixed in the atmosphere, leading to strong spatial and temporal heterogeneities (Myhre et al., 2013). Therefore, their global warming impact highly depends on the region of emissions and are very short-term, which makes their aggregation to a CO_2e unit challenging (Levasseur et al., 2016a). This regional variability increases uncertainties associated with emission metrics for near-term climate forcers. However, their contribution to the rate of temperature increase in the short-term has been shown to be important (Allen et al., 2020; Fuglestvedt et al., 2008).

Another issue related to the evaluation of the contribution of CO_2 and other GHGs to global warming is the consideration of the timing of emissions and uptakes. Some emissions and uptakes within the life cycle of biofuels might occur several years or decades before or after the fuel is produced and used. The two most discussed cases in the literature regard carbon uptake by growing biomass, which can occur over several years after harvesting for forest or long-rotation crops (Lamers and Junginger, 2013; McKechnie et

---

[7] GWP* is an alternative application of GWPs where the CO_2-equivalence of short-lived climate pollutant emissions is predominantly determined by changes in their emission rate. See https://iopscience.iop.org/article/10.1088/1748-9326/ab6d7e.

al., 2011; Zanchi et al., 2012), and upfront LUC emissions for short-rotation crops, which are usually amortized over several years of biomass production (Fargione et al., 2008; Levasseur et al., 2010). Upfront LUC emissions, incurred immediately but occurring both in the short- and long-term, and delayed carbon uptakes both lead to a so-called carbon debt.

Using common accounting approaches, a given CO_2 emission is considered compensated by the uptake of an equivalent amount of CO_2, no matter when they occur. However, if the emission occurs several years before the uptake, the CO_2 released will contribute to global warming on the short-term, and the time required to reach a net-zero warming effect can be several decades or centuries (Levasseur et al., 2012). Moreover, biogenic carbon emissions from biofuel combustion are usually disregarded in LCA, as they are systematically considered compensated by equivalent uptakes, which could lead to potential accounting errors (Searchinger et al., 2009). Given the urgency of climate change and the ambitious binding targets for GHG emission reduction set by climate policies such as the Paris Agreement (i.e., net zero CO_2 emissions in 2050), the increase in GHG emissions in the short-term could be an issue even if they are compensated by equivalent uptakes later on.

Different approaches have been proposed to address the timing issue of GHGs in LCA. Some of them are applicable to very specific cases. For instance, the PAS 2050 carbon footprint standard (British Standards Institute, 2008) and the EU International Life Cycle Data System (ILCD) Handbook (European Commission, 2010) proposed methods to account for the climate benefits associated with temporary carbon storage in long-lived products. O'Hare et al. (2009) and Kendall et al. (2009) proposed two different approaches to amortize upfront LUC emissions for short-rotation crops, while considering their contribution on global warming. Cherubini et al. (2011) proposed a new metric to assess biofuel combustion emissions while accounting for the delay of carbon uptake, depending on the biomass rotation period. Levasseur et al. (2010) proposed the broader dynamic LCA approach in order to account for the timing of every GHG emission and uptake in LCA. The method develops a temporally differentiated inventory, then calculates associated radiative forcing over time based on the models used to calculate GWP values. A similar approach has been proposed and applied to the assessment of bioenergy systems, using the temperature change as an indicator instead of radiative forcing (Ericsson et al., 2013).

The use of any of these approaches implies the selection of a time horizon beyond which global warming impacts are disregarded. Therefore, an emission occurring the first year will lead to a higher impact than an emission occurring 25 years later, as this second emission will contribute to global warming for a period of 75 years instead of 100 years, given that a 100-year time horizon is selected. For methods proposing new metrics for specific cases (e.g., Cherubini et al., 2011; Kendall et al., 2009; O'Hare et al., 2009), the time horizon is fixed by the developers. In contrast, methods based on the dynamic LCA approach provide time-dependent impacts (radiative forcing or temperature change), and it is up to the decision-maker to select one or more time horizons for the comparison of fuels. Dynamic LCA results are more complex to understand for non-experts compared to approaches proposing specific metrics, but they allow decision-makers to compare the global warming contribution of different fuels on the short-, mid-, and long-terms as different time horizon can be selected for the analysis.

Conclusion 6-4: Several metrics in addition to global warming potential for 100 years are now available with differing emphases such as short-term, long-term, or cumulative impacts.

Different metrics capture different climate change impacts, depending on their characteristics (e.g., cumulative or instantaneous, time horizon). It is recognized that the most appropriate metric depends on which aspects of climate change are most important to a particular application. Some biofuel pathways lead to a so-called carbon debt, increasing global warming impacts in the short-term. Given the urgency of climate change and the ambitious binding targets for GHG emission reduction set by climate policies such as the Paris Agreement (i.e., net zero CO_2 emissions in 2050), the increase in GHG emissions in the short-term could be an issue even if they are compensated by equivalent uptakes later on. The most recent research, including the recently published IPCC Sixth Assessment Report, present different metrics to aggregate GHG emissions into a single unit. Different metrics can be used depending on the context of application (e.g., national-level inventories, LCA), and the type of climate change impacts to address (e.g., extreme weather events or species adaptation related to the short-term temperature peak, sea-level rise associated with the longer-term stabilization temperature). Some studies reviewed different GHG metrics, such as Brandão et al. (2019). There is a clear consensus in the literature that the sole use of GWP_{100} does not capture the full range of climate change impacts.

Low-carbon fuels could also affect the climate through mechanisms other than GHG and near-term climate forcer emissions. The production of biomass often leads to land cover changes (e.g., transformation of natural lands to crops, forest harvesting), which could then affect the albedo, i.e., the proportion of solar radiation reflected by the surface. Researchers have estimated the potential warming or cooling effect of albedo modifications due to different land cover changes for bioenergy production (Cai et al., 2016; Cherubini et al., 2012; Darvin et al., 2014; Holtsmark, 2015). Some have shown that, in some cases, the order of magnitude of the climate effect due to albedo modifications can be as high as that of associated carbon stock changes (Betts, 2000; Jones et al., 2013). This is the case for harvesting forests at high latitudes because of the reflectivity of snow cover in the winter (Bernier et al., 2011). Although potentially very important, climate impacts from albedo changes are difficult to quantify in LCA because they are site-specific and rely on solar irradiance measurements (Sieber et al., 2019). Some approaches have been proposed by researchers for their integration in LCA, but they have not yet been implemented (Muñoz et al., 2010; Sieber et al., 2020).

Biofuels could affect the climate through mechanisms such as modifications in surface albedo due to land cover changes. Although potentially very important, climate impacts from changes in albedo, evapotranspiration, or other biogeophysical changes are difficult to quantify in LCA because they are site-specific and rely on solar irradiance measurements. Researchers have proposed different approaches to account for the timing of GHG emissions and uptakes. However, all these approaches rely on the subjective choice of a time horizon beyond which impacts are disregarded. There is currently no consensus in the literature regarding the valuation of delayed emissions in LCA.

> **Recommendation 6-6:** Use of more than one climate change metric should be considered in the analysis of low-carbon fuel policies.
>
> **Recommendation 6-7:** Further research should be conducted to better understand the suitability of different GHG metrics for LCA.
>
> **Recommendation 6-8:** Further research should be conducted to develop a framework to include albedo effects from land cover change, and near-term climate forcers, in LCA of low-carbon fuels.
>
> **Recommendation 6-9:** Further research should be conducted to better understand the climate implications of increased GHG emissions on the short-term (carbon debt) to support the selection of an appropriate approach to account for the timing of GHG emissions and uptakes in LCA.

VEHICLE–FUEL COMBINATIONS AND EFFICIENCIES

Life-cycle GHG emissions of transportation fuels can be compared on a per-unit-energy basis (e.g., emissions could be measured per MJ of a fuel's energy), but such a comparison can be incomplete or misleading without also considering how much energy is needed to propel a vehicle with each type of fuel as well as how much energy is required, and how much emissions are created in the production and maintenance of each type of vehicle. Efficiency and production emissions can vary widely both within and across vehicle fuel type technologies, making fair comparisons with single point estimates challenging. This section discusses issues specific to the vehicles that convert transportation fuels into transportation services.

To compare emissions from transportation fuels, rather than using an energy-based functional unit, a more meaningful functional unit might be based on the transportation services delivered. Common functional units for passenger transportation in the United States are vehicle-mile or passenger-mile, and a common functional unit for transportation of goods in the United States is ton-mile. Using a functional unit based on transportation services provided requires knowing or estimating the efficiency of the vehicle and, in some cases, the number of passengers or weight of cargo transported. The service-based functional unit could be reported in addition to an energy-based functional unit.

The life-cycle implications of transportation fuels depend on the vehicle efficiency as well as emissions associated with production, maintenance, and disposal of vehicles.

Conclusion 6-5: To make a meaningful comparison of the LCA of transportation fuels, the vehicles that use those fuels should be considered.

Recommendation 6-10: LCA of transportation fuels may include analysis using functional units based on the transportation service provided, such as passenger-mile or ton-mile, or otherwise be based on comparison of comparable transportation services. This may be reported in addition to an energy-based functional unit. LCAs should clearly describe their assumptions for the energy- and service-based functional units, such as through vehicle efficiency, market share, or other factors.

The efficiency of the vehicle that will use the fuel is typically unknown to the fuel producer and heterogeneous across fuel customers. The CA-LCFS policy handles this by introducing energy efficiency ratios. This section will first discuss issues of LCA of transportation fuels and then discuss how energy efficiency ratios are used in CA-LCFS policy.

Efficiency

Vehicle efficiency can vary widely within and across fuel types, as well as across driving conditions; individual vehicles also have variations based on the level of maintenance (e.g., tire pressure). Models that use a single "representative" vehicle to represent each fuel type can mask this heterogeneity. This section summarizes and quantifies the efficiency of vehicles available today that use each fuel type and summarizes how much the efficiency of each fuel can change depending on conditions that vary regionally and across drivers, such as weather, terrain, driving style, regional energy sources, and other factors.

Variation in Vehicle Design

Figure 6-3 summarizes the energy consumption rates of vehicles by fuel type. Energy consumption rates (kWh/mile) are used here rather than efficiency (mile/kWh) because emissions implications are proportional to consumption rates (and inversely proportional to efficiency) (Larrick and Soll, 2008). Figure 6-3 shows a range of new vehicles available for sale today, with efficiency values taken from fueleconomy.gov, including (1) the least efficient mass-market vehicle for each fuel type (excluding low-volume luxury and sport models, such as Rolls-Royce and Lamborghini designs); (2) the most efficient mass-market vehicle for each fuel type; and (3) one of the most popular vehicles for each fuel type sold in the United States. In addition, for passenger cars the figures include the efficiency value used in GREET. There are no mass-market hydrogen vehicles today, so the three available (low-volume) models are used to establish the range. All energy values are converted to kWh, and consumption rates are reported in kWh per 100 miles. Plug-in hybrid electric vehicles (PHEVs), which use a combination of grid electricity and gasoline, are not shown, and E85 vehicles represent the efficiency of flex fuel vehicles when operating on E85 (with the exception of the GREET data point, which represents a dedicated E85 vehicle). For comparison, the energy efficiency ratios (inverted to map to relative consumption rates) used in CA-LCFS are also plotted on a secondary axis.

Conclusion 6-6: If an LCA uses a single point estimate for efficiency of each vehicle type, its conclusions may vary substantially depending on which vehicle design (make-model-trim) is used to represent each fuel type.

Recommendation 6-11: When comparing life-cycle emissions of different transportation fuels, LCA studies that assess or inform policy should consider the range of vehicle efficiencies within each fuel type to ensure that the comparisons are made on comparable transportation services, such as passenger capacity, payload capacity, and performance.

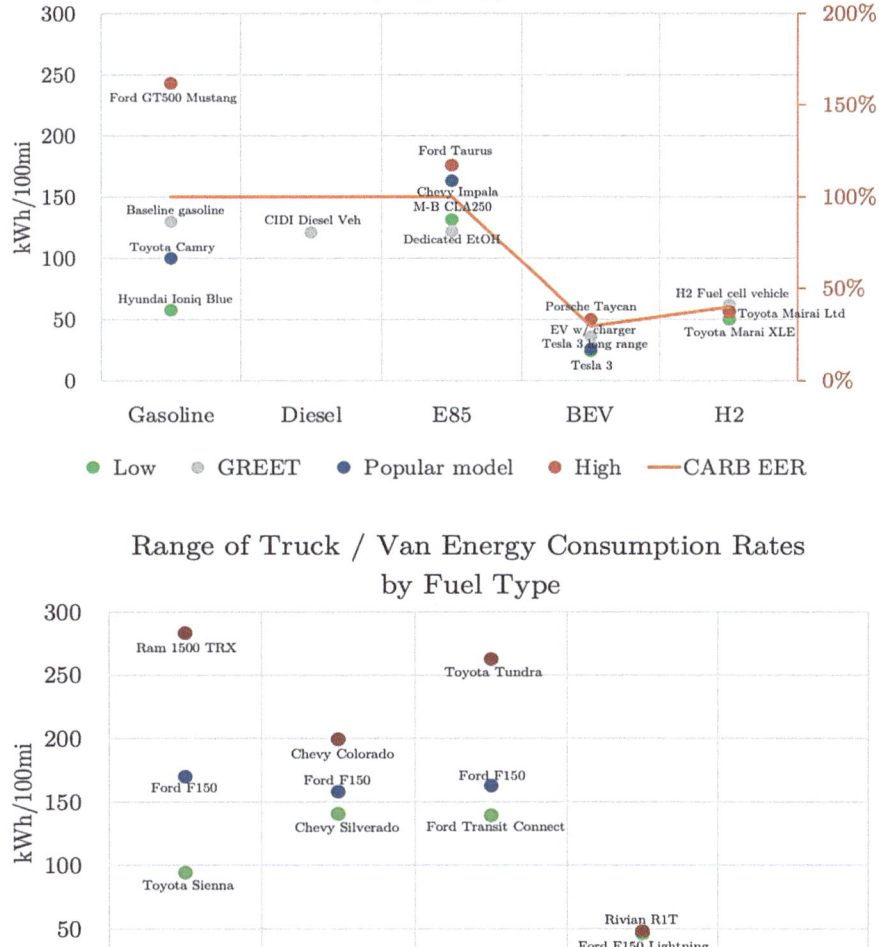

FIGURE 6-3 Range of vehicle energy consumption rates (inverse of efficiency) by fuel type. NOTE: Low = vehicle with lowest energy consumption rate among all currently available new mass-market U.S. vehicles; High = vehicle with highest energy consumption rate among all currently available new mass-market U.S. vehicles; Popular model = consumption rate of one of the most popular U.S. make-model options currently available; GREET = efficiency value used in the GREET model; CARB EER = inverse of energy efficiency ratio assumed in CA-LCFS.

Variation in Use Conditions

The energy consumption rates presented previously are based on driving cycles performed in a laboratory, where each vehicle is placed on a dynamometer (like a treadmill for a vehicle), driven to follow a prescribed sequence of vehicle speeds, and monitored to assess fuel consumption and emissions. The advantage of such a test is standardization—all vehicles can be compared in the same conditions, ensuring an apples-to-apples comparison. However, such a test masks important variation in real-world effects that can differ across fuel types.

Real-world driving conditions, including speed, acceleration, frequency and duration of stops, driving distance, precipitation, temperature, humidity, and other factors affect the efficiency of all vehicles, but the effect can be larger for some fuel types than others (Karabasoglu et al., 2013).

Climate and weather conditions also affect vehicle efficiency for different fuel types differently. Cold weather generally makes all vehicles less efficient, but it has a larger effect on electric vehicles (EVs) than gasoline vehicles. Gasoline vehicles are less efficient when cold for a variety of reasons, including increased viscosity of engine lubricants, but gasoline vehicles can use waste heat from the engine to heat the passenger cabin. In contrast, EVs are less efficient when cold both because batteries are less efficient when it is cold and because energy from the battery must be used to heat the cabin. Figure 6-4 shows energy consumption rates for an internal combustion engine vehicle (ICEV), hybrid electric vehicle (HEV), PHEV and battery electric vehicle (BEV) normalized to fair weather efficiency. Electrified vehicle efficiency is more sensitive to ambient temperature than gasoline vehicle efficiency. Note that Figure 6-4 is shown relative to fair-weather efficiency. In absolute terms, electrified vehicles are typically more efficient.

Variation of fuel consumption rates can be larger within fuel type than across fuel type, especially for gasoline, for which fuel consumption rates vary by more than a factor of 4. Some of this variation is due to vehicle class (e.g.: sports car vs. compact car), but there is substantial variation within vehicle class.

Figure 6-5 shows the implications of these differences in efficiency for energy consumption in different parts of the United States, based on temperature data across a typical year. The same vehicle with the same driving patterns consumes about 20 percent more energy in regions with extreme heat and cold than in regions with consistently mild weather.

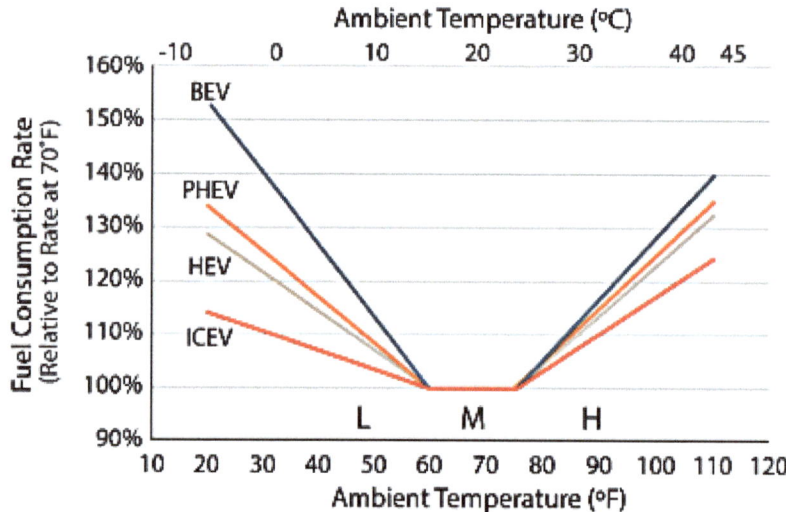

FIGURE 6.4 Relationship between fuel consumption rate and ambient temperature. 100 percent is indexed to the baseline fuel consumption rate for each powertrain. NOTES: BEV = battery electric vehicle; PHEV = plug-in hybrid electric vehicle; HEV = hybrid electric vehicle; ICEV = internal combustion engine vehicle. L= low; M = medium; H = high. SOURCE: Wu et al. (2019, p. 10563). Reprinted with permission from American Chemical Society, see https://pubs.acs.org/doi/10.1021/acs.est.9b00648.

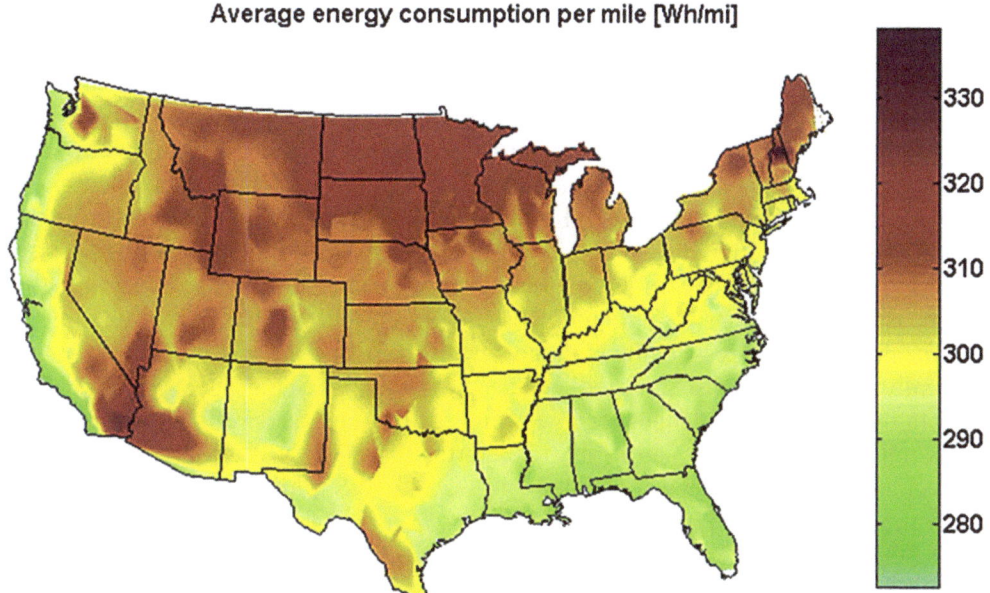

FIGURE 6-5 Estimated average energy consumption (watt hour) per mile for a Nissan Leaf in different regions of the United States based on regional ambient temperature over the year. SOURCE: Yuksel et al. (2015). Reprinted with permission from Yuksel, T. and Michalek, J.J. (2015). Effects of regional temperature on electric vehicle efficiency, range, and emissions in the United States. Environmental Science & Technology, 49(6), pp.3974-3980. Copyright 2015 American Chemical Society.

Simultaneous Variation in Vehicle Design and Multiple Use Conditions

When considering all of these factors together, using regional weather information, regional marginal emission factors, and regional differences in driving style (assuming city driving in urban counties, highway driving in rural counties, and combined driving in suburban counties), the comparison of life-cycle GHG emissions from gasoline and electric vehicles depends not only on which specific vehicle models are being compared but also on regional differences in climate, grid mix, and driving patterns (Yuksel et al., 2016).

While specific estimates of life-cycle GHG emissions for each vehicle type change over time and may vary across different LCA studies, depending on what assumptions are used and which vehicle models are used to represent each fuel type, the main point here is that there is so much variation across vehicle designs within each fuel type and across use conditions that using a single efficiency point estimate to represent each fuel type can substantially affect findings of an LCA, depending which vehicle design and use conditions are assumed. Use conditions, including driving style, climate, and energy source, all of which vary regionally, affect the emissions of vehicles for different fuel types differently and can affect which fuel type is estimated to have higher or lower life-cycle GHG emissions.

> **Conclusion 6-7:** If an LCA uses a single point estimate for efficiency of each vehicle type, its conclusions may vary substantially depending on which use conditions are assumed.

> **Recommendation 6-12:** When comparing life-cycle emissions of different transportation fuels, LCA studies should avoid relying on a single point estimate for efficiency of each vehicle fuel type and instead consider the range of vehicle efficiencies within each fuel type across vehicles and common or likely operating conditions.

Fuel Octane Rating

Fuel octane ratings are standardized measurements of the ability of a fuel to withstand compression without detonation. Higher octane fuels can withstand higher compressions without premature combustion. Operating engines at higher compression can improve power and efficiency, although in modern engines this is just one variable of many.[8,9] Fuel properties affect internal combustion engine performance, and engine design and controls affect fuel needs. The combined effects of both result in measurable differences in performance, fuel consumption, and tailpipe emissions. Premature fuel combustion and engine knock were observed in gasoline internal combustion engines more than a hundred years ago, and various solutions have been implemented. Early solutions focused more on fuel composition, while more recent ones have focused on engine control, especially with the advent of controlling spark advance mechanically, and later electronically. This section discusses fuel octane rating as one key fuel specification with implications for fuel LCA. In order to ensure adequate octane ratings of finished fuels refiners add aromatic hydrocarbons (benzene, toluene, and xylene combined often referred to as BETX, plus others), which are molecules with high octane ratings. Ethanol, when blended into the fuel, has a high octane rating and therefore can dilute and substitute aromatics in gasoline. There has been a long history of using additives to increase octane rating by changing fuel composition. All of these have had positive impacts on engine performance but usually were found to have significant detrimental effects on health and the environment, which led to progressive banning of these additives. Most notoriously, the first additive in large-scale use, tetraethyl lead introduced in 1924, has been found to cause significant health damage, especially in children. It was removed gradually and finally banned in 1996. Methyl tertiary-butyl ether was introduced in the 1990s and was removed gradually over the next two decades due to ground water contamination issues and replaced with ethanol.[10]

Ethanol blending in gasoline increased rapidly between 1997 (0.31 percent) and 2016 (9.57 percent), driven by a combination of methyl tertiary-butyl ether replacement, Volumetric Ethanol Excise Tax Credit (VEETC) blending tax credits, the requirements of the RFS1 in 2005 and RFS2 in 2007, and changing economics of gasoline blending.[11] By 2016, almost all U.S. gasoline was blended with 10 percent ethanol.[12] The increased octane of ethanol allowed refiners to reduce the aromatic content of gasoline, which fell from 25 percent to 19 percent over the same time frame (EPA Fuel Trends Report 2017).

Fuels with high octane ratings allow vehicle manufacturers to increase the compression ratio in an engine, which enables that engine to extract more mechanical energy from a given mass of air–fuel mixture due to its higher thermal efficiency.[13] This has attracted engine and emissions researchers to study and develop engines that utilize higher octane fuels (Costenoble and de Groot, 2020; DOE, 2017; Schifter et al., 2020; Storey et al., 2016; West et al., 2018; Yang et al., 2019). The reviewed studies show that optimized higher octane fuel engines may at least partially or more than fully compensate for ethanol's lower volumetric fuel economy (due to its lower heating value) and result in increased energy economy ratio, which is defined as the energy consumption in British thermal unit (joule) of the conventional E10 vehicle divided by that of the alternative fuel (Unnasch and Browning, 2000). For example, Oak Ridge National Laboratory research finds that high-octane fuel can provide "an improvement in vehicle fuel efficiency in vehicles designed and dedicated to use the increased octane" (Theiss et al., 2016). Consequences including LUC implications may be assessed by CLCA. The findings related to high-octane fuel in this section describe the efficiency of the optimized fuel engine system rather than the broader issues associated with the lifecycle of ethanol, which are described at length in the rest of this report.

---

[8] See https://www.eia.gov/energyexplained/gasoline/octane-in-depth.php.
[9] See https://www.astm.org/d2700-21.html.
[10] See https://archive.epa.gov/mtbe/web/html/faq.html.
[11] EPA fuel trends report, see https://nepis.epa.gov/Exe/ZyPDF.cgi?Dockey=P100T5J6.pdf.
[12] See EIA https://www.eia.gov/todayinenergy/detail.php?id=26092.
[13] The compression ratio is defined as the ratio between the volume of the cylinder with the piston in the bottom position and in the top position. The higher this ratio, the greater will be the power output from a given engine.

Conclusion 6-8: Specifically formulated high-octane fuels in combination with dedicated fuel engine technologies can provide efficiency improvements in fuel combustion that affect LCA results.

Recommendation 6-13: LCAs of high-octane fuels should consider the impact of fuel octane on vehicle efficiency, but for the purpose of broad policy assessment LCA should be based on the actual and anticipated vehicle fleet, and following common practice for fuel vehicle assessments include only combinations that reflect reality.

Vehicle Production Emissions

For each transportation fuel, there are emissions associated with producing, maintaining, and managing the end of life of the vehicle that converts the fuel into transportation services. Emissions from producing and maintaining gasoline and ethanol vehicles are similar, but emissions from producing BEVs tend to be larger because of the large battery packs needed to store enough energy to deliver a vehicle range comparable to gasoline vehicles (Elgowainy et al., 2018; Karabasoglu and Michalek, 2013; Ma et al., 2012; Nealer and Hendrickson, 2015; Yuksel et al., 2016). This does not imply, however, that emissions from BEVs are higher on a per mile or per ton-mile basis, as emissions productions are allocated over the service life of vehicles in LCA. Estimates of the emissions from producing BEVs vary widely, and the technology is still in flux, with lithium ion chemistries varying in the amount of cobalt, nickel, magnesium, aluminum, iron, and other materials used, among other factors, including the type and origin of these metals and lithium (Deng et al., 2020; DOE, 2020; NASEM, 2021; Schmuch et al., 2018; Transportation Research Board and National Research Council, 2015).

Conclusion 6-9: Ignoring vs. including vehicle production emissions in an LCA could affect its conclusion about which transportation fuels have the lowest carbon emission implications per unit of transportation services delivered.

Recommendation 6-14: For regulatory impact assessment, LCA of transportation fuels and transportation fuel policy should consider a range of estimates for possible changes in the emissions of vehicle production required to convert transportation fuels into transportation services, and the resulting changes in vehicle fleet composition.

Logistics

An additional issue, particularly relevant for LCA of electric trucks, is that some fuel types require powertrain designs that take up more space or weight, reducing the amount of cargo that can be transported. In particular, electric Class 8 trucks sized for long-distance range require large, heavy battery packs, reducing payload capacity, and requiring more truck trips to deliver the same quantity of goods (Figure 6-6; Sripad and Viswanathan, 2017). Use of a per-ton-mile functional unit can partially account for these differences when comparing transportation fuels, though logistical constraints can be complicated and nonlinear.

Different fuel types imply different logistical constraints and may impose changes on travel demand. In particular, BEVs substantially reduce payload for Class 8 trucks, and more truck travel is required to move the same amount of goods.

Conclusion 6-10: A per-vehicle-mile functional unit is, on its own, not fully informative for comparing transportation fuels for weight-constrained or space-constrained applications, such as Class 8 trucks.

Recommendation 6-15: LCA comparing transportation fuels for weight-constrained applications should present a per-ton-mile functional unit and/or explicitly model the logistical implications of payload effects by fuel type.

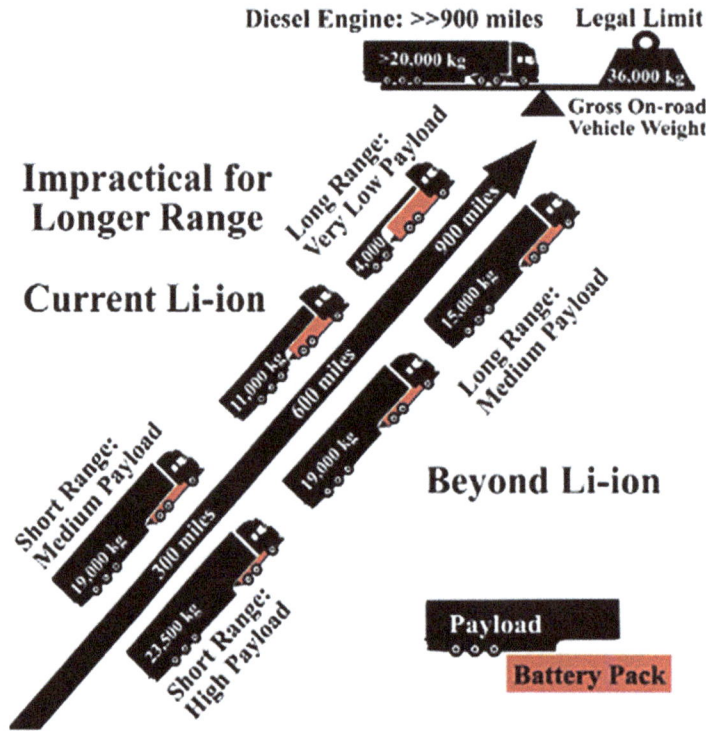

FIGURE 6-6 A comparison between current and beyond Li-ion batteries for electrifying semi trucks. Trucks present a depiction of the ration of payload to battery pack size of vehicles with particular range, noted on the arrow. SOURCE: Sripad and Viswanathan (2017, p. 1669). Reprinted with permission from Sripad, S. and V. Viswanathan. 2017. Performance Metrics Required of Next Generation Batteries to Make a Practical Electric Semi Truck. *ACS Energy Lett.* 2017, 2, p. 1669. Copyright © 2017 American Chemical Society.

REFERENCES

Allen, R. J., S. Turnock, P. Nabat, D. Neubauer, U. Lohmann, D. Olivié, N. Oshima, M. Michou, T. Wu, J. Zhang, T. Takemura, M. Schulz, K. Tsigaridis, S. E. Bauer, L. Emmons, L. Horowitz, V. Naik, T. v. Noije, T. Bergman, J.-F. Lamarque, P. Zanis, I. Tegen, D. M. Westervelt, P. Le Sager, P. Good, S. Shim, F. O'Connor, D. Akritidis, A. K. Georgoulias, M. Deushi, L. T. Sentman, J. G. John, S. Fujimori, and W. J. Collins. 2020. Climate and air quality impacts due to mitigation of non-methane near-term climate forcers. *Atmospheric Chemistry and Physics* 20:9641–9663.

ASTM (American Society for Testing and Materials). 2021. *Standard Test Method for Motor Octane Number of Spark-Ignition Engine Fuel.* https://www.astm.org/d2700-21.html (accessed March 10, 2022).

Berndes, G., S. Ahlgren, P. Börjesson, and A. Cowie. 2013. Bioenergy and land use change—state of the art. *WIREs Energy and Environment* 2:282–303.

Bernier, P. Y., R. L. Desjardins, Y. Karimi-Zindashty, D. Worth, A. Beaudoin, Y. Luo, and S. Wang. 2011. Boreal lichen woodlands: A possible negative feedback to climate change in eastern North America. *Agricultural and Forest Meteorology* 151(4):521–528.

Betts, R. A. 2000. Offset of the potential carbon sink from boreal forestation by decreases in surface albedo. *Nature* 408(6809):187–190.

Brandão, M., A. Levasseur, M. U. Kirschbaum, B. P. Weidema, A. L. Cowie, S. V. Jørgensen, M. Z. Hauschild, D. W. Pennington, and K. Chomkhamsri. 2013. Key issues and options in accounting for carbon sequestration and temporary storage in life cycle assessment and carbon footprinting. *The International Journal of Life Cycle Assessment* 18(1):230–240.

Brandão, M., M. U. F. Kirschbaum, A. L. Cowie, and S. V. Hjuler. 2019. Quantifying the climate change effects of bioenergy systems: Comparison of 15 impact assessment methods. *GCB Bioenergy* 11:727–743. https://doi.org/10.1111/gcbb.12593.

Bredeson, L., R. Quiceno-Gonzalez, X. Riera-Palou, and A. Harrison. 2010. Factors driving refinery CO_2 intensity, with allocation into products. *International Journal of Life Cycle Assessment* 15:817–826. https://doi.org/10.1007/s11367-010-0204-3.

Buchholz, T., S. Prisley, G. Marland, C. Canham, and N. Sampson. 2014. Uncertainty in projecting GHG emissions from bioenergy. *Nature Climate Change* 4:1045–1047. https://doi.org/10.1038/nclimate2418.

Cai, H., J. Wang, Y. Feng, M. Wang, Z. Qin, and J. B. Dunn. 2016. Consideration of Land Use Change-Induced Surface Albedo Effects in Life-Cycle Analysis of Biofuels. *Energy & Environmental Science* 9(9):2855–67. https://doi.org/10.1039/C6EE01728B.

Cain, M., J. Lynch, M. R. Allen, J. S. Fuglestvedt, D. J. Frame, and A. H. Macey. 2019. Improved calculation of warming-equivalent emissions for short-lived climate pollutants. *Climate and Atmospheric Science* 2:29.

CARB. 2018. *Carbon Capture and Sequestration Protocol under the Low Carbon Fuel Standard.* 8/13/2018. https://ww2.arb.ca.gov/sites/default/files/2020-03/CCS_Protocol_Under_LCFS_8-13-18_ada.pdf.

CARB. 2020. *Comment Log for Low Carbon Fuel Standard Public Workshop to Discuss Potential Regulation Revisions.* https://www.arb.ca.gov/lispub/comm2/bccommlog.php?listname=lcfs-wkshp-oct20-ws.

CARB. 2021. *LCFS Pathway Certified Carbon Intensities.* https://ww2.arb.ca.gov/resources/documents/lcfs-pathway-certified-carbon-intensities.

Cherubini, E., D. Franco, G. M. Zanghelini, and S. R. Soares. 2018. Uncertainty in LCA case study due to allocation approaches and life cycle impact assessment methods. *International Journal of Life Cycle Assessment* 23:2055–2070. https://doi.org/10.1007/s11367-017-1432-6.

Cherubini, F., Bright, R.M. and Strømman, A.H., 2012. Site-specific global warming potentials of biogenic CO_2 for bioenergy: contributions from carbon fluxes and albedo dynamics. *Environmental Research Letters* 7(4), p.045902.

Cherubini, F., J. Fuglestvedt, T. Gasser, A. Reisinger, O. Cavalett, M. A. J. Huijbregts, D. J. A. Johansson, S. V. Jørgensen, A. Raugei, G. Schivley, A. H. Strømman, K. Tanaka, and A. Levasseur. 2016. Bridging the gap between impact assessment methods and climate science. *Environmental Science & Policy* 64:129–140.

Cherubini, F., G. P. Peter, T. Berntsen, A. H. Strømman, and E. Hertwich. 2011. CO_2 emissions from biomass combustion for bioenergy atmospheric decay and contribution to global warming. *GCB Bioenergy* 3(5):413–426.

Cintas, O., G. Berndes, A. L. Cowie, G. Egnell, H. Holmström, and G. I. Ågren. 2016. The climate effect of increased forest bioenergy use in Sweden: evaluation at different spatial and temporal scales. *Wiley Interdisciplinary Reviews: Energy and Environment* 5(3):351–369.

Collins, W.J., D. J. Frame, J. S. Fuglestvedt, and K. P. Shine. 2020. Stable climate metrics for emissions of short and long-lived species – combining steps and pulses. *Environmental Research Letters* 15:024018.

Colorado State University. n.d. DayCent: Daily Century Model. https://www2.nrel.colostate.edu/projects/daycent/ (accessed February 10, 2022).

Communication from the Commission on the practical implementation of the EU biofuels and bioliquids sustainability scheme and on counting rules for biofuels, 2010/C 160/02 § (n.d.).

Costenoble, O., and T. de Groot. 2020. *Engine tests with new types of biofuels and development of biofuel standards*. Report by the Netherlands Standardization Institute (NEN); Delft, The Netherlands. 2020-01-24. https://advancedbiofuelsusa.info/ortwin-costenoble-on-future-biofuels-standards-in-the-eu/.

Cowie, A. L., G. Berndes, N. S. Bentsen, M. Brandão, F. Cherubini, G. Egnell, B. George, L. Gustavsson, M. Hanewinkel, Z. M. Harris, and F. Johnsson. 2021. Applying a science-based systems perspective to dispel misconceptions about climate effects of forest bioenergy. *GCB Bioenergy* 13(8):1210–1231. https://onlinelibrary.wiley.com/doi/epdf/10.1111/gcbb.12844.

Cowie, A., G. Berndes, and T. Smith. 2013. *On the Timing of Greenhouse Gas Mitigation Benefits of Forest-Based Bioenergy* (Vol. 4). IEA Bioenergy Executive Committee Statement. https://www.ieabioenergy.com/publications/on-the-timing-of-greenhouse-gas-mitigation-benefits-of-forest-based-bioenergy.

Darvin, E. L., S. I. Seneviratne, P. Ciais, A. Olioso, and T. Wang. 2014. Preferential cooling of hot extremes from cropland albedo management. *Proceedings of the National Academies of Sciences of the United States of America* 111(27):9757–9761.

De Kleine, R. D., T. J. Wallington, J. E. Anderson, and H. C. Kim. 2017. Commentary on "Carbon balance effects of US biofuel production and use" by DeCicco et al. (2016). *Climatic Change* 144:111–119. https://doi.org/10.1007/s10584-017-2032-y.

DeCicco, J. M. 2017. Author's response to commentary on "Carbon balance effects of U.S. biofuel production and use." *Climatic Change* 144:123–129. https://doi.org/10.1007/s10584-017-2026-9.

DeCicco, J.M., and W. H. Schlesinger. 2018. Reconsidering bioenergy given the urgency of climate protection. *Proceedings of the National Academy of Sciences*, 115(39), pp.9642-9645.

DeCicco, J. M., D. Y. Liu, J. Heo, R. Krishnan, A. Kurthen, and L. Wang. 2016. Carbon balance effects of US biofuel production and use. *Climate Change* 138:667–680.

Deng, J., C. Bae, A. Denlinger, and T. Miller. 2020. Electric vehicles batteries: requirements and challenges. *Joule* 4(3):511–515. https://www.sciencedirect.com/science/article/pii/S254243512030043X.

Denicoff, M. R., M. E. Prater, and P. Bahizi. 2014. *Soybean Transportation Profile*. USDA. https://www.ams.usda.gov/sites/default/files/media/Soybean%20Transportation%20Profile.pdf (accessed February 10, 2022).

Directive (EU). 2018. 2018/2001 of the European Parliament and of the Council of 11 December 2018 on the promotion of the use of energy from renewable sources, Pub. L. No. 32018L2001, OJ L 328 (2018). http://data.europa.eu/eli/dir/2018/2001/oj/eng.

DOE (US Department of Energy). 2017. *U.S. Department of Energy, Office of Energy Efficiency and Renewable Energy, Vehicle Technologies Office*. Advanced Combustion Systems and Fuels 2017 Annual Progress Report. https://www.energy.gov/sites/default/files/2018/05/f51/AdvCombSysAndFuels_FY2017_APR_Final.pdf (accessed February 11, 2022).

DOE. 2020. *2019 Annual Progress Report – Batteries*. https://www.energy.gov/eere/vehicles/downloads/2019-annual-progress-report-batteries (accessed April 23, 2022).

Dwivedi, P., M. Khanna, A. Sharma, and A. Susaeta. 2016. Efficacy of Carbon and Bioenergy Markets in Mitigating Carbon Emissions on Reforested Lands: A Case Study from Southern United States. *Forest Policy and Economics* 67:1–9.

Dynarski, K., D. A. Bossio, and K. M. Scow. 2020. Dynamic Stability of Soil Carbon: Reassessing the "Permanence" of Soil Carbon Sequestration. *Frontiers in Environmental Science* 8. https://doi.org/10.3389/fenvs.2020.514701.

EIA (US Energy Information Administration). *Gasoline Explained*. Replaced by ASTM. https://www.eia.gov/energyexplained/gasoline/octane-in-depth.php (accessed February 10, 2022).

Elgowainy, A., J. Han, J. Ward, F. Joseck, D. Gohlke, A. Lindauer, T. Ramsden, M. Biddy, M. Alexander, S. Barnhart, and I. Sutherland. 2018. Current and future United States light-duty vehicle pathways: Cradle-to-grave lifecycle greenhouse gas emissions and economic assessment. *Environmental Science & Technology* 52(4):2392–2399.

Emery, I., S. Mueller, Z. Qin, and J. B. Dunn. 2017. Evaluating the potential of marginal land for cellulosic feedstock production and carbon sequestration in the United States. *Environmental Science & Technology*, 51(1):733–741. https://doi.org/10.1021/acs.est.6b04189.

EPA (Environmental Protection Agency). 2010. *Renewable Fuel Standard Program (RFS2) Regulatory Impact Analysis*. U.S. Environmental Protection Agency. EPA-420-R-10-006. https://19january2017snapshot.epa.gov/sites/production/files/2015-08/documents/420r10006.pdf.

EPA. 2014. *Framework for Assessing Biogenic CO_2 Emissions from Stationary Sources, Office of Air and Radiation Office of Atmospheric Programs, Climate Change Division*. United States Environmental Protection Agency, Washington DC. https://yosemite.epa.gov/Sab/Sabproduct.nsf/0/3235DAC747C16FE985257DA90053F252/$File/Framework-for-Assessing-Biogenic-CO2-Emissions+(Nov+2014).pdf).

EPA. 2017. *Office of Transportation and Air Quality*. Fuel Trends Report: Gasoline 2006 – 2016 EPA-420-R-17-005. https://nepis.epa.gov/Exe/ZyPDF.cgi?Dockey=P100T5J6.pdf (accessed February 10, 2022).

EPA. n.d. *Methyl Tertiary Butyl Ether (MTBE)*. https://archive.epa.gov/mtbe/web/html/faq.html.

Ericsson, N., C. Porsö, S. Ahlgren, A. Nordberg, C. Sundberg, and P. A. Hansson. 2013. Time-dependent climate impact of a bioenergy system – methodology development and application to Swedish conditions. *GCB Bioenergy* 5(5):580–590.

European Commission. 2010. *International Reference Life Cycle Data System (ILCD) Handbook – General guide for life cycle assessment – Detailed guidance. Joint Research Centre – Institute for Environment and Sustainability*. EUR24708 EN, Publications Office of the European Union, Luxembourg.

Fargione, J., J. Hill, D. Tilman, S. Polasky, and P. Hawthorne. 2008. Land clearing and the biofuel carbon debt. *Science* 319(5867):1235–1238. https://www.science.org/doi/full/10.1126/science.1152747.

Field, J. L., H. Cai, M. S. Laser, D. S. LeBauer, S. P. Long, L. R. Lynd, K. Paustian, Z. Qin, T. L. Richard, J. J. Sheehan, P. Smith, E. A. H. Smithwick, and M. Q. Wang. 2020. Robust paths to net greenhouse gas mitigation and negative emissions via advanced biofuels. *PNAS* 117:21968–21977.

Forster, P., T. Storelvno, K. Armour, W. Collins, J. L. Dufresne, D. Frame, D. J. Lunt, T. Mauritsen, M. D. Palmer, M. Watanabe, M. Wild, and H. Zhang. 2021. The Earth's Energy Budget, Climate Feedbacks, and Climate Sensitivity. In *Climate Change 2021: The Physical Science Basis*. Contribution of Working Group I to the Sixth Assessment Report of the Intergovernmental Panel on Climate Change. V. Masson-Delmotte, P. Zhai, A. Pirani, S. L. Connors, C. Péan, S. Berger, N. Caud, Y. Chen, L. Goldfarb, M. I. Gomis, M. Huang, K. Leitzell, E. Lonnoy, J. B. R. Matthews, T. K. Maycock, T. Waterfield, O. Yelekçi, R. Yu, B. Zhou, eds. Cambridge University Press.

Fuglestvedt, J., T. Berntsen, G. Myhre, K. Rypdal, and R. B. Skeie. 2008. Climate forcing from the transport sector. *Proceedings of the National Academy of Sciences of the United States of America* 105(2):454–458.

Giuntoli, J., S. Searle, R. Jonsson, A. Agostini, N. Robert, S. Amaducci, L. Marelli, and A. Camia. 2020. Carbon accounting of bioenergy and forest management nexus. A reality-check of modeling assumptions and expectations, *Renewable and Sustainable Energy Reviews* 134. https://doi.org/10.1016/j.rser.2020.110368.

GRAS (Global Risk Assessment). n.d. Sustainable, transparent & deforestation-free supply chains. https://www.gras-system.org/.

Haberl, H., D. Sprinz, B. Marc, P. Cocco, Y. Desaubies, M. Henze, O. Hertel, R. K. Johnson, U. Kastrup, P. Laconte, E. Lange, P. Novak, J. Paavola, A. Reenberg, S. van den Hove, T. Vermeire, P. Wadhams, and T. Searchinger. 2012. Correcting a fundamental error in greenhouse gas accounting related to bioenergy. *Energy Policy* 45:18–23.

Heijungs, R., K. Allacker, E. Benetto, M. Brandão, J. B. Guinee, S. Schaubroeck, T. Schaubroeck, and A. Zamagni. 2021. System expansion and substitution in LCA: A lost opportunity of ISO 14044 amendment 2. *Frontiers in Sustainability* 2.

Hicks Pries, C. E., C. Castanha, R. C. Porras, and M. S. Torn. 2017. The whole-soil carbon flux in response to warming. *Science* 355(6332):1420–1423. https://www.science.org/doi/abs/10.1126/science.aal 1319. FN 21.

Holtsmark, B. 2015. A comparison of the global warming effects of wood fuels and fossil fuels taking albedo into account. *Global Change Biology Bioenergy* 7(5):984–997.

ICF International. 2015. *Waste, Residue and By-Product Definitions for the California Low Carbon Fuel Standard.* Washington, DC: International Council on Clean Transportation, 2015. https://theicct.org/sites/default/files/publications/ICF_LCFS_Biofuel_Categorization_Final_Report_011816-1.pdf.

IPCC (Intergovernmental Panel on Climate Change). 1990. *Climate Change: The IPCC Scientific Assessment.* Contribution of Working Group I to the First Assessment Report of the Intergovernmental Panel on Climate Change. J.T. Houghton, G.H. Jenkins, J.J. Ephraums, eds. Cambridge University Press, United Kingdom, 365 p.

IPCC. 2019. *Refinement to the 2006 IPCC Guidelines for National Greenhouse Gas Inventories* (Vol 4). Agriculture, Forestry and Other Land Use. https://www.ipcc-nggip.iges.or.jp/public/2019rf/vol4.html (accessed February 10, 2022).

IPCC. n.d. *Task Force on National Greenhouse Gas Inventories.* http://www.ipcc-nggip.iges.or.jp/ (accessed February 10, 2022).

ISO (International Organization for Standardization). 2006. *Environmental Management: Life Cycle Assessment; Requirements and Guidelines.* ISO Geneva.

Jolliet, O., A. Antón, A. M. Boulay, F. Cherubini, P. Fantke, A. Levasseur, T. E. McKone, O. Michelsen, L. Milá i Canals, M. Motoshita, S. Pfister, F. Verones, B. Vigon, and R. Frischknecht. 2017. Global guidance on environmental life cycle impact assessment indicators: Impacts of climate change, fine particular matter formation, water consumption and land use. *The International Journal of Life Cycle Assessment* 23:2189–2207.

Jones, A. D., W. D. Collins, J. Edmonds, M. S. Torn, A. Janetos, K. V. Calvin, A. Thomson, L. P. Chini, J. F. Mao, X. Y. Shi, P. Thornton, G. C. Hurtt, and M. Wise. 2013. Greenhouse gas policy influences climate via direct effects of land-use change. *Journal of Climate* 26(11):3657–3670.

Jonker, J. G. G., M. Junginger, and A. Faaij. 2014. Carbon payback period and carbon offset parity point of wood pellet production in the South-eastern United States. *GCB Bioenergy* 6(4):371–389. https://onlinelibrary.wiley.com/doi/full/10.1111/gcbb.12056.

Karabasoglu, O., and J. Michalek. 2013. Influence of driving patterns on life cycle cost and emissions of hybrid and plug-in electric vehicle powertrains. *Energy Policy* 60:445–461.

Kendall, A., B. Chang, and B. Sharpe. 2009. Accounting for time-dependent effects in biofuel life cycle greenhouse gas emissions calculations. *Environmental Science & Technology* 43:7142–7147.

Khanna, M., W. Wang, and M. Wang. 2020. Assessing the additional carbon savings with biofuel, *BioEnergy Research* 13:1082–1094. https://doi.org/10.1007/s12155-020-10149-0.

Lamers, P., and M. Junginger. 2013. The 'debt' is in the detail: A synthesis of recent temporal forest carbon analyses on woody biomass for energy. *Biofuels, Bioproducts and Biorefining* 7(4):373–385.

Lan, K., L. Ou, S. Park, S. S. Kelley, P. Nepal, H. Kwon, H. Cai, and Y. Yao. 2021. Dynamic life-cycle carbon analysis for fast pyrolysis biofuel produced from pine residues: implications of carbon temporal effects. *Biotechnology Biofuels* 14:191. https://doi.org/10.1186/s13068-021-02027-4.

Larrick, R. P., and J. B. Soll. 2008. The MPG illusion. *Science* 320(5883):1593–1594.

Ledo, A., J. Hillier, P. Smith, E. Aguilera, S. Blagodatskiy, F. Q. Brearley, A. Datta, E. Diaz-Pines, A. Don, M. Dondini, J. Dunn, D. M. Feliciano, M. A. Liebig, R. Lang, M. Llorente, Y. L. Zinn, N. McNamara, S. Ogle, Z. Qin, P. Rovira, R. Rowe, J. L. Vicente-Vicente, J. Whitaker, Q. Yue, and A. Zerihun. 2019. A global, empirical, harmonised dataset of soil organic carbon changes under perennial crops. *Science Data* 6:57. https://doi.org/10.1038/s41597-019-0062-1.

Leifeld, J., and S. G. Keel. 2022. Quantifying negative radiative forcing of non-permanent and permanent soil carbon sinks, *Geoderma* 423:115971.

Levasseur, A., O. Cavalett, J. S. Fuglestvedt, T. Gasser, D. J. A. Johansson, S. V. Jørgensen, M. Raugei, A. Reisinger, G. Schivley, A. Strømman, K. Tanaka, and F. Cherubini. 2016b. Enhancing life cycle impact assessment from climate science: Review of recent findings and recommendations for application to LCA. *Ecological Indicators* 71:163–174.

Levasseur, A., A. de Schryver, M. Hauschild, Y. Kabe, A. Sahnoune, K. Tanaka, and F. Cherubini. 2016a. Greenhouse gas emissions and climate change impacts. In R. Frischknecht, and O. Jolliet, eds. *Global Guidance for Life Cycle Impact Assessment Indicators - Volume 1* (pp. 60–79): United Nations Environment Programme.

Levasseur, A., P. Lesage, M. Margni, M. Brandão, and R. Samson. 2012. Assessing temporary carbon sequestration and storage projects through land use, land-use change and forestry: Comparison of dynamic life cycle assessment with ton-year approaches. *Climatic Change* 115:759–776.

Levasseur, A., P. Lesage, M. Margni, L. Deschênes, and R. Samson. 2010. Considering time in LCA: Dynamic LCA and its application to global warming impact assessments. *Environmental Science & Technology* 44:3169–3174.

Ma, H., F. Balthasar, N. Tait, X. Riera-Palou, and A. Harrison. 2012. A new comparison between the life cycle greenhouse gas emissions of battery electric vehicles and internal combustion vehicles. *Energy policy* 44:160–173. https://www.sciencedirect.com/science/article/abs/pii/S030142151200 0602#f0025.

McKechnie, J., S. Colombo, J. Chen, W. Mabee, and H. L. MacLean. 2011. Forest bioenergy or forest carbon? Assessing trade-offs in greenhouse gas mitigation with wood-based fuels. *Environmental Science & Technology* 45(2):789–795.

Mueller, S., G. Dennison, and S. Liu. 2021. An assessment on ethanol-blended gasoline/diesel fuels on cancer risk and mortality. *International Journal of Environmental Research* 18:6930. *Public Health*. https://doi.org/10.3390/ijerph18136930.

Muñoz, I., P. Campra, and A. R. Fernández-Alba. 2010. Including CO_2-emission equivalence of changes in land surface albedo in life cycle assessment. Methodology and case study on greenhouse agriculture. *The International Journal of Life Cycle Assessment* 15:672–681.

Myhre, G., D. Shindell, F. M. Bréon, W. Collins, J. Fuglestvedt, J. Huang, D. Koch, J. F. Lamarque, D. Lee, B. Mendoza, T. Nakajima, A. Robock, G. Stephens, T. Takemura, and H. Zhang. 2013. Anthropogenic and Natural Radiative Forcing. In T.F Stocker, D. Qin, G.K. Plattner, M. Tignor, S.K. Allen, J. Boschung, A. Nauels, Y. Xia, V. Bex, and P.M. Midgley, eds. *Climate Change 2013: The Physical Science Basis. Contribution of Working Group I to the Fifth Assessment Report of the Intergovernmental Panel on Climate Change.* Cambridge University Press, Cambridge, United Kingdom and New York, NY.

NASEM (National Academies of Sciences, Engineering, and Medicine). 2019. *Negative Emissions Technologies and Reliable Sequestration: A Research Agenda.* Washington, DC: The National Academies Press. https://doi.org/10.17226/25259.

NASEM. 2021. *Assessment of Technologies for Improving Light-Duty Vehicle Fuel Economy—2025-2035.* Washington, DC: The National Academies Press. https://doi.org/10.17226/26092.

Nealer, R., and T. P. Hendrickson. 2015. Review of recent lifecycle assessments of energy and greenhouse gas emissions for electric vehicles. *Current Sustainable/Renewable Energy Reports* 2:66–73. https://doi.org/10.1007/s40518-015-0033-x.

O'Hare, M., R. J. Plevin, J. I. Martin, A. D. Jones, A. Kendall, and E. Hopson. 2009. Proper accounting for time increases crop-based biofuels' greenhouse gas deficit versus petroleum. *Environmental Research Letters* 4:024001.

Parton, W. J. 1996. The CENTURY model. In D.S. Powlson, P. Smith, J.U. Smith. eds. *Evaluation of Soil Organic Matter Models.* NATO ASI Series (Series I: Global Environmental Change), Vol 38. Springer, Berlin, Heidelberg. https://doi.org/10.1007/978-3-642-61094-3_23.

Peñaloza, D., F. Røyne, G. Sandin, M. Svanström, and M. Erlandsson. 2019. The influence of system boundaries and baseline in climate impact assessment of forest products. *International Journal of Life Cycle Assessment* 24:160–176. https://doi.org/10.1007/s11367-018-1495-z.

Qin, Z., J. B. Dunn, H. Kwon, S. Mueller, and M. M. Wander. 2014. Soil carbon sequestration and land use change associated with biofuel production: Empirical evidence. *GCB Bioenergy* 8:66–80. https://doi.org/10.1111/gcbb.12237.

Red Trail Energy. 2019. *Red Trail Energy Low Carbon Fuel Standard (LCFS) Design-Based Pathway Application – Carbon Capture and Storage Integrated with Ethanol Production. Red Trail Energy.* Richardton, ND.

Schifter, I., L. Díaz, G. Sánchez-Reyna, C. González-Macías, U. González, and R. Rodríguez. 2020. Influence of gasoline olefin and aromatic content on exhaust emissions of 15% ethanol blends. *Fuel* 265:116950.

Schmuch, R., R. Wagner, G. Hörpel, T. Placke, and M. Winter. 2018. Performance and cost of materials for lithium-based rechargeable automotive batteries. *Nature Energy* 3:267–278. https://doi.org/10.1038/s41560-018-0107-2.

Schrijvers, D. L., P. Loubet, and G. Sonnemann. 2016. Developing a systematic framework for consistent allocation in LCA. *International Journal of Life Cycle Assessment* 21:976–993. https://doi.org/10.1007/s11367-016-1063-3.

Searchinger, T. D. 2010. Biofuels and the need for additional carbon. *Environmental Research Letters* 5:24007.

Searchinger, T., S. P. Hamburg, J. Melillo, W. Chameides, P. Havlik, D. M. Kammen, G. E. Likens, R. N. Lubowski, M. Obersteiner, M. Oppenheimer, G. P. Robertson, W. H. Schlesinger, and G. D. Tilman. 2009. Fixing a critical climate accounting error. *Science* 326:527–528.

Searchinger, T., R. Heimlich, R. A. Houghton, F. Dong, A. Elobeid, J. Fabiosa, S. Tokgoz, D. Hayes, and T. H. Yu. 2008. Use of US croplands for biofuels increases greenhouse gases through emissions from land-use change. *Science* 319(5867):1238–1240. https://www.science.org/doi/10.1126/science.1151861.

Sieber, P., N. Ericsson, T. Hammar, and P. A. Hansson. 2020. Including albedo in time-dependent LCA of bioenergy. *GCB Bioenergy* 12(6):410–425.

Sieber, P., N. Ericsson, and P.A. Hansson. 2019. Climate impact of surface albedo change in life cycle assessment: Implications of site and time dependence. *Environmental Impact Assessment Review* 77:191–200.

Seber, G., R. Malina, M. N. Pearlson, H. Olcay, J. I. Hileman, and S. R. Barrett. 2014. Environmental and economic assessment of producing hydroprocessed jet and diesel fuel from waste oils and tallow. *Biomass and Bioenergy* 67:108–118.

Shine, K. P. P., R. P. P. Allan, W. J. J. Collins, and J. S. S. Fuglestvedt. 2015. Metrics for linking emissions of gases and aerosols to global precipitation changes. *Earth System Dynamics* 6(2):525–540.

Shine, K., J. Flugestvedt, K. Hailemariam, and N. Stuber. 2005. Alternatives to the global warming potential for comparing climate impacts of emissions of greenhouse gases. *Climatic Change* 68:281–302.

Smith, M. A., M. Cain, and M. R. Allen. 2021. Further improvement of warming-equivalent emissions calculation. *Climate and Atmospheric Science* 4:19.

Sripad, S., and V. Viswanathan. 2017. Performance metrics required of next generation batteries to make a practical electric semi-truck. *ACS Energy Letters* 2:1669−1673. https://pubs.acs.org/doi/pdf/10.1021/acsenergylett.7b00432 (accessed February 10, 2022).

Storey, J., M. DeBusk, S. Huff, S. Lewis, F. Li, J. Thomas, and M. Eibl. 2016. *Characterization of GDI PM during start-stop operation with alcohol fuel blends. Oak Ridge National Laboratory.* Presented at Effects of Fuel Composition on PM Health Effects Institute Workshop, December 8, 2016. Chicago, IL. https://www.healtheffects.org/sites/default/files/Storey-Characterization_of_GDI_PM.pdf. (accessed February 11, 2022).

Theiss, T. J., T. Alleman, A. Brooker, A. Elgowainy, G. Fioroni, Gina, J. Han, Jeongwoo, S. P. Huff, C. Johnson, M. D. Kass, P. N. Leiby, R. U. Martinez, R. McCormick, K. Moriarty, E. Newes, G. A. Oladosu, J. P. Szybist, J. F. Thomas, M. Wang, and B. H. West. 2016. *Summary of High-Octane Mid-Level Ethanol Blends Study.* ORNL/TM-2016/42, 1286966. Oak Ridge National Laboratory. https://doi.org/10.2172/1286966.

Tillman, A.-M. 2000. Significance of decision-making for LCA methodology. *Environmental Impact Assessment Review* 20(1):113–123. https://doi.org/10.1016/S0195-9255(99)00035-9.

Transportation Research Board and National Research Council. 2015. *Overcoming Barriers to Deployment of Plug-in Electric Vehicles.* The National Academies Press, Washington, DC. https://doi.org/10.17226/21725.

U.K. Department for Transport. *Renewable Transport Fuel Obligation Guidance Part Two Carbon and Sustainability.* London, UK: UK Department for Transport. 2021. https://assets.publishing.service.gov.uk/government/uploads/system/uploads/attachment_data/file/947710/rtfo-guidance-part-2-carbon-and-sustainability-2021.pdf.

U.K. Department for Transport. RTFO Guidance Year 10: *Carbon & Sustainability Guidance.* London, U.K.: U.K. Department for Transport. 2018. https://assets.publishing.service.gov.uk/government/uploads/system/uploads/attachment_data/file/604627/rtfo-guidance-part-2-carbon-sustainability-guidance-year-10.pdf.

USDA (US Department of Agriculture). 2020.Agricultural Marketing Service; 7 CFR Part 205. https://www.govinfo.gov/content/pkg/FR-2020-08-05/pdf/2020-14581.pdf.

Unnasch, S., and L. Browning. 2000. *Fuel Cycle Energy Conversion Efficiency Analysis.* California Energy Commission and Air Resources Board, California, viewed, 11, 07-12 with follow in analysis in Unnasch, S. and Chan, M. (2007) Full Fuel Cycle Assessment Tank to Wheels Emissions and Energy Consumption. California Energy Commission Report CEC-600-2007-003-D.

Verity Tracking. 2022. Measurement, Reporting, Verification and Value. https://www.veritytracking.com/.

Wang, M., J. Kelly, H. Kwon, X. Liu, Z. Lu, L. Ou, P. Sun, O. Winjobi, H. Xu, E. Yoo, G. Zaimes, and G. Zang. 2020. *Greenhouse Gases, Regulated Emissions, and Energy Use in Technologies Model ® (2020 Excel).* Argonne National Laboratory (ANL), Argonne, IL, United States. https://doi.org/10.11578/GREET-EXCEL-2020/DC.20200912.1.

Wang, M., J. Han, J. B. Dunn, H. Cai, and A. Elgowainy. 2012. Well-to-wheels energy use and greenhouse gas emissions of ethanol from corn, sugarcane and cellulosic biomass for US use. *Environmental Research Letters*, 7:045905.

Wang, M., H. Hong, and S. Arora. 2011. Methods of dealing with co-products of biofuels in life-cycle analysis and consequent results within the U.S. context. *Energy Policy* 39(10):5726–36. https://doi.org/10.1016/j.enpol.2010.03.052.

Weidema, B. 1999. *System Expansions to Handle Co-Products of Renewable Materials* 4, 45–48 in Presentation Summaries of the 7th LCA Case Studies Symposium SETAC-Europe. https://lca-net.com/files/casestudy99.pdf (accessed February 10, 2022).

West, B., S. Huff, L. Moore, M. DeBusk, and S. Sluder. 2018. *Effects of High-Octane E25 on Two Vehicles Equipped with Turbocharged, Direct-Injection Engines.* ORNL/TM-2018/814; September 2018. https://info.ornl.gov/sites/publications/Files/Pub109556.pdf (accessed February 10, 2022).

Wiloso, E. I., R. Heijungs, G. Huppes, and K. Fang. 2016. Effect of biogenic carbon inventory on the life cycle assessment of bioenergy: Challenges to the neutrality assumption. *Journal of Cleaner Production* 125:78–85. https://www.sciencedirect.com/science/article/pii/S0959652616301676#bib21.

Wu, D., F. Guo, F. R. Field III, R. D. De Kleine, H. C. Kim, T. J. Wallington, and R. E. Kirchain. 2019. Regional heterogeneity in the emissions benefits of electrified and lightweighted light-duty vehicles. *Environmental Science & Technology* 53(18):10560–10570. https://pubs.acs.org/doi/10.1021/acs.est.9b00648.

Xu, H., H. Sieverding, H. Kwon, D. Clay, C. Stewart, J. M. Johnson, Z. Qin, D. L. Karlen, and M. Wang. 2019. A global meta-analysis of soil organic carbon response to corn stover removal. *GCB Bioenergy* 11(10):1215–1233. https://doi.org/10.1111/gcbb.12631.

Yang, J., P. Roth, T. D. Durbin, M. M. Shafer, J. Hemming, D. S. Antkiewicz, A. Asa-Awuku, and G. Karavalakis. 2019. Emissions from a flex fuel GDI vehicle operating on ethanol fuels show marked contrasts in chemical, physical and toxicological characteristics as a function of ethanol content. *Science of the Total Environment* 683:749–761. https://www.energy.gov/sites/default/files/2018/05/f51/AdvCombSysAndFuels_FY2017_APR_Final.pdf.

Yuksel, T., and J. J. Michalek. 2015. Effects of regional temperature on electric vehicle efficiency, range, and emissions in the United States. *Environmental Science & Technology* 49(6):3974–3980.

Yuksel, T., M. Tamayao, C. Hendrickson, I. Azevedo, and J. J. Michalek. 2016. Effect of regional grid mix, driving patterns and climate on the comparative carbon footprint of electric and gasoline vehicles. *Environmental Research Letters* 11(4):044007.

Zanchi, G., N. Pena, and N. Bird. 2012. Is woody bioenergy carbon neutral? A comparative assessment of emissions from consumption of woody bioenergy and fossil fuel. *GCB Bioenergy* 4(6):761–772.

Zhang, Y., J. M. Lavallee, A. D. Robertson, R. Even, S. M. Ogle, K. Paustian, and M. F. Cotrufo. 2021. Simulating measurable ecosystem carbon and nitrogen dynamics with the mechanistically defined MEMS 2.0 model, *Biogeosciences* 18:3147–3171. https://doi.org/10.5194/bg-18-3147-2021.

Part III

Specific Fuel Issues for Life-Cycle Analysis

7
Fossil and Gaseous Fuels for Road Transportation

LIQUID FOSSIL FUELS

Liquid fossil fuels account for the majority of all transportation fuels today. The United States produces about 20 percent of global liquid fossil fuels, and in 2020, petroleum products accounted for about 90 percent of the total U.S. transportation sector energy use. (Biofuels contributed about 5 percent; natural gas accounted for about 3 percent; and electricity provided less than 1 percent) (EIA, 2020). Worldwide, fossil fuels supply about 84 percent of world energy (Ritchie and Roser, 2020), and petroleum and liquid fossil fuels are globally traded fungible commodities, hence no one country's fossil fuels market is truly independent of global market supply and demand considerations.

The U.S. product mix from refining crude oil is more gasoline-heavy than most other large economies, where middle distillates predominate. The most recent data on international petroleum products produced from crude oil, from the Energy Information Administration (EIA, 2022), are shown in Figure 7-1.

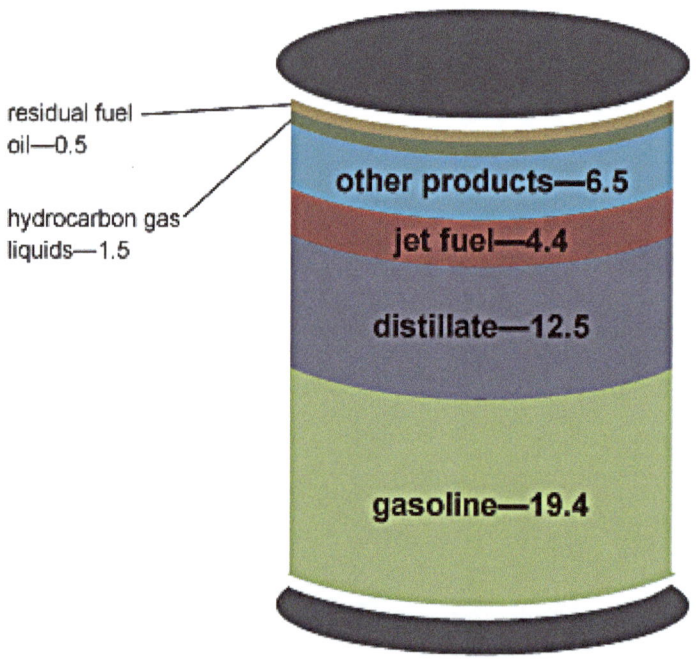

Note: A 42-gallon (U.S.) barrel of crude oil yields about 45 gallons of petroleum products because of refinery processing gain. The sum of the product amounts in the image may not equal 45 because of independent rounding.

Source: U.S. Energy Information Administration, *Petroleum Supply Monthly*, June 2021, preliminary data

FIGURE 7-1 Petroleum products made from a barrel of crude oil, worldwide, 2020 (in gallons). SOURCE: U.S. EIA (2022) see https://www.eia.gov/energyexplained/oil-and-petroleum-products/.

In addition to conventional gasoline, reformulated gasoline was mandated in the 1990 Clean Air Act amendments. The first phase of reformulated gasoline began in 1995 and the second phase in 2000. Reformulated gasoline now comprises about 25 percent of all gasoline sold in the United States, and it is required in cities with high smog levels as defined by the U.S. Environmental Protection Agency (EPA), and is it is optional elsewhere. Reformulated gasoline is currently used in 16 states and the District of Columbia. In addition to the federal reformulated gasoline standards, California has its own reformulated gasoline formulation (currently, CaRFG3), and while these regulations do not require the use of ethanol there is a separate oxygen content requirement.[1] Ethanol as a gasoline substitute or as an additive has a long history but consumption of ethanol, mainly corn ethanol, increased in the United States for a variety of factors, including but not limited to blending mandates under the U.S. Renewable Fuel Standard (RFS) established in 2005 (RFS1) and in 2007 (RFS2), volumetric ethanol excise tax credits, and oxygenate demand following the phase out of methyl tertiary-butyl ether in 2006.

Several distillate fuels are used for transportation in the United States. Number 1 diesel fuel is a relatively light oil (550°F at the 90 percent recovery) defined in the American Society for Testing and Materials (ASTM) D975[2] that is used in variable speed diesel engines such as those in city buses. Number 2 diesel fuel has a 90 percent distillation temperature of 640°F (also specified in ASTM D975). It is used in more uniform speed diesel engines, such as locomotives, trucks, and automobiles. Distillate fuels (1 through 6) are also used as heating oil and in numerous industrial applications. Since the United States has always used proportionally more gasoline than distillate fuels, U.S. refineries have always been configured for more gasoline production. Still, the United States has been exporting significant quantities of distillate fuels (more than 27 percent of production, or more than 1.2 MMb/d).

The United States consumes approximately 30 percent of worldwide jet fuel production (about 1.8 MMb/d pre-pandemic). Kerosene jet fuel (Jet A, A-1, JP-5, and JP-8) has a carbon number of C8 to C16, whereas the lighter naphtha-based jet fuel (Jet B and JP-4) are seldom used, other than for special applications. Other specialty jet fuels (such as JP1 and JP2) are either obsolete or used for very specific applications.

Synthetic Fuels

Synthetic fuels ("synfuels") are also being proposed by some as potential low-carbon fuel substitutes or additives. Synfuels are liquid fuels that can be produced from coal, natural gas, or biomass through direct liquefaction, or more commonly through gasification of the source hydrocarbons to synthesis gas (mainly carbon monoxide and hydrogen), followed by cleanup and shift reaction and then synthesized into liquid or gaseous products, such as synthetic gasoline, distillates, lubricants, jet fuel, synthetic natural gas, and so on. There are two major routes to liquid fuels from synthesis gas, through the production of methanol followed by conversion of methanol over a zeolite catalyst to liquid products, or through Fischer-Tropsch synthesis. Synfuels can also be produced using water and a carbon input (e.g., flue gases or captured carbon dioxide [CO_2]) by using renewable electricity to supply energy for electrolysis; subsequently, synthetic fuels can be generated through Fischer–Tropsch synthesis or methanation (Schaaf et al., 2014). Current production of renewable synthetic fuels is negligible as of 2021.

The economics of synfuels are challenged by the intensive chemical reactions needed, which result in low thermal efficiencies and in the production of CO_2. Beyond issues of petroleum scarcity, some synfuel projects were justified as ways to monetize stranded natural gas, as demonstration projects, or as a way to produce high-value byproducts (such as high-quality waxes or synthetic lubricant blending stocks). In general, the products of synthetic fuel plants can be of very high quality, free of trace pollutants, and tailored to precise specifications. There are proposals to couple such projects with carbon capture utilization and storage (CCUS), to produce high quality fuel with less of an impact on greenhouse gas (GHG) emissions.

---

[1] For details, see https://www.epa.gov/gasoline-standards/reformulated-gasoline.
[2] ASTM D975 is the Standard Specification for Diesel Fuel Oils, see https://www.astm.org/d0975-21.html.

Calculation of Greenhouse Gas Emissions for Liquid Fossil Fuels

Though the majority of emissions from liquid fossil fuels come from their final combustion, emissions attributable to upstream extraction and processing may still be significant; emissions from petroleum systems in the United States alone totaled 47.2 Mtonne CO_2e in 2019 (EPA, n.d.). Furthermore, there is substantial variability in the upstream emissions attributable to extraction and processing prior to final use, influenced by crude oil quality, region, and refinery configuration.

Liquid fossil fuels have been extensively analyzed in the literature, as well as incorporated into various life-cycle models. The Oil Production Greenhouse Gas Emissions Estimator (OPGEE) model estimates the life-cycle GHG emissions from the production, processing, and transport of petroleum.[3] The model's system boundary is from crude oil exploration through to the refinery gate. The model is open-source and allows users to input their own data on various parameters of crude oil production, including energy consumption, fugitive emissions, flaring, and qualities of the crude oil itself.

Downstream of oil extraction, refinery emissions are another important contributor to the life-cycle GHG emissions attributable to liquid fossil fuels. Refinery emissions include hydrogen production emissions used for hydroprocessing, emissions from energy inputs, and onsite emissions (both point source and fugitive emissions) at refineries. The Petroleum Refinery Life Cycle Inventory Model (PRELIM)[4] is a life-cycle tool that can be used in conjunction with data on crude oil inputs to estimate refinery emissions across an array of crude oil and refinery configuration combinations. Further, as refinery outputs have different market values, energy densities and end uses, PRELIM may be used to attribute a share of overall emissions to specific outputs using different allocation assumptions. Generally, heavier and more sour crudes require additional energy and hydrogen to process, resulting in higher refinery emissions; also, additional refinery stages like cracking can increase emissions further. Generally, the end products with the highest share of emissions at the refinery tend to be low-sulfur diesel and gasoline.

There are significant uncertainties and significant variations in the current calculations of life-cycle GHG emissions from petroleum fuels (Masnadi et al., 2018). Data on flaring of methane (CH_4) is available from satellite nighttime radiometry. Direct venting or leakage of CH_4 has a considerably higher contribution to GHG emissions than combusting that CH_4 through flaring due to the high global warming potential (GWP) of CH_4; several studies estimate the contribution of these emissions (Lauvaux et al., 2022). Many oil fields with above-average petroleum production carbon intensities (CIs) have significant flaring.[5] CH_4 fugitive emissions and venting from oil and gas facilities are poorly detected, measured, and monitored, and are one of the significant sources of uncertainty in GHG life-cycle calculations for petroleum fuels. Improved reporting and transparency on oil sector emissions is needed to reduce uncertainty in life-cycle GHG emissions calculation for petroleum fuels.

The refining of petroleum produces multiple products: gasoline, diesel, naptha and other products. Emissions are assigned to the co-products using either allocation or displacement. Co-product displacement is typically used in U.S. analyses of petroleum fuels. It should be noted that allocation at the aggregate refinery level compared with allocation at the refining process level can result in different energy and emissions results, as may the allocation method (e.g., mass, energy-content, market value) at the refining process level (Wang et al., 2004).

Enhanced oil recovery, practiced at older wells, also increases the energy requirements and GHG emissions of petroleum, particularly when steam flooding is used. Heavy oils have higher energy requirements and GHG emissions for both extraction and processing. Renewable electricity can be used for extraction and processing, as is currently practiced in North Sea offshore fields. Utilization of CH_4 as natural

---

[3] See https://eao.stanford.edu/opgee-oil-production-greenhouse-gas-emissions-estimator.
[4] See PRELIM: The Petroleum Refinery Life Cycle Inventory Model, See https://www.ucalgary.ca/energy-technology-assessment/open-source-models/prelim (accessed March 7, 2022).
[5] See Carnegie Endowment, (n.d.). Oil Climate Index, https://oci.carnegieendowment.org/#analysis (accessed March 7, 2022).

gas avoids the need for flaring and can provide additional energy for onsite use or distribution if natural gas supply chain infrastructure is available.

Due to variation in CI across different sources of crude oil, changes in aggregate demand for crude oil may result in non-linear changes in the average CI of crude oil at the margins of production. Econometric analysis suggests that changes in demand may not be uniformly distributed across the mix of crude oils on the global market; for example, higher CI crudes such as heavy oil sands may be the first oils displaced by small demand reductions (Masnadi et al., 2021). In contrast, more profitable, lighter crudes with a higher market share are more likely to maintain production; however, these effects would dissipate with larger demand shocks.

The OPGEE model has been used to estimate the upstream crude oil intensity for the California low-carbon fuel standard (CA-LCFS), which incorporates annual updates of the CI of its crude oil mix to estimate deficit generation within the policy. The California Air Resources Board (CARB) uses OPGEE to estimate the weighted average CI of crude oil consumed in California,[6] based on the volumes of crude recorded as either produced or imported into the state, and the sources of those crude oils. The CI of each crude is estimated based on site-specific parameters, and as noted previously, the inclusion of specific CI values in this report does not imply the committee's endorsement of those values. The average CI for the production and transportation of crudes consumed in California in 2020 was estimated to be 13.39 gCO_2e/MJ, following a steadily rising trend over the lifetime of the LCFS. To this, 92 gCO_2e/MJ is added, representing the carbon emissions from refining, distribution, and combustion of the fuel, for a total of 105.39 gCO_2e/MJ for 2020. These values are recalculated every year and California's most recent reported value for crude oil is 100.8 g CO2e/MJ, as of 2021.[7]

The International Civil Aviation Organization's Carbon Offsetting and Reduction Scheme for International Aviation (CORSIA) estimates a weighted average value (Prussi et al., 2021) for the average global jet fuel, based on global petroleum production and consumption, of 89 gCO_2e/MJ. This value includes crude oil recovery, transportation and refining, jet fuel transportation, and jet fuel combustion, calculated with methodology consistent with the CA-LCFS calculations discussed above. However, while the CA-LCFS baseline values are evaluated annually and updated as petroleum fuels life-cycle emissions change, the CORSIA baseline reflects an estimated future global life-cycle emissions average and remains fixed over the course of the period of the scheme.

The U.S. RFS2 developed a 2005 baseline estimate of 93.08 g CO_2e/MJ for gasoline and 91.9 gCO_2e/MJ for diesel, with this baseline value remaining fixed throughout the regulatory period.

In sum, GHG emissions calculations for petroleum fuels include emissions from production, transportation, refining, and combustion. Emissions vary depending on the crude oil, transportation, and refining. Some low-carbon and renewable fuel policies, such as CORSIA and U.S. RFS2, have adopted a fixed value for baseline fossil fuel emissions; others, like CA-LCFS, use annual data to calculate new values annually.

Recommendation 7-1: Policymakers may consider recognizing the variation in GHG emissions across different petroleum fuel pathways, and include mechanisms to reduce these emissions in fuel policies.

Conclusion 7-1: Additional data, reporting, and transparency is needed for petroleum sector operations, including improved information on venting and flaring of methane.

---

[6] See: https://ww2.arb.ca.gov/sites/default/files/classic/fuels/lcfs/crude-oil/2020_crude_average_ci_value_final.pdf.
[7] See: CARB (n.d.) https://ww2.arb.ca.gov/resources/documents/lcfs-pathway-certified-carbon-intensities (accessed March 7, 2022).

GASEOUS FOSSIL FUELS AND HYDROGEN

The discussion of "gaseous" fuels in this section covers fuels derived from compounds that are gases at ambient conditions, rather than the state at which they are stored or used in internal combustion engines or fuel cells. The main gaseous fuels considered in this section are natural gas derived (mostly compressed natural gas and liquefied natural gas [LNG]), propane and hydrogen, and hydrogen derivatives such as ammonia.

It is estimated that there are nearly 30 million natural gas vehicles using LNG, compressed natural gas, and biomethane in the world, mostly in Asia (Iran, Pakistan, China and other countries), followed by lesser incidence in Europe (Italy, and gaining in other countries) and Latin America (Argentina, Brazil, and Colombia). LNG is used in large volumes as a way of transporting natural gas over long distances, where long distance pipelines are infeasible or disadvantaged, from major natural gas-producing areas such as the Middle East, Australia, several African countries, and increasingly the United States to markets mostly in Asia Pacific (China, Japan, Korea, Taiwan) and in Europe. In this case, LNG is mostly regasified and used for electric power production, with smaller amounts used for transportation.

The "hydrogen economy" is a proposed set of comprehensive alternative ways to produce, use, and store energy through hydrogen and its derivatives. While hydrogen is the most common of the known elements in the universe, it does not exist in usable quantities in nature and it needs to be produced from hydrocarbons (mostly through steam reforming of natural gas), or through electrolysis of water.

Hydrogen is used extensively in industry, for example for the manufacture of ammonia and methanol, and in petroleum refining. More than 95 percent of hydrogen is made today via steam reforming, mostly of natural gas, and is called "grey" hydrogen. Six percent of global natural gas production and 2 percent of global coal production is used for hydrogen production, resulting in about 830 million tonnes of CO_2 emissions annually. Hydrogen is also a fuel component of rocket fuel, and has other uses such as a fuel in fuel cells. Proposed "blue" hydrogen is also to be produced mostly from natural gas through steam reforming, with the associated CO_2 emissions captured, stored, or used productively. "Green" hydrogen is usually considered to be hydrogen produced through electrolysis with power derived from renewable sources. Less than 1 percent of global hydrogen production is produced from low-carbon sources today. "Turquoise" hydrogen production refers to a novel conversion process wherein CH_4 is treated with thermal decomposition (i.e., CH_4 pyrolysis) to generate hydrogen and solid carbon; this conversion process has not been commercialized (Sanchez-Bastardo et al., 2021). Figure 7-2 summarizes the color-based terms used to describe hydrogen produced with different technologies from different feedstocks.

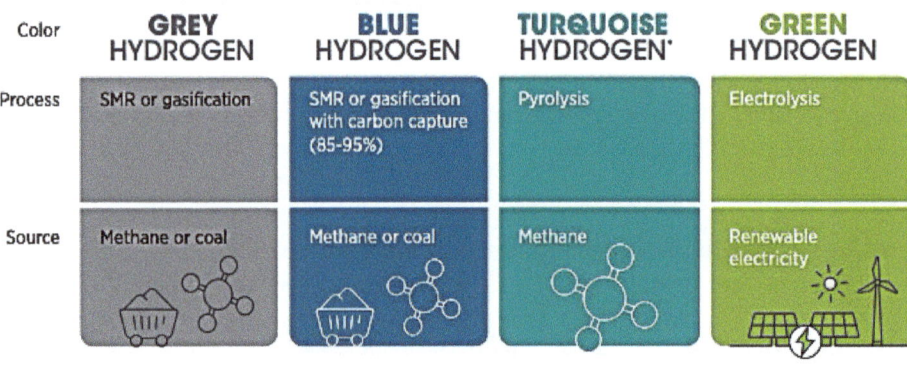

FIGURE 7-2 Hydrogen produced from different technologies and feedstocks. NOTE: SMR = steam-methane reforming; not all colors of hydrogen are depicted in this figure. SOURCE: IRENA (2020). © 2020 IRENA - International Renewable Energy Agency. All Rights Reserved.

There are many proposals to use hydrogen for transportation and also in the context of a more ambitious "hydrogen economy", that is, using hydrogen to store renewable energy (such as offshore wind energy); and using hydrogen for the production of "green" steel, ammonia, and concrete, among other products. This discussion will assess the current state, focusing mostly on the United States, but referencing countries with significant hydrogen proposals, namely the European Union (EU), Japan, and China, and provide a context of some history of past hydrogen economy efforts. Figure 7-3 shows the increased global production of hydrogen since 1975.

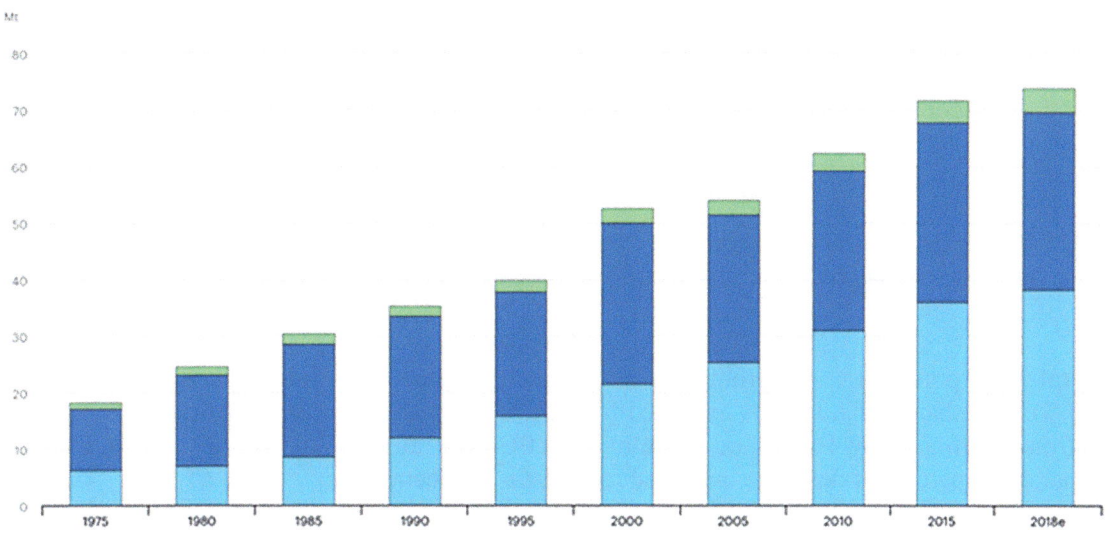

FIGURE 7-3 Global demand for pure hydrogen, 1975-2018. NOTES: Aqua = refining; blue = ammonia; green = other. SOURCE: IEA (2019) World Energy Outlook. All rights reserved.

Because today hydrogen is most commonly produced via steam reforming of natural gas, the methodological issues in life-cycle assessments (LCAs) of these two fuels are closely related. Natural gas can be used as a transportation fuel in its own right and is a common input to other fuel supply chains. LCA methodological issues for evaluating natural gas systems can therefore have wide-ranging implications for the understanding of low-carbon fuels.

Natural Gas

Global natural gas production has been increasing at a higher rate than oil production, and while there has been a pandemic-caused decline in 2020, it is picking up again and similarly to global oil production, it is projected to keep increasing through this decade. In many major markets, including the United States, natural gas has displaced coal as a major electrical energy production source. Figure 7-4 shows domestic natural production since 1971[8] The IEA estimates that the oil and gas sector emitted 82 million tonnes of CH_4 (around 2.5 gigatons of CO_2 equivalent) in 2019.

---

[8] See https://iea.blob.core.windows.net/assets/52f66a88-0b63-4ad2-94a5-29d36e864b82/KeyWorldEnergyStatistics 2021.pdf (accessed on 04/21/2022).

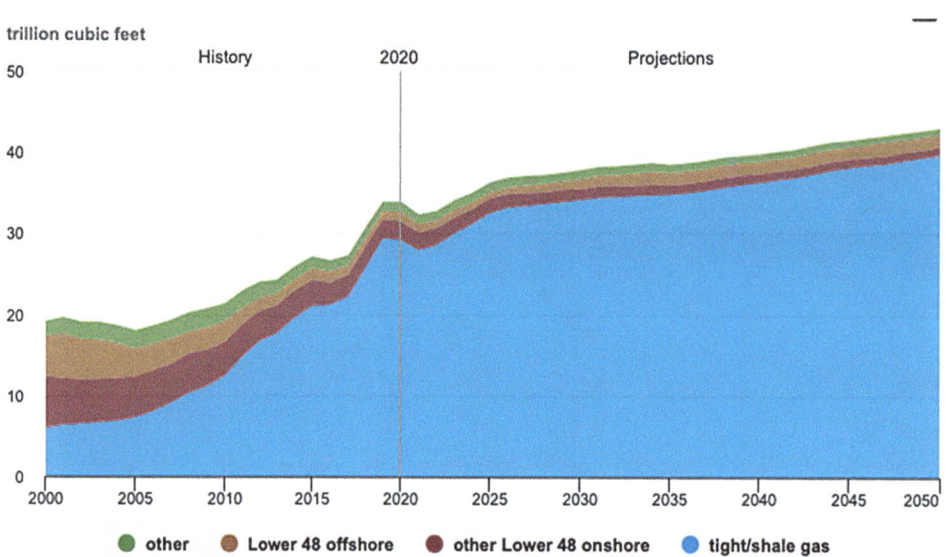

FIGURE 7-4 U.S. Dry natural gas production, 2000-2050 (trillion cubic feet). SOURCE: EIA (2021, February), see https://www.eia.gov/energyexplained/natural-gas/where-our-natural-gas-comes-from.php.

Natural gas is produced mostly from shale and tight gas resources from shale plays across the country (Figure 7-5). The share of natural gas extracted from shale plays has increased significantly over the last decade, from approximately 165 billion cubic meters (BCM) in 2010 to 805 BCM in 2020, comprising approximately 78 percent of domestic marketed natural gas production.

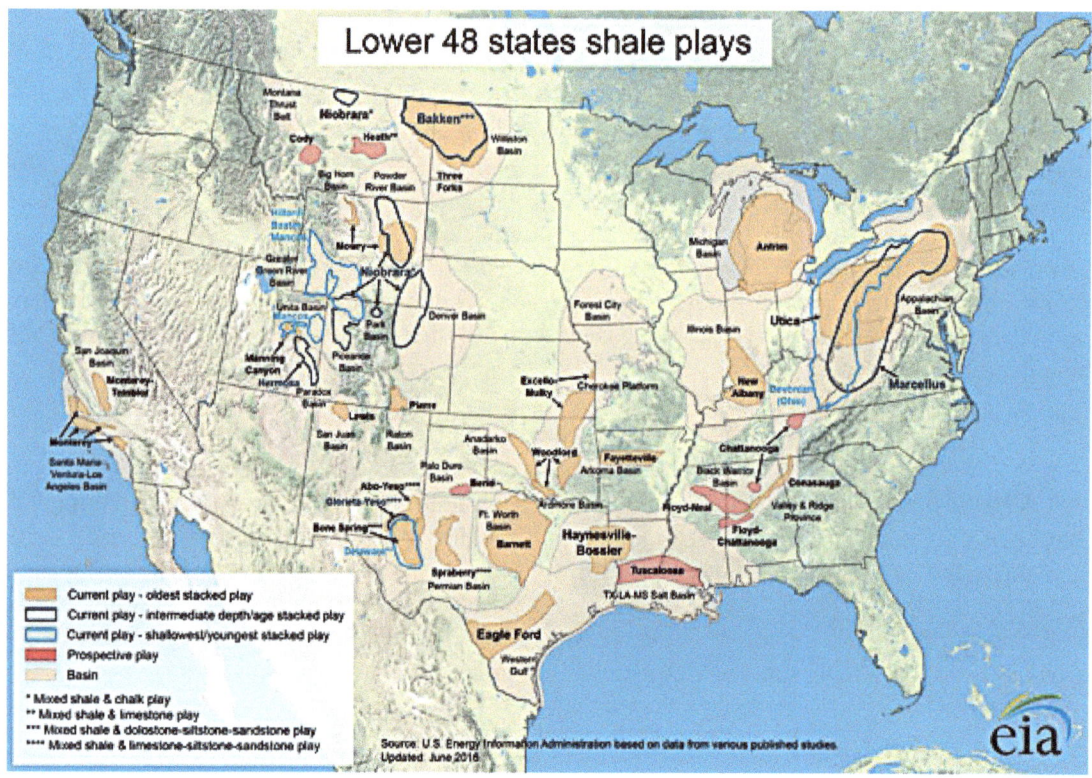

FIGURE 7-5 Location of shale gas plays in the continental United States. SOURCE: EIA (2016, June).

There are multiple steps involved in natural gas recovery and delivery to the point-of-use (see Figure 7-6). In natural gas LCAs, it is important to consider direct CH_4 emissions from each step of this supply chain, which may derive from venting, flaring, or leaking (Burnham et al., 2011).

One of the first steps in producing natural gas is to drill a well. During well completion, water, hydrocarbon liquids, and natural gas all flow up from the well (Allen et al., 2013). These components of the flowback are separated. Tanks that contain the water and hydrocarbons are vented to the atmosphere. The gas may be flared or retained for sales. The configuration of the equipment that manages the flowback, which Allen et al. (2013) grouped into five different categories, can affect the amount of fugitive CH_4 emissions from the operation along with the duration of the flowback period. CH_4 emissions from completion at 27 different sites exhibited a striking range from 0.01 megagrams (Mg) to 17 Mg. It is this type of site-to-site variation, which holds for completion along with other stages in the natural gas supply chain, that leads to differing results among natural gas LCA, which may not necessarily show uncertainty.

Operating wells also have several different CH_4 emission sources. Over time in a natural gas well, the pressure from the reservoir declines, and liquids can accumulate that reduce the amount of gas flowing to the surface. A liquid unloading procedure removes these liquids. Well blowdown is one type of liquids unloading procedure. In it, the well is closed, pressure builds up, and then the well is vented to the atmosphere, releasing the problematic liquids along with CH_4. The choice of unloading procedure and the well's characteristics determine the level of CH_4 emissions that will occur during liquids unloading. As a well ages, it is likely that it will require repair, or workovers, which can involve venting the well and, accordingly, CH_4 emissions.

Fugitive CH_4 can escape from oil or condensate tanks, pneumatic controllers and pumps, and fittings, flanges, and valves. The rates of these emissions are highly dependent upon the adoption of technology to reduce these emissions such as replacement of high-bleed valves with low-bleed valves and using LiDAR-based programs to detect and replace leaky system components.

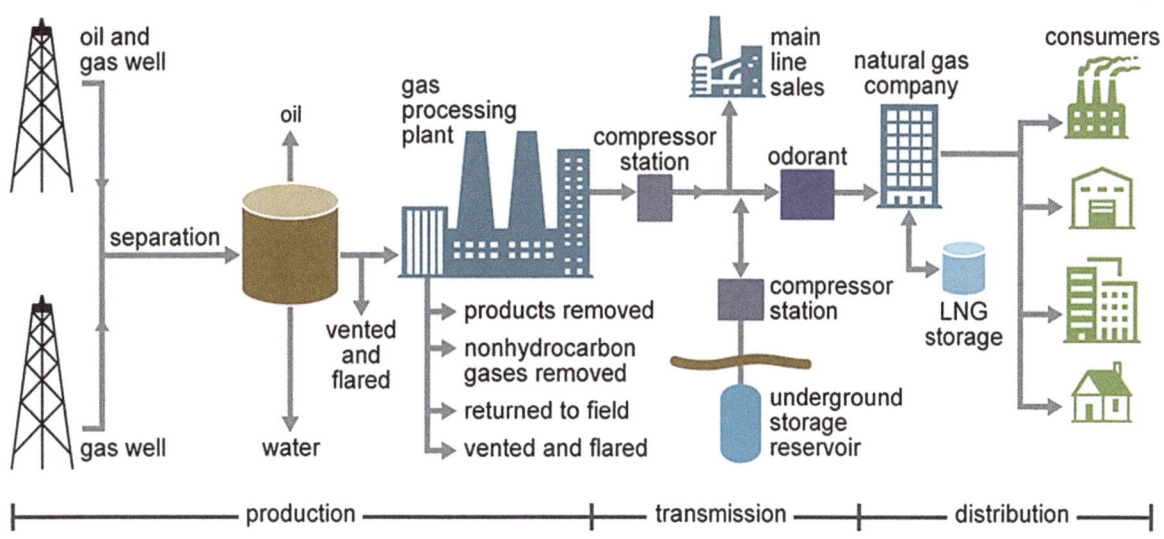

FIGURE 7-6 Steps in the natural gas supply chain from recovery to delivery. SOURCE: EIA (2022, February).

Flaring of natural gas emits CO_2 and CH_4. There are commitments and targets to reduce flaring, but these are not always met. Rates of flaring can depend on several factors including the number of nearby wells. When other wells are nearby, the economics of building pipelines to transport gas for use instead of flaring on site improve; flaring rates may be lower (Willyard, 2019).

The last step in natural gas production is the processing step. During this step, the acid gas removal process vents CO_2, which is another notable emissions source in the overall natural gas supply chain. After processing, natural gas enters the transmission and distribution system to the point of end use. CH_4 leakage throughout this system contributes to life-cycle GHG emissions of natural gas.

There is variability in CH_4 emissions throughout the natural gas supply chain depending on the location of natural gas production and the extent to which emissions reductions measures are in place (Alvarez et al., 2018). One method to estimate these emissions that could inform parameter selection in LCAs is developing emissions factors from measurements on representative samples of equipment and then multiplying these factors by activity levels (e.g., numbers of equipment and associated throughput). This method is called "bottom-up" and is summarized in Figure 7-7 in the context of Rutherford et al. (2021).

Another approach, called "top down," is to measure CH_4 concentrations in the atmosphere near natural gas infrastructure, which may be done, for example, with low-flying aircraft. Recently, a top-down estimate was developed for the Permian Basin using satellite observations paired with atmospheric inverse modeling that reported that 3.7 percent of the CH_4 produced in the Permian Basin is emitted (Zhang et al., 2020). A second study reported CH_4 emissions to be 9.4 percent of gross natural gas production in the portion of the Permian Basin in New Mexico (Chen et al., 2022). On the other hand, Alvarez et al. report a national average value of 2.3 percent as a national average CH_4 leakage rate from natural gas processing based on bottom-up methods of emissions estimation (Alvarez et al., 2018). They report relative agreement with other top-down studies. Alvarez et al. hypothesized that differences between bottom-up and top-down emissions estimates can often be explained by the latter's inclusion of abnormal, high-emitting events.

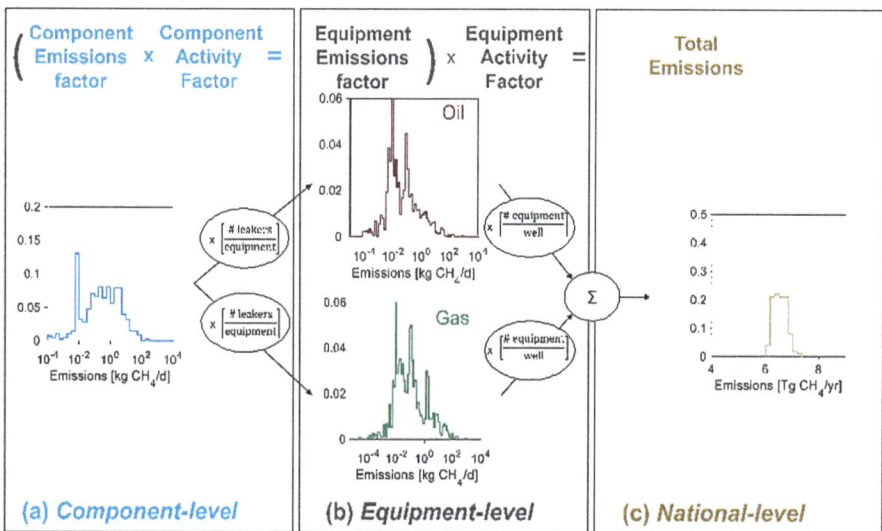

Fig. 1 Schematic of this study's bottom-up CH_4 emissions estimation tool. Calculation of total CH_4 emissions involves multiplication of emission factors (e.g., emissions per valve) by activity factors (e.g., number of valves per wellhead). Two sequential extrapolations are performed using an iterative bootstrapping approach. First, our database of component-level (e.g., valve, connector) emissions measurements (a) is extrapolated using component-level activity factors to generate equipment-level (e.g., wellhead, separator) emission factors (b). Second, these equipment-level emission factor distributions are extrapolated using equipment-level activity factors to generate a 2015 United States oil and natural gas production-segment CH_4 emissions estimate. This extrapolation is performed 100 times to generate a distribution of national-level CH_4 emissions (c) and estimate a 95% confidence interval (CI).

FIGURE 7-7 Bottom-up methane (CH_4) emissions estimation tool. SOURCE: Rutherford et al. (2021). Copyright © 2021, Jeffrey S. Rutherford et al. CC BY license.

Top-down estimates can capture irregular CH_4 leakage such as unintentional emissions during storage and equipment leakage (Rutherford et al., 2021) that may be excluded from bottom-up estimates. Rutherford et al. (2021) developed a new model for CH_4 emissions in the natural gas production supply chain based on existing inventory data (3700 measurements from six studies). It employed a bottom-up approach (Figure 7-7) and a new statistical technique (bootstrap resampling) to these data that aimed to capture these unintentional emissions from so-called super emitters. Emissions estimates based on the model exhibited strong agreement with site-level measurements, indicating that bottom-up approaches can be robust and provide reliable emissions estimates. Rutherford et al. (2021) compared their results with those of the EPA Greenhouse Gas Emissions Inventory (GHGI), which is a commonly-used data source for understanding natural gas emissions. The emissions estimates in the EPA GHGI, however, are based on data from the 1990s. As with previous studies (Alvarez et al., 2018), Rutherford et al. generated estimates of CH_4 emissions in the natural gas supply chain that greatly exceeded (~80 percent) the EPA GHGI's estimate. Breaking down the reason behind this discrepancy, Rutherford et al. determined that one driving factor was emissions from tanks. Calculated from engineering models in the EPA GHGI but based on measurements in Rutherford et al.'s work, these emissions were significantly underestimated in the EPA GHGI. Rutherford et al. noted the lack of tank emissions data as being a significant data gap that requires addressing. Overall, Rutherford et al. noted that updated component (e.g., valves, tanks) counts in the natural gas system in the United States requires updating from the 1990s inventory that is commonly used, including in the EPA GHGI and Rutherford et al.'s work. There is also a pressing, ongoing need to increase CH_4 emissions collection at the component level. Building inventory data will cut data gaps and improve sample size and representativeness to enable enhanced statistical techniques for inventory building.

The Greenhouse gases, Regulated Emissions and Energy use in Technologies (GREET) model (Argonne National Laboratory, n.d.) is widely used in transportation fuel LCAs and updates the natural gas parameters it contains annually. The GREET model's use of different parameters for upstream natural gas leakage is discussed to illustrate their impact on the overall emissions for natural gas, not as an endorsement of the model's validity or to provide a central estimate for natural gas emissions. The 2021 GREET release uses a combination of data from Rutherford et al. (2021) and Alvarez et al. (2018) as its default values for natural gas pathways, rather than values from the EPA GHGI, because of the growing recognition that the latter data source underestimates CH_4 emissions in the natural gas supply chain. Changing the CH_4 leakage assumptions in the GREET model can illustrate the overall impact in natural gas LCA results when using parameters from EPA GHGI or from the literature (Rutherford et al., 2021; Alvarez et al., 2018). The differences cause a ~40 percent difference between the two emissions estimates.

Together, these estimates suggest CH_4 emissions rate is one of the most critical parameters in an LCA of natural gas and products or fuels made from it. Judicious, transparent selection with a well-documented rationale based on the most recent inventory data is essential. As shale gas grows to comprise a larger share of domestic natural gas production, additional data collection may be necessary to characterize its contribution to natural gas life-cycle emissions more broadly.

Another complicating aspect of natural gas LCA is co-product handling. Natural gas is often co-produced with oil and natural gas liquids in ratios that are highly regionally dependent. For example, in the Eagle Ford Shale, wells can range from nearly entirely dry (i.e., producing almost entirely natural gas) to producing a mix of the three co-products (Chen et al., 2019).

The above-described emissions must be allocated among these products. One option for allocation is energy allocation because all products are used as energy carriers; this may create some inconsistency for instances where a co-product is a chemical input (see Allocation section, Chapter 6). Market value allocation would also be possible but is complicated by substantial fluctuations in the market values of the various co-products. Another option is to allocate each supply chain element's emissions to each product depending on whether that element is necessary to produce that co-product. In this approach, for example, emissions from tanks storing condensate oil would only be assigned to the oil. Furthermore, emissions could be evaluated at scales ranging from individual gas processing plants to regions within a basin, and at the basin-level. Chen et al. (2019) investigated the effects of co-product handling and choice of scale in these systems in the Eagle Ford Shale and reported that differences in GHG emissions per MJ of energy product could be up to 25 percent different depending on choice of co-product allocation technique. Accordingly, there is a need for consistent and transparent methodologies and metrics in co-product allocation approaches in natural gas LCA. For example, if emissions are assigned only to the natural gas product as opposed to across all co-products within a basin, natural gas emissions will be reported as significantly higher (again, depending on the amount of co-products) than if emissions are allocated among all co-products (Allen et al., 2021).

As Grubert and Brandt (2019) illustrate, choice of GWP in LCAs of natural gas can have a profound effect on results. The choice of GWP for CH_4 can have a large impact on the final CO_2e emissions impact of natural gas, even when applied to relatively small quantities of CH_4 leakage. First, the estimate of CH_4 GWP has evolved over the years in Intergovernmental Panel on Climate Change (PICC) reports based on enhanced scientific understanding (Figure 7-8). Further, it is possible to use a GWP in calculations that account for climate–carbon feedback. Finally, the choice of using a 20-year or 100-year GWP has a significant impact on results. Choice of GWP must be clearly documented in an LCA of natural gas and, ideally, results will be reported for both 20- and 100-year GWPs to show results' sensitivity to this choice.

Limited supply chain GHG emissions from natural gas systems are relatively straightforward to identify. Extended supply chain and market-mediated effects that influence GHG emissions are more challenging. Increased production of ethane as a co-product of natural gas has led to this natural gas liquid's main use as a feedstock for ethylene. This has some effect on petroleum refinery configuration and emissions. Oil produced from natural gas fields could serve as a tie-over as the nation transitions to increased electrification and could prolong dependence on liquid fuels made from fossil carbon. The natural gas market is tied closely to heating and power generation; the increased use of natural gas as a transportation fuel could drive up prices leading heating and power generation to turn to other energy sources, which could include biomass or coal in addition to renewables.

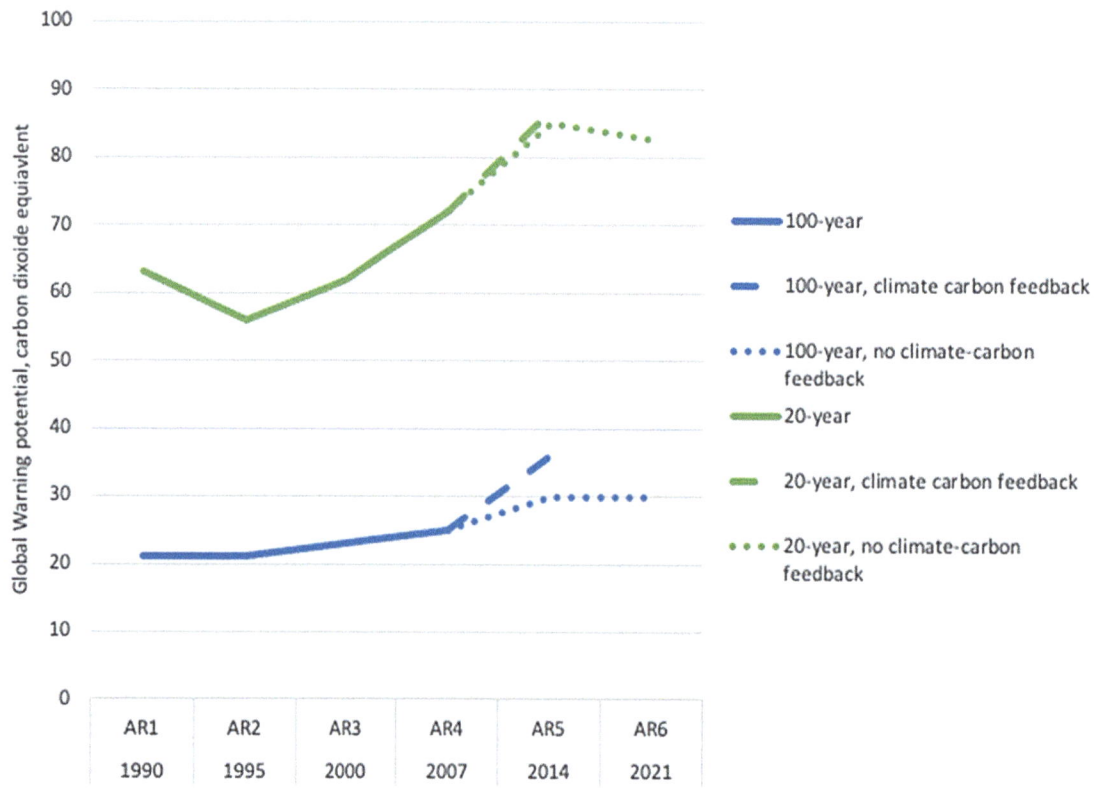

FIGURE 7-8 Evolution of the IPCC estimate of methane global warming potential from 1990 to 2021.
SOURCE: Reproduced from data in the IPCC Sixth Assessment Report.

Several factors complicate natural gas LCA, including spatial variability in the amounts and types of co-products produced and CH_4 emissions from venting, flaring, and leaking. These facets of natural gas systems will evolve over time as wells age, new wells are drilled, and industry adopts emissions-reducing practices.

Conclusion 7-2: More emissions inventory data from natural gas systems are needed, particularly regarding emissions from storage tanks.

Conclusion 7-3: The share of natural gas extracted from shale as a share of overall domestic consumption in the United States has increased rapidly and additional research and data collection will be necessary to better understand its production process and climate implications.

Top-down emissions estimates for the natural gas industry collected using satellites and field measurements often capture irregular or unintended emissions of CH_4 along the supply chain that may be excluded from bottom-up assessments. As more component-level emissions measurements are collected, there may be increasing ability to reconcile top-down and bottom-up emissions estimates.

Additional research and data collection are necessary for the LCA community to inform the judicious and consistent selection of natural gas LCA parameters and methods.

Conclusion 7-4: Assumptions on co-product handling methods have broad implications on natural gas LCAs. Additional research and data collection on industry practices can assist in the understanding of choice of co-product handling for natural gas production.

Conclusion 7-5: The selection of methane emissions leakage rate within an LCA has profound impacts on the overall estimated climate impact of natural gas production. Additional research and data collection is necessary to identify representative leakage rates for the natural gas industry and is essential to enable comparison of natural gas LCA across studies.

Recommendation 7-2: Further research should be done on the key parameters used to assess the climate impacts of natural gas production, such as methane leakage rates. These parameters will evolve as technology advances, data availability increases, and statistical methods may be used to translate the additional data into improved emissions estimates.

Recommendation 7-3: Further research on the climate impacts of natural gas production should draw upon real world activity data in part supplied by the natural gas industry and in part from independent studies using satellite and remote sensing technology to improve methane emissions rate estimates; these should be revisited frequently— at least every five years.

As described previously, hydrogen can be made from steam-methane reforming or authothermal reforming of CH_4 with (blue) or without (grey) CCUS. Currently, global CCUS capacity is limited to a few relatively small demonstration projects. In the case of blue and grey hydrogen, the LCA issues pertaining to natural gas as outlined in the previous section can significantly influence hydrogen LCA results.

For the proposed blue hydrogen schemes, the approach to handling CCUS is critical. It should be noted that CO_2 emissions are in both the steam-methane reforming and authothermal reforming product stream and flue stream. CO_2 is more difficult to capture in the latter case because it is less pure. Therefore, the rate of CO_2 capture should be specified explicitly for each stream that contains CO_2 so that a holistic rate of CO_2 capture is reported. This should hold true regardless of the location of CO_2 capture (e.g., prior to or at the flue gas outlet). Further, the ability of technology to capture CO_2 is improving with rates predicted to reach 90 percent in the future. Currently, rates at existing blue hydrogen facilities, which number only four, range from 29 percent to 43 percent (Gorski et al., 2021). The level of CO_2 capture from emissions streams accordingly will affect life-cycle GHG emissions of blue hydrogen. The use of present-day measured CO_2 capture rates rather than assumed future values has a significant impact on the net GHG savings associated with blue hydrogen (Howarth and Jacobsen, 2021). Other emission sources in steam-methane reforming and authothermal reforming include combustion of fuel to provide energy to the facilities, including, potentially, for operation of CCUS processes. The source of energy used for steam-methane reforming or authothermal reforming influences the life-cycle GHG emissions of blue hydrogen.

Other relevant LCA issues include the permanence of sequestered CO_2 in carbon capture and storage systems. Given that carbon capture and storage systems are relatively young, it is unclear how much leakage may occur in the future. Furthermore, treatment of sequestered carbon in carbon capture and storage systems has come under great scrutiny in the LCA community (Corsten et al., 2013). For example, it has been suggested that no emissions avoidance credit should be taken by the product system that uses CCUS for the CO_2 that is stored underground (Singh, 2021).

If the captured CO_2 in a blue hydrogen facility is converted into a second product (e.g., algae-derived fuel, polyol, or other chemical), LCA issues pertaining to carbon capture and utilization must come into play as described in a previous report (NASEM, 2018) and by von der Assen et al. (2013). In this instance, a co-product allocation method has to be selected to manage the division of energy and emissions burdens between the co-produced hydrogen and product derived from carbon capture and utilization.

Green hydrogen is to be primarily made using electrolysis powered by renewable electricity. The emissions attributable to this pathway take into account any energy consumption and associated emissions for delivering and purifying the water. One key uncertainty in the LCA of green hydrogen is whether the renewable electricity that is assumed to be used in its production is "additional." That is, it is desired that demand from electrolyzers for renewable electricity does not divert that electricity from other uses. If that occurred, the overall demand for electricity would rise. Accordingly, increased generation, potentially from fossil fuels like coal and natural gas, would be necessary. This renewable electricity additionality is a requirement in the European Union Hydrogen Strategy, which specifically states that additionality of the renewable energy used to make green hydrogen should be verified (Pototschnig, 2021). In the absence of additionality requirements or measures to ensure that renewable electricity is used to produce green hydrogen, the upstream GHG emissions attributable to fossil fuels in the grid would be attributed to the hydrogen and can therefore greatly increase its assessed emissions.[9]

There are several methods to demonstrate the additionality of renewable electricity used to produce green hydrogen. One way to ensure that renewable electricity is used to produce green hydrogen is to physically connect electrolyzers to solar or wind power as their sole power source. A second is to use a power purchase agreement or similar certification to link the production of green hydrogen and the generation of renewable electricity. However, the electrolyzers could consume electricity at a time when no electricity produced from renewable resources are provided to it. This type of certification could have varying levels of effectiveness; for example, the eligibility may require that the renewable electricity is not incentivized by any incentives beyond the power purchase agreement held by the green hydrogen producer. A final option is to produce green hydrogen during hours when only renewable electricity is produced as the marginal electricity source. In any case, proving additionality of renewable electricity that is used in green hydrogen production is complex (Bracker, 2017), and no best practice has been established. In LCA, a consequential framework would be best able to handle shifts in electricity generation in response to increased demand from electrolyzers. Attributional LCAs may adopt an average grid CI or assume renewable electricity is used in the production of green hydrogen as a means of comparing the CI of green hydrogen— assuming it is indeed produced using renewable electricity— against other fuels.

The main factors that influence blue hydrogen GHG emissions are the energy source for steam-methane reforming or authothermal reforming and the carbon capture efficiency, along with parameters used to model upstream natural gas production. The main factors that influence green hydrogen emissions are the additionality of the electricity used for green hydrogen production.

---

[9] EIA, 2020. Green Hydrogen: A Guide for Policymaking. See https://www.irena.org/-/media/Files/IRENA/Agency/Publication/2020/Nov/IRENA_Green_hydrogen_policy_2020.pdf.

Conclusion 7-6: The life-cycle emissions attributable to green hydrogen are sensitive to assumptions on the upstream source of electricity used for electrolysis, as the difference in emissions between hydrogen produced from renewable electricity and even grid-average electricity is substantial.

Recommendation 7-4: To ensure renewable electricity is supplied via the grid to produce green hydrogen in the context of an LCFS, certification is necessary to ensure that the source of the electricity and its additionality.

Recommendation 7-5: In the context of an LCFS, LCAs of hydrogen should be well documented with choices of key parameters supported with facility-measured data or well-supported citations from the literature. These key parameters include the choice of energy source for steam-methane reforming or authothermal reforming, the carbon capture level from the waste gaseous stream, source of upstream electricity, and the rate of methane or CO_2 leakage. Where relevant, the approach to quantifying emissions of upstream natural gas production should align with those used elsewhere in an LCFS for other fuels produced from natural gas.

REFERENCES

Allen, D. T., Q. Chen, and J. B. Dunn. 2021. Consistent metrics needed for quantifying methane emissions from upstream oil and gas operations. *Environmental Science & Technology Letters* 8(4):345-349. https://pubs.acs.org/doi/10.1021/acs.estlett.0c00907.

Allen, D. T., V. M. Torres, J. Thomas, D. W. Sullivan, M. Harrison, A. Hendler, S. C. Herndon, C. E. Kolb, M. P. Fraser, A. D. Hill, and B. K. Lamb. 2013. Measurements of methane emissions at natural gas production sites in the United States. *Proceedings of the National Academy of Sciences* 110(44):17768-17773.

Alvarez, R. A., D. Zavala-Araiza, D. R. Lyon, D. T. Allen, Z. R. Barkley, A. R. Brandt, K. J. Davis, S. C. Herndon, D. J. Jacob, A. Karion, and E. A. Kort. 2018. Assessment of methane emissions from the US oil and gas supply chain. *Science* 361(6398):186-188. https://www.science.org/doi/10.1126/science.aar7204.

Bracker, J. 2017. *An outline of sustainability criteria for synthetic fuels used in transport.* Oko Institute. Frieburg, Germany.

CARB (California Air Resources Board). 2021. Calculation of 2020 Crude Average Carbon Intensity Value. https://ww2.arb.ca.gov/sites/default/files/classic/fuels/lcfs/crude-oil/2020_crude_average_ci_value_final.pdf (accessed March 07, 2022).

CARB. n.d. LCFS Pathway Certified Carbon Intensities. https://ww2.arb.ca.gov/resources/documents/lcfs-pathway-certified-carbon-intensities (accessed March 07, 2022).

Chen, Q., J. B. Dunn, and D. T. Allen. 2019. Aggregation and allocation of greenhouse gas emissions in oil and gas production: Implications for life-cycle greenhouse gas burdens. *ACS Sustainable Chemistry & Engineering* 7(20):17065-17073. 10.1021/acssuschemeng.9b03136.

Chen, Y., E. D. Sherwin, E. S. Berman, B. B. Jones, M. P. Gordon, E. B. Wetherley, E. A. Kort, and A. R. Brandt. 2022. Quantifying Regional Methane Emissions in the New Mexico Permian Basin with a Comprehensive Aerial Survey. *Environmental Science & Technology.* https://pubs.acs.org/doi/10.1021/acs.est.1c06458.

Corsten, M., A. Ramirez, L. Shen, J. Koornneef, and A. Faaij. 2013. Environmental impact assessment of CCS chains – lessons learned and limitations from LCA literature. *International Journal of Greenhouse Gas Control* 13:59–71.

EIA (Energy Information Administration). 2022. Oil and petroleum products explained. https://www.eia.gov/energyexplained/oil-and-petroleum-products/.

EIA. 2022. Natural Gas Explained. Where our natural gas comes from. https://www.eia.gov/energyexplained/natural-gas/where-our-natural-gas-comes-from.php#:~:text=Natural%20gas%20is%20pro duced%20from%20onshore%20and%20offshore,10%25%20greater%20than%20U.S.%20total%20natural%20gas%20consumption (accessed September 2022).

EIA. 2020. Use of energy explained: Energy use for transportation https://www.eia.gov/energyexplained/use-of-energy/transportation.php (accessed March 07, 2022).

Elgowainy, A., J. Han, H. Cai, M. Wang, G. S. Forman, and V. B. DiVita. 2014. "Energy efficiency and greenhouse gas emission intensity of petroleum products at US refineries." *Environmental Science & Technology* 48(13):7612-7624.

Gorski, J. 2021. *Carbon intensity of blue hydrogen production: Accounting for technology and upstream emissions.* https://www.pembina.org/pub/carbon-intensity-blue-hydrogen-production

Gorski, J., T. Jutt and K. T. Wu. 2021. *Carbon Intensity of Blue Hydrogen Production: Accounting for Technology and Upstream Emissions.* Technical paper, Pembina Institute. https://www.pembina.org/reports/carbon-intensity-of-blue-hydrogen-revised.pdf.

Grubert, E. A., and A. R. Brandt. 2019. Three considerations for modeling natural gas system methane emissions in life cycle assessment. *Journal of Cleaner Production* 222:760-767.

IRENA (International Renewable Energy Agency). 2020. *Green Hydrogen: A guide to policy making.* International Renewable Energy Agency, Abu Dhabi.

Lauvaux, T., C. Giron, M. Mazzolini, A. d'Aspremont, R. Duren, D. Cusworth, D. Shindell, and P. Ciais. 2022. Global assessment of oil and gas methane ultra-emitters. *Science* 375(6580):557-561.

Masnadi, M. S., G. Benini, H. M. El-Houjeiri, A. Milivinti, J. E. Anderson, T. J. Wallington, R. De Kleine, V. Dotti, P. Jochem, and A. R. Brandt. 2021. Carbon implications of marginal oils from market-derived demand shocks. *Nature* 599:80–84. https://doi.org/10.1038/s41586-021-03932-2.

Masnadi, M. S., H. M. El-Houjeiri, D. Schunack, Y. Li, J. Englander, A. Badahdah, J. Monfort, J. E. Anderson, T. J. Wallington, J. A. Bergerson, D. M. Gordon, J. G. Koomey, S. V. Przesmitzki, I. M. Azevedo, X. T. Bi, J. E. Duffy, G. A. Heath, G. A. Keoleian, C. Mcglade, D. N. Meehan, S. Yeh, F. You, M. Wang, and A. R. Brandt. 2018. Global carbon intensity of crude oil production. *Science* 361:851-853.

NASEM (National Academies of Sciences, Engineering, and Medicine). 2018. *Developing a Research Agenda for Utilization of Gaseous Carbon Waste Streams.* Washington, DC: The National Academies Press.

Pototschnig, A. 2021. *Policy Brief Renewable hydrogen and the additionality requirement: why making it more complex than is needed?* Florence School of Regulation Issue 2021/35. September 2021.

Prussi, M., U. Lee, M. Wang, R. Malina, H. Valin, F. Taheripour, C. Velarde, M. D. Staples, L. Lonza, and J. I. Hileman. 2021. CORSIA: The first internationally adopted approach to calculate life-cycle GHG emissions for aviation fuels.

Ritchie, H., and M. Roser. 2020. Energy. Published online at OurWorldInData.org. https://ourworldindata.org/energy. https://ourworldindata.org/fossil-fuels#:~:text=In%202019%2C%20around%2084%25%20of,from%20coal%2C%20oil%20and%20gas.

Rutherford, J.S., E. D. Sherwin, A. P. Ravikumar, G. A. Heath, J. Englander, D. Cooley, D. Lyon, M. Omara, Q. Langfitt, and A. R. Brandt. 2021. Closing the methane gap in US oil and natural gas production emissions inventories. *Nat Commun* 12:4715. https://doi.org/10.1038/s41467-021-25017-4.

Sánchez-Bastardo, N., R. Schlögl, and H. Ruland. 2021. Methane pyrolysis for zero-emission hydrogen production: A potential bridge technology from fossil fuels to a renewable and sustainable hydrogen economy. *Industrial & Engineering Chemistry Research* 60(32):11855–81. https://doi.org/10.1021/acs.iecr.1c01679.

Sasol. 2021. Sasol Limited–Climate Change Report for the Year Ended 30 June 2021. https://www.sasol.com/sites/default/files/financial_reports/Sasol%20Climate%20Change%20Report_2021_22Sep21.pdf.

Schaaf, T., J. Grünig, M. R. Schuster, T. Rothenfluh, and A. Orth. 2020. Methanation of CO_2 - storage of renewable energy in a gas distribution system. *Energy, Sustainability and Society*, 4, 2. https://doi.org/10.1186/s13705-014-0029-1.

Singh, U., and L.M. Colosi. 2021. The case for estimating carbon return on investment (CROI) for CCUS platforms. *Applied Energy* 285:116394. https://doi.org/10.1016/j.apenergy.2020.116394.

von der Assen, N., J. Jung, and A. Bardow. 2013. Life-cycle assessment of carbon dioxide capture and utilization: Avoiding the pitfalls. *Energy & Environmental Science* 6:2721-2734. https://pubs.rsc.org/en/content/articlelanding/2013/ee/c3ee41151f.

Wang, M., H. Lee, and J. Molburg. 2004. "Allocation of Energy Use in Petroleum Refineries to Petroleum Products – Implications for Life-Cycle Energy Use and Emission Inventory of Petroleum Transportation Fuels," *International Journal of Life-Cycle Assessment* 9(1):34–44.

Willyard, K. A. 2020. A license to pollute? Opportunities, incentives, and influences on oil and gas venting and flaring in Texas. *Energy Research & Social Science* 62:101381.

Zhang, Y., R. Gautam, S. Pandey, M. Omara, J. D. Maasakkers, P. Sadavarte, D. Lyon, H. Nesser, M. P. Sulprizio, D. J. Varon, and R. Zhang. 2020. Quantifying methane emissions from the largest oil-producing basin in the United States from space. *Science Advances* 6(17):5120.

8
Aviation and Maritime Fuels

AVIATION FUELS

The life-cycle climate impacts of aviation fuels have been evaluated in the academic literature and as part of regulatory assessments for several fuel policies. There has been analysis of both conventional, petroleum-derived jet fuel and of a variety of alternative fuels produced through a wide array of conversion processes (i.e., alternative aviation fuels). In this section, the term "alternative aviation fuels" (AAFs) is used to refer to alternatives to conventional fossil aviation fuel. The term "sustainable aviation fuels" has been commonly used to describe alternative (non-petroleum) aviation fuels and is used in some policy contexts to refer to aviation pathways that satisfy certain sustainability criteria. However, this term in a general context may not necessarily indicate environmental benefits. Therefore, the term "sustainable aviation fuels" is not used in this report because it suggests an endorsement of the environmental benefits for all non-petroleum aviation fuels. For the purposes of evaluating the climate impacts of AAFs within fuels policies, this section discusses several key areas that may require special consideration beyond the approaches used for alternative fuels used in other sectors: (1) the non-carbon dioxide (CO_2) effects of aviation fuels when combusted at high altitudes, (2) the impacts of alternative fuels on airplane efficiency, and (3) the impact of a flexible product slate on the life-cycle emissions calculations for aviation fuels. Some of these effects are discussed further in the subsequent sections.

The standards organization American Society for Testing and Materials (ASTM) International has certified seven types of AAFs under its ASTM D7566 standard; this certification ensures the physical and chemical characteristics of fuels and their operational performance up to a specific blend level for each fuel (Prussi et al., 2021). Key criteria include composition, volatility, and stability; these fuels are suitable for commercial use as "drop-in" fuels when blended with conventional jet fuel up to their maximum blend level. ASTM International certification does not determine the technology-readiness level or sustainability of certified fuels. An overview of the ASTM International–approved and pending approval upcoming AAF pathways and their likely feedstocks is provided in Table 8-1. In addition to liquid fuels, energy supplied by electricity and hydrogen to alternative airframe designs can also be considered an AAF, though these technologies do not go through the ASTM International liquid fuel certification process (Viswanathan et al., 2022). Given that each pathway may utilize different feedstocks with varying environmental implications, a life-cycle analysis (LCA) is used to evaluate the impacts of each pathway relative to petroleum jet fuel, as well as to assess the impacts of different feedstocks of the production systems within each pathway.

Prior LCAs of AAFs have used a variety of analytical approaches and scopes, with a wide variation in results based on the authors' assumptions and methodology. Across the literature, a major source of variation in the emissions estimates for AAFs is associated with the types of feedstocks; within a given conversion pathway, there are often some variations in emissions outcomes as across different conversion pathways. As shown in Table 8-2, most analyses have taken a primarily attributional approach, estimating on the energy and emissions directly attributable to feedstock production through to final use, without attributing market-mediated effects to AAFs. We note that the existing literature largely consists of attributional analyses and relatively few consequential assessments of the impact of AAFs; this is a limitation within the literature and is not meant to imply that market-mediated emissions or the consequential LCA (CLCA) approach in general are not an important consideration in the assessment of AAFs. The ratio of attributional LCA (ALCA) to CLCA approaches in Table 8-2 may not represent the importance or significance of the CLCA approach for AAFs. Across the literature, allocation approaches vary considerably by study; in some cases, studies utilize process-level allocation, wherein the allocation approach may vary depending on the life-cycle stage. For example, the impacts of producing biomass co-products such as soy oil and soy meal may be allocated on a mass basis, whereas the products from a bio-refinery may be allocated on an energy basis. Some studies, such as Han et al. (2017), utilized a system expansion approach to attribute impacts from co-products with a different performance metric than energy, such as for the dried distillers' grains with solubles co-products of corn alcohol to jet production. Many studies described in Table 8-2 include a sensitivity analysis to illustrate the impact of allocation assumptions on their emissions estimates. The Allocation section (Chapter 6) provides additional background on the methodology of allocation and its role in LCA.

Carbon Offsetting and Reduction Approaches

Several fuels policies also incorporate the LCAs of aviation fuel climate impacts, which are examined in the following sections.

TABLE 8-1 Summary of Approved and Pending Alternative Aviation Fuel Production Pathways

Fuel	Blend Level	Typical Feedstocks	Status
Hydroprocessed esters and fatty acids synthetic paraffinic kerosene (HEFA-SPK)	50%	Vegetable oils; waste fats, oils and greases	Approved in 2011
Hydroprocessed fermented sugars to synthetic isoparaffins (HFS-SIP)	10%	Sugar crops	Approved in 2014
Fischer-Tropsch synthetic paraffinic kerosene with aromatics (FT-SPK/A)	50%	Lignocellulosic crops, residues and wastes	Approved in 2015
Alcohol to jet synthetic paraffinic kerosene (ATJ-SPK)	50%	Starchy and sugary crops; lignocellulosic crops, residues and wastes; industrial flue gases	Approved in 2016
Co-processing in petroleum refinery	N/A	Vegetable oils; waste fats, oils and greases	Approved in 2018
Fischer-Tropsch synthetic paraffinic kerosene (FT-SPK)	50%	Lignocellulosic crops, residues and wastes	Approved 2019
Catalytic hydrothermolysis synthesized kerosene (CH-SK, or CHJ)	50%	Vegetable oils; waste fats, oils and greases	Approved in 2020
Integrated hydropyrolysis and hydroconversion (HC-HEFA-SPK)	10%	Lignocellulosic crops, residues and wastes	Approved in 2020
High freeze point hydroprocessed esters and fatty acids synthetic kerosene (HFP HEFA-SK or HEFA+)	10%	Vegetable oils; waste fats, oils and greases	In progress
Hydro-deoxygenation synthetic aromatic kerosene (HDO-SAK)	N/A	Starchy and sugary crops; lignocellulosic crops, residues and wastes	In progress
Alcohol-to-jet synthetic kerosene with aromatics (ATJ-SKA)	N/A	Starchy and sugary crops; lignocellulosic crops, residues and wastes	In progress

TABLE 8-2 Sample of Published LCAs of Aviation Fuels (inclusion in this list does not imply endorsement by this committee)

Study	Fuel Pathways	Feedstocks	Region	LCA Methodology	Co-Product Handling Methods
Stratton et al. (2010)	Conventional jet fuel, HEFA-SPK, FT-SPK	Crude oil (average, ultra-low sulfur, oil sands, oil shale), soy, palm, rapeseed, jatropha, algae, Salicornia, coal, natural gas, switchgrass	United States	Process-based attributional	Primarily energy allocation. Sensitivity analysis of other methods.
Elgowainy et al. (2012)	HEFA-SPK, FT-SPK, pyrolysis-to-jet	Crude oil (conventional, oil sands, average mix), natural gas, coal, soy, corn stover, algae	United States	Process-based attributional	Process-level allocation, primarily energy allocation with system expansion for some co-products.
Han et al. (2013)	HEFA-SPK, FT-SPK, pyrolysis-to-jet	Crude oil, coal, jatropha, rapeseed, camelina, soy, palm, corn stover	United States	Process-based attributional	Primarily energy allocation. Sensitivity analysis of other methods.
Cox et al. (2014)	HEFA-SPK, HFS-SIP	Algae, pongamia, sugarcane molasses	Australia	Process-based attributional	Economic allocation, with system expansion as a sensitivity analysis.
Staples et al. (2014)	ATJ-SPK HFS-SIP	Corn, sugarcane, switchgrass	United States	Process-based attributional	Economic allocation, with system expansion as a sensitivity analysis.
Seber et al. (2014)	HEFA-SPK	Used cooking oil, tallow	United States	Process-based attributional; comparison of tallow as a by-product vs. a co-product via sensitivity analysis	Primarily energy allocation. Sensitivity analysis of other methods.
Moreira et al. (2016)	HFS-SIP	Sugarcane	Brazil	Process-based attributional with consequential ILUC added.	System expansion
DeJong et al. (2017)	HEFA-SPK, FT-SPK, ATJ-SPK, HFS-SIP, Pyrolysis-to-Jet, Hydrothermal Liquefaction,	Used cooking oil, jatropha, camelina, willow, poplar, corn stover, forestry residues, corn, sugar cane	European Union	Process-based attributional	Primarily energy allocation with hybrid displacement approach. Sensitivity analysis of other methods.
Han et al. (2017)	ATJ-SPK STJ	Corn, corn stover	United States	Process-based attributional	Hybrid approach (energy and system expansion)
Suresh et al. (2018)	FT-SPK	MSW	United States	Primarily process-based attributional; includes indirect emissions for avoided methane at landfills	Primarily energy allocation. Sensitivity analysis of other methods.

continued

TABLE 8-2 continued

Study	Fuel Pathways	Feedstocks	Region	LCA Methodology	Co-Product Handling Methods
O'Connell et al. (2019)	HEFA-SPK, FT-SPK, pyrolysis-to-jet	Rapeseed, sunflower, soybean, palm, forest residues, short-rotation forest, wheat straw	European Union	Primarily process-based attributional; use of marginal production values and discussion of ILUC	Primarily energy allocation. Sensitivity analysis of other methods.
GREET (2021)	HEFA-SPK, FT-SPK, ATJ-SPK, pyrolysis jet, catalytic sugars-to-hydrocarbons	Soy, palm, canola, jatropha, camelina, corn oil, algae, corn, agricultural residues, forest residues, lignocellulosic energy crops, coal, natural gas, biomethane, electricity	United States	Process-based attributional; consequential option for ILUC emissions via CCLUB	Process-level allocation. Option for energy, mass, market value and displacement.
Prussi et al. (2021)	HEFA-SPK, FT-SPK, ATJ-SPK, HFS-SIP	Soy, palm, canola, camelina, corn oil, algae, agricultural residues, forest residues, MSW, lignocellulosic energy crops	United States, European Union, Brazil, Southeast Asia	Process-based attributional analysis with consequential ILUC assessment	Energy allocation

NOTES: Inclusion in this list does not imply endorsement by this committee. ATJ-SPK = alcohol-to-jet synthetic paraffinic kerosene; CCLUB = Carbon Calculator for Land Use and Land Management; FT-SPK = Fischer-Tropsch synthetic paraffinic kerosene; HEFA-SPK = hydroprocessed esters and fatty acids synthetic paraffinic kerosene; HFS-SIP = hydroprocessing of fermented sugars—synthetic iso-paraffinic kerosene; ILUC= indirect land use change; MSW = municipal solid waste; STJ = sugar-to-jet.

The Carbon Offsetting and Reduction Scheme for International Aviation

The Carbon Offsetting and Reduction Scheme for International Aviation (CORSIA) was developed and adopted by the International Civil Aviation Organization (ICAO) to reduce and offset a portion of the growth in greenhouse gas (GHG) emissions for international aviation. This offsetting scheme includes an assessment of the life-cycle emissions of a selection of various qualifying AAFs (ICAO, 2019; Prussi et al., 2021). These default values are intended to provide an accounting method for tracking some GHG reductions from petroleum displacement through the use of AAF, and are not presented as reflecting the view of this committee. In CORSIA, GHG emissions reductions may be generated by subtracting the life-cycle emissions for AAFs from the fossil fuel baseline, calculated as 89 gCO_2e for jet fuel and 95 gCO_2e/MJ for aviation gasoline. More information on the variation in upstream carbon intensity (CI) of fossil fuels is in Chapter 7.

The methodological approach for LCA in CORSIA is summarized by Prussi et al. (2021) and presented in detail in ICAO (2019). The assessment takes a primarily attributional approach using energy allocation, in addition to a consequential induced land use change (ILUC) (as described by CORSIA) assessment for each crop-based fuel pathway. Emissions estimates from each pathway are broken into emissions values for "core LCA values" and "ILUC LCA value." The core-LCA reflects the attributional emissions for each fuel, from its feedstock production through to final combustion. The ILUC assessment is based on a set of pathway-specific demand shocks input into the Global Trade Analysis Project BIO (GTAP-BIO) and the Global Biosphere Management (GLOBIOM) economic models, which reflect policy demand for a mix of alternative jet and road transportation biofuels co-products. Added together, the default core-LCA and ILUC emission factors for pathways can be compared to the fossil fuel baseline to determine a fuel's eligibility relative to the 10 percent GHG reduction threshold for CORSIA eligibility.

CORSIA has defined a set of sustainability criteria. According to these criteria, an eligible AAF cannot be produced from feedstock "made from biomass obtained from land converted after 1 January 2008 that was primary forest, wetlands, or peat lands and/or contributes to degradation of the carbon stock in primary forests, wetlands, or peat lands as these lands all have high carbon stocks" (CORSIA, 2021). CORSIA also obligates AAF producers to also determine direct land use change (LUC) emissions for eligible land conversions that occurred after January 2008 and replace ILUC emissions with direct LUC emissions in cases where the estimated direct LUC is larger than ILUC. This approach, though implemented as a safeguard, may nevertheless ignore some ILUCs.

On the other hand, CORSIA suggests that certain land types, land management practices, and innovative agricultural practices can be considered to contribute to low risk for land area change. As a result, aviation feedstocks produced from these lands, upon check and verification, can receive a value of zero for ILUC in the LCA of a batch of fuels.

CORSIA specifies two approaches for low LUC risk aviation fuel feedstock production: the Yield Increase Approach, and the Unused Land Approach. For the Unused Land Approach, CORSIA (ICAO 2019, 10-11) states: "Eligible lands for the unused land approach could include, among others, marginal lands, underused lands, unused lands, degraded pasture lands, and lands in need of remediation." In order to qualify as sustainable aviation fuel feedstock under the low land use risk category, certification is required by one of the CORSIA approved certification schemes. The certification schemes, in turn, work with CORSIA on the technical implementation of the policy. As a result, technical documents have been submitted to CORSIA to evaluate which types of land could in practice qualify under the low LUC risk land category. For example, a report by the University of Illinois at Chicago and Southern Illinois University Edwardsville (Mueller et al., 2021) explored qualifying reclaimed coal mining land for the carbon credits under this category. The low LUC risk categories as implemented within CORSIA are policy instruments intended to promote low LUC GHG values of a fuel produced within this framework, though their precise impact on emission remains uncertain.

CORSIA is unique as it is a global rather than national or regional low-carbon transportation policy effort. Therefore, when it came to ILUC modeling its guiding technical committee, the Committee on Aviation Environmental Protection (CAEP), had to oversee and negotiate the use of a larger range of possible models and input parameters than was the case for past national efforts. On the one hand, multiple models can reduce methodological uncertainties. The CORSIA "Life Cycle Assessment Methodology Document" states: "Because no model can pretend completeness of the representation, the comparison of different model results can help address model design uncertainty." On the other hand, differing results and parameters had to be reconciled as part of the CORSIA LCA process. CORSIA's ILUC modeling is based on two economic models, the GTAP-BIO developed at Purdue University and the GLOBIOM model developed by the International Institute for Applied Systems Analysis in Austria. These models represent two very different modeling approaches. The results from both models were reviewed by ICAO's Fuels Task Group experts and reconciled into default ILUC values. The alignment between the two models varied across the 17 pathways reported by Prussi et al. (2021). In most cases, the differences were relatively small and close across the two models. Therefore, the average of the model's results for each of these pathways has been adopted as the default ILUC value. However, for other pathways results were significantly different across the two models. For these pathways, after a review process, the Fuels Task Group assigned the lower of the two ILUC values from the modeling and added an adjustment factor of 4.45 gCO_2e/MJ.

The U.S. Renewable Fuel Standard

The U.S. Environmental Protection Agency (EPA) has also assessed the LCA emissions of various AAFs pathways to determine their eligibility under the Renewable Fuel Standard (RFS). Approved pathways include hydroprocessed esters and fatty acid fuels produced from soy, algae, waste fats, oils and greases, camelina, corn oil, and sorghum oil (EPA, n.d.). Within the RFS, these fuels' life-cycle emissions have been assessed by EPA and meet, according to EPA's methods, a 50 percent GHG reduction threshold in 2022 relative to the diesel comparator of 94 gCO_2e/MJ, including the impacts of ILUC; they are eligible for D4 and D5 Renewable Identification Numbers (see Chapter 3) as either biomass-based diesel or advanced biofuel, respectively.

The California Low-Carbon Fuel Standard

The California low-carbon fuel standard (CA-LCFS) was amended in 2019 to include aviation fuels. Unlike petroleum fuels used in the road sector, jet fuel is not subject to the CA-LCFS regulation and does not generate deficits. Rather, the production and blending of AAFs can generate credits as an "opt-in" fuel; credits are generated by subtracting the AAF's CI from benchmarks calculated by the California Air Resources Board (CARB). Starting with a baseline CI of 87 gCO_2e/MJ that reflects the average CI of jet fuel consumed in California, the jet fuel benchmark remains fixed at the 2010 baseline CI for conventional jet fuel, with a zero percent reduction in each year, until the benchmark for diesel substitutes declines below the CI baseline for jet fuel in 2023 (CARB, 2020). From 2023 onward, the CI decline for jet fuel will move in parallel with that for diesel fossil fuel.

Non-CO_2 Effects of Aviation Fuels

Beyond the climate forcing contributions of conventional GHGs such as CO_2, methane (CH_4) and nitrous oxide (N_2O) throughout the life cycle of aviation fuels, jet fuel combustion at altitude may also contribute to climate change through other forms of radiative forcing. Incomplete combustion of jet fuel may result in nitrogen oxides (NO_x) formation, aerosolized sulfates and soot, contrail formation, and contribution to cirrus cloud formation (IPCC, 1999). These emissions, in conjunction with NO_x, sulfates, and water vapor, may contribute to contrails and increased cirrus cloud formation, together called aviation-induced cloudiness. Notably, the magnitude and sign of these effects differs substantially—for example, sulphate aerosols are a negative radiative forcer whereas contrails provide a net warming effect. Figure 8-1 illustrates the possible routes that incomplete aviation fuel combustion may contribute to radiative forcing. These impacts may warrant consideration in LCA approaches for aviation fuels, as they are partly attributable to the types of fuels combusted; therefore, a portion of these impacts may be attributable to fuels. This raises issues of both how to set a baseline for aviation fuels and how to estimate the difference in overall climate impact between conventional aviation fuels and AAFs.

A recent analysis of the cumulative contribution of aviation to the climate through 2018 estimates with greater precision the contribution of aviation-induced cloudiness. Lee et al. (2021) estimates that non-CO_2 effects comprise approximately 2/3 of aviation's current radiative forcing. Figure 8-2 separates out the effective radiative-forcing impacts of AAFs into several discrete categories in milliWatts per square meter (mW/m^2). Each component, which may have either a cooling or a warming effect, is shown with a best estimate and a confidence interval. NO_x emissions at high altitudes can have a variety of different effects that may both warm and cool the atmosphere. Summed together, these impacts add up to 17.5 mW/m^2. However, the largest contributor to the total impact is contrails and increased cirrus cloud formation, with an impact of over 50 mW/m^2. When taking into account non-CO_2 effects, the CO_2 combustion impact of aviation alone declines to approximately 34 percent of the total. In particular, the impacts of NO_x, soot, and contrail formation are estimated to add significantly to the overall climate impact of aviation.

The aromatic content of jet fuel is strongly correlated to the soot emissions. The aromatic content of jet fuel is an important factor in determining its operational performance, as it affects density, boiling point, smoke point, and freeze point for the fuel. Aromatic content is limited to a maximum of 25 percent and minimum of 8 percent by volume as part of the ASTM International certification for conventional jet fuel. Within that range, there may be substantial variation in aromatic content for fossil jet fuel (Hemighaus et al., 2006). Reduced aromatic content can generate a decrease in ice crystal numbers in contrails, but also an increase in water vapor. The net impact of the change in the fuels' aromatic content remains highly uncertain and requires additional testing and measurement, particularly to quantify the impact of sustainable aviation fuel blending. Early research suggests that reduced aromatic content in fuel could lead to a reduction in optical depth of the contrails, shorter contrail lifetimes, and overall decreased radiative forcing. (Bräuer et al., 2021). Klower et al. (2021) parameterized the impact of low aromatic AAFs on contrail cirrus formation as a function of the square root of the fossil fuel share, implying that a 50 percent reduction of jet fuel aromatic content via AAF blending reduces contrail cirrus radiative forcing by about 30 percent.

Reducing aromatic content through lower-aromatic conventional fossil jet fuel or added AAF content could therefore reduce contrail cirrus radiative forcing.

The uncertainty of non-CO_2 contributions to the climate impact of aviation may be significantly higher than that of CO_2 from jet fuel combustion; Lee et al. (2021) estimate that non-CO_2 forcing terms contribute about eight times more than CO_2 to the uncertainty in aviation's net effective radiative forcing. The non-CO_2 effects of aviation fuels raise several important questions for LCA of aviation fuels and in turn, for aviation fuel policy.

Though existing LCAs of aviation fuels have generated a set of estimates for the emissions of conventional GHGs released from the well-to-wake production and use of aviation fuels, compared to the quantity of existing literature on the life-cycle GHG emissions for AAF production, the data are more sparse on the non-CO_2 effects of AAFs. Preliminary research suggests that lower aromatic content in aviation fuels may result in reduced contrail formation (Voight et al., 2021). The magnitude and direction of some effects may differ based on the use case, route, and altitude of the flight in question; for example, non-CO_2 emissions at stratospheric altitudes from supersonic airplanes would differ from the impacts of subsonic flights consuming the same fuel at lower altitudes.

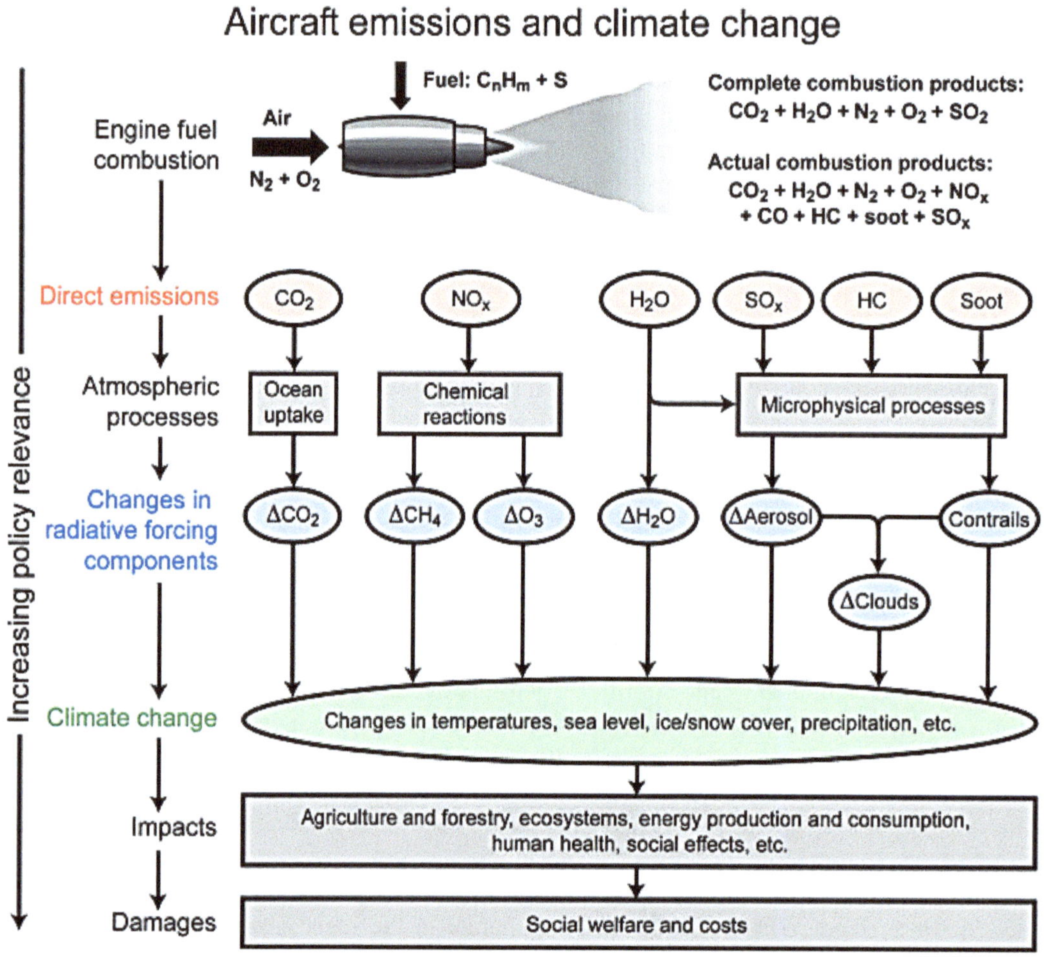

FIGURE 8-1 The principal emissions from aviation operations and the atmospheric processes that lead to changes in radiative forcing components. SOURCE: Reproduced from Lee et al. (2009). Reprinted from Atmospheric Environment, Elsevier.

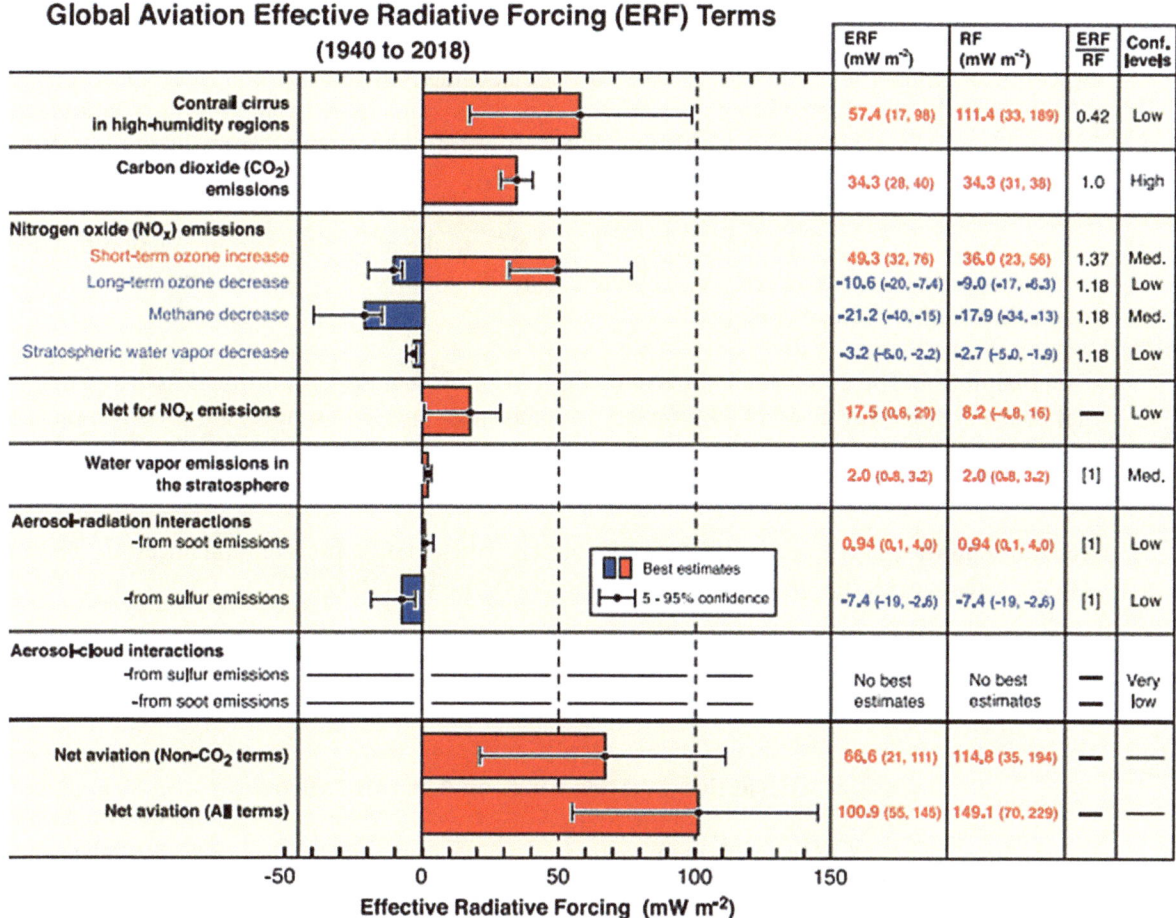

FIGURE 8-2 Radiative forcing components in milliwatts per square meter from global aviation as evaluated from preindustrial times until 2018. SOURCE: Reproduced from Lee et al. (2021, Fig. 3). Reprinted from Atmospheric Environment, Elsevier.

It is challenging to present the non-CO_2 impacts of aviation fuel combustion on a consistent basis alongside the climate impacts of the fuel cycle, largely due to the different time scales. The fuel cycle is dominated by CO_2 emissions and thus is not sensitive to assumptions of time horizon; in contrast, the impact of non-CO_2 emissions is much more sensitive to the assumptions of the time horizon and operating conditions (altitude and atmospheric conditions). Non-CO_2 effects have the highest warming impact at short time-scales, with the CO_2 impact overshadowing other warming effects on longer time scales (Fahey et al., 2016). Alternative metrics such as global temperature potential (GTP) and average temperature response (ATR) suffer from the same problem, as they remain sensitive to assumptions of time horizon (EASA, 2020).

Combusting aviation fuels in flight releases CO_2 in addition to a mix of other pollutants such as soot, water vapor, sulfates, and NO_x capable of generating climate impacts by interacting with the atmosphere at high altitudes.

Conclusion 8-1: Non-CO_2 effects from aviation fuels may be high but remain uncertain. The largest non-CO_2 impact from aviation fuel combustion may be aviation-induced cloudiness, with the remaining contributions being much smaller.

Conclusion 8-2: Reduced aviation fuel aromatic content, whether through processing of fossil fuels or blending of alternative aviation fuels may have beneficial climate effects on non-CO_2 emissions. However, additional research is necessary to more accurately assess the contribution of non-CO_2 effects from the fuel cycle on a consistent basis with existing LCA of alternative aviation fuels. Due to the non-linearity of these effects, additional testing is necessary to evaluate the effect of alternative aviation fuel blending on non-CO_2 emissions.

Conclusion 8-3: The overall addition of CO_2 and NO_x to the atmosphere from aviation fuel combustion is well-characterized; however, there is substantial uncertainty on the emissions of sulfur, soot, and aviation-induced cloudiness.

Conclusion 8-4: The combustion emissions from aviation fuel are proportional to the quantity of fuel consumed. However, non-CO_2 impacts are non-linear and do not necessarily correspond proportionally to fuel switching. Furthermore, changes in airplane routing, such as location, altitude, and time of day may also influence non-CO_2 impacts of aviation.

Recommendation 8-1: Because the non-CO_2 effects from aviation fuels remain uncertain, research should be done to clarify the magnitude and direction of these effects.

Conclusion 8-5: Though there is evidence that fuel blending can mitigate the impact of some non-CO_2 climate forcing; attributing these impacts to fuel switching in policies may result in inaccurate crediting of these fuels.

Impacts of Alternative Aviation Fuels on Aircraft Efficiency

Using alternative fuels or electrification technologies (e.g., batteries or fuel cells) for aviation will alter aircraft efficiency if they impact the total weight of the aircraft, which is particularly important during takeoff. For example, Bills et al. (2020) illustrated the impact of batteries on aircraft weight in their 2020 article (see Figure 8-3). During the climb phase of the flight, energy consumption per unit time approximately doubles relative to the cruise phase of the flight. Because of restrictions on takeoff weight, increasing or decreasing the weight of the aircraft through the use of alternative fuels or electrification technologies will impact the payload an aircraft is capable of carrying and the maximum range for long flights. The length of a given flight will also affect the overall energy use and life-cycle emissions per passenger-kilometer or tonne-kilometer transported; shorter flights will have larger impacts per unit cargo-kilometer because of the outsized impact that the climb phase will have on total trip-level energy use.

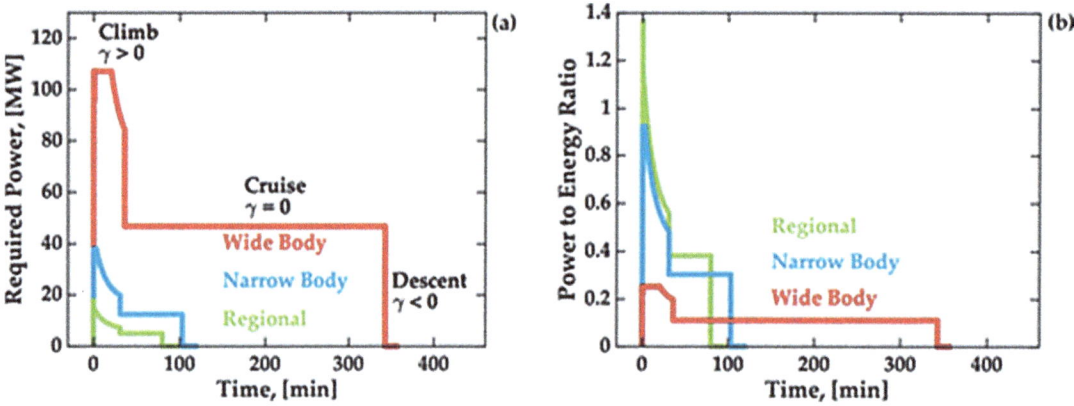

FIGURE 8-3 Aircraft power profiles and power to energy ratio by flight segment. SOURCE: Bills et al. (2020). Reprinted with permission from ACS Energy Letters, American Chemical Society.

So far, there is no standard set of aircraft design and flight duration assumptions used by researchers or LCA practitioners for comparative analysis of AAFs. For liquid fuels that can be combusted in conventional jet engines, the Breguet Range Equation can be used to estimate the impact of more or less energy-dense fuels on total aircraft efficiency, given a set of specific aircraft specifications and trip length, as illustrated in Baral et al. (2019). In most cases, the fuel savings associated with a more energy-dense liquid fuel will be small because these fuels will be blended with conventional petroleum jet fuels. Given the small expected differences in aircraft efficiency, comparing emissions savings on a per-MJ higher heating value content basis for most AAFs is appropriate. Table 8-3 provides an overview of different fuels and energy carriers used in aviation and provides their energy density and density.

Assigning appropriate functional units when comparing across battery-electric aircraft, fuel cell electric aircraft, and aircraft with jet engines running on liquid fuels is more complex. Such cross-comparisons have been done in the scientific literature for passenger vehicles by using 1 km or mile traveled by a typical sedan as a functional unit (e.g., see Yuksel and Michalek, 2015). Chester and Horvath (2009) similarly explored vehicle-km and passenger-km as different functional units, noting that ridership impacts the latter unit on mass transportation modes where ridership may be well under maximum capacity in some cases. This approach works well for research applications, where introducing additional complexity is more acceptable when that complexity advances the goal of providing the most accurate comparison possible.

TABLE 8-3 Comparison of Physical Properties across Different Energy Carriers Used for Aviation

Energy Carrier/Storage	Energy Stored per Unit Mass		Density (kg/liter)
	Higher heating value (MJ/kg)	Net heat of combustion (MJ/kg)	
Jet A		> 42.8	0.775-0.84
Jet A-1		> 42.8	0.775-0.84
Jet B		> 42.8	0.75-0.801
Li-ion Battery Pack (current technology)			
	0.936 (Gray et al. 2021)		2.81 (Gray et al. 2021)
Hydrogen Fuel Cell (compressed gaseous H_2), including storage tank			
	5 (Mukhopadhaya and Rutherford, 2022)		0.04 (Mukhopadhaya and Rutherford, 2022)
Hydrogen Fuel Cell (liquid H_2), including storage tank			
	8.5 (Mukhopadhaya and Rutherford, 2022)		0.07 (Mukhopadhaya and Rutherford, 2022)
Alternative Liquid Fuels			
Hydroprocessed esters and fatty acids (HEFA) synthetic paraffinic kerosene (HEFA-SPK)		43.9 (Huq et al., 2021)	0.73-0.77 (Van Dyk and Saddler, 2021)
Alcohol-to-jet SPK (ATJ-SPK)		43.9 (Huq et al. 2021)	0.73-0.77 (Van Dyk and Saddler, 2021)
Fischer Tropsch-SPK (FT-SPK/FT-SPK-A)		43.7-44.1 (Huq et al., 2021)	0.73-0.8 (Van Dyk and Saddler, 2021)
Limonane	46.32 (Baral et al., 2019)	43.41 (Baral et al., 2019)	0.804 (Baral et al., 2019)
Bisabolane	46.66 (Baral et al., 2019)	43.76 (Baral et al., 2019)	0.814 (Baral et al., 2019)
Epi-isozizaane	44.89 (Baral et al., 2019)	42.33 (Baral et al., 2019)	0.929 (Baral et al., 2019)
RJ-4	45.06 (Baral et al., 2019)	42.59 (Baral et al., 2019)	0.92 (Baral et al., 2019)
Dimethylcyclooctane		43.82 (Rosenkoetter et al., 2019)	0.827 (Rosenkoetter et al., 2019)

Conclusion 8-6: The blending of alternative aviation fuels with different energy densities, as well as the introduction of alternative technologies such as electric-drive or hydrogen-powered airframes, may change the operating weight and efficiency of aircraft. In order to compare the emissions of these alternative fuels to conventional jet fuel on a consistent basis, these changes in efficiency must be taken into consideration.

Recommendation 8-2: Alternative fuels and airframe combinations, particularly those with large density differences such as battery electric technology and hydrogen, may impact airplane efficiency and thus influence overall emissions. The comparative LCA of these technologies should use functional units based on the transportation service provided or otherwise be based on comparison of consistent transportation services.

Effects of a Mixed Product Slate

Many AAF production pathways generate a variety of hydrocarbons as part of their product slate; in some cases, jet fuel is not even the largest share of the end product (either on an energy or market value basis) (Pearlson et al., 2013; Stratton et al. 2010). For some AAF processes, such as hydroprocessed esters and fatty acids synthetic paraffinic kerosene (HEFA-SPK) and Fischer–Tropsch synthetic paraffinic kerosene (FT-SPK), it may be impossible to generate a 100 percent jet product slate. Though these processes are currently optimized to maximize the share of middle distillates in their product slate, it may be possible in some cases to increase the jet fraction, while simultaneously increasing the share of less-valuable light ends such as naphtha and propane (Pearlson et al., 2013). Increasing the jet fraction requires additional hydrogen and may therefore increase the overall energy consumption and emissions for a biorefinery.

For the purposes of an LCA of jet fuel, the operating parameters of a biorefinery that produces jet fuel as a part of an overall product slate may have important implications on its climate impact. For example, a middle distillate-optimized biorefinery may have a different overall energy consumption and product slate than a jet-optimized biorefinery, changing overall emissions and the relative shares of co-products.

For AAF production pathways that generate both an aviation and a road fuel co-product, maximizing the share of AAF output requires additional energy and may reduce overall biofuel yields, increasing emissions attributable to AAF production.

Conclusion 8-7: There are some variations in the life-cycle emissions attributable to alternative aviation fuels at facilities for some fuel pathways, depending on whether they are configured to maximize alternative aviation fuel output or to maximize yields of other co-products, such as middle distillates.

Recommendation 8-3: Alternative aviation fuel LCA estimates developed for fuel policy should reflect existing practices at facilities or the expected behavior in response to future policies.

MARITIME FUELS

International goods movement on ocean-going vessels carried 90 percent of international merchandise trade (11 billion tons) in 2016 (Li, 2020). In transporting this large amount of goods, marine vessels consume about half of fuel oil demand (IEA, 2020). Fuels for these vessels are undergoing notable changes mandated in the International Convention for the Prevention of Pollution from Ships (MARPOL) Annex VI treaty. These rules require cuts in fuel sulfur levels from 3.5 percent to 0.5 percent m/m for vessels operating outside of designated emission control areas, which have a different set of more stringent emissions limits. They went into effect in 2020 and affect the consumption of predominant marine fuels including higher sulfur fuel oil. In 2018, the International Maritime Organization adopted a resolution on the initial strategy to reduce GHG emission from ships. It has set targets of 20 percent GHG emissions reductions from ships in 2020, 40 percent by 2030, and 50 percent by 2050. The method for achieving these targets remains under development. In the meantime, MARPOL Annex VI undoubtedly will influence life-cycle GHG emissions of marine fuels and goods transport as the marine sector adopts technology to comply with it.

One main MARPOL Annex VI compliance strategy is using low sulfur fuels. Oil-derived options include very low sulfur fuel oil and marine gas oil. Refineries are adjusting to the increased demand for very low sulfur fuel oil, which is expected to dominate the marine fuel market (IEA, 2020). To produce lower sulfur fuels, refineries may install hydrodesulfurization technologies that consume energy and hydrogen beyond current levels or produce more distillates to blend with higher-sulfur fuels (Van et al., 2019). These changes would influence the life-cycle GHG emissions of oil-derived marine fuels and would vary from refinery to refinery. Installing on-board scrubbers that remove sulfur oxides (SOx) from ship exhaust streams is a second compliance option. These scrubbers take up space and, if releasing scrubber waste streams to the ocean in open loop systems, can have negative environmental consequences. A third option is to use liquefied natural gas (LNG), which has a higher calorific value (~20 percent) than liquid fuels, but requires vessels to be retrofitted for its use. Current trends indicate existing vessels are more likely to switch to lower-sulfur, oil-derived fuels or install scrubbers whereas new vessels may be built to use LNG as a fuel (Li, 2020). Using other marine fuels is also an option: see Table 8-4.

In sum, future climate regulations from the International Maritime Organization may further influence the mix of fuels supplied to the marine sector. Sulfur regulations such as MARPOL Annex VI are already driving changes in the production and use of fuels in the marine sector, including in refineries that are experiencing increasing demand for very low sulfur fuel oil. Importantly, marine fuels have similar supply chains to other transportation fuels (e.g., aviation, road transport). Unique aspects of their life cycle that are relevant to quantifying their emissions primarily come in the operations stage, such as CH_4 slip from LNG combustion in marine engines. CH_4 slip from LNG engines will vary based on the engine type and the load profile of the engine (Ushakov, 2019). In general, LCA methodological considerations for marine fuels are similar to those for other transportation fuels (Tan et al., 2021).

Conclusion 8-8: Estimating the life-cycle GHG emissions of very low sulfur fuel oil will depend upon information about individual refinery choices in meeting marine fuel sulfur requirements.

TABLE 8-4 Marine Fuels That Could Be Included in an LCFS

Fuel	Production Routes and Notes
Higher sulfur fuel oil	Produced from oil refining. Shifts in refinery operations to produce more very low sulfur fuel oil that affect production emissions. Higher sulfur fuel oil requires scrubbing, which may affect in-use emissions depending on energy consumed for scrubbing operations.
Very low sulfur fuel oil	
Marine gas oil	
Liquefied natural gas (LNG; see gaseous fuels)	Produced from natural gas extraction and processing. Methane slip from marine vessel engines could contribute to use-phase GHG emissions.
Dimethyl ether	Can be produced from various feedstocks including natural gas, other fossil fuels, and biomass.
Methanol	Produced from natural gas. Could be produced from biomass or coal
Hydrogen (see gaseous fuels)	Produced primarily from natural gas.
Ammonia	Produced primarily from natural gas.
Biodiesel (see biofuels)	Primarily produced from vegetable oils, fats, oils, greases. Renewable diesel or Fischer-Tropsch diesel are also possible biomass-derived fuels.
Ethanol (see biofuels)	Predominantly produced from cornstarch fermentation.
Biogas (see biofuels)	Can be produced from anaerobic digestion, landfills, and other sources.

Recommendation 8-4: LCA of oil-derived marine fuels should use new data as available for the feedstock conversion life-cycle stage. A body such as the International Maritime Organization should strive to collect data that will enable reliable marine fuel LCAs.

Recommendation 8-5: The baseline life-cycle GHG emissions for marine fuels should reflect current industry trends stemming from MARPOL Annex VI and potentially be updated after several years' time once the industry adjusts more fully to the new regulations through, for example, deployment of more liquefied natural gas-fueled vessels.

Recommendation 8-6: Marine fuel pathways should be evaluated with methods that are consistent with on-road and aviation fuels while considering unique factors in the oil refining and use phase aspects of a marine fuel's life cycle.

REFERENCES

Baral, N. R., O. Kavvada, D. Mendez-Perez, A. Mukhopadhyay, T. S. Lee, B. A. Simmons, and C. D. Scown. 2019. Techno-economic analysis and life-cycle greenhouse gas mitigation cost of five routes to bio-jet fuel blendstocks. *Energy & Environmental Science* 12(3):807–824.

Bills, A., S. Sripad, W. L. Fredericks, M. Singh, and V. Viswanathan. 2020. Performance metrics required of next-generation batteries to electrify commercial aircraft. *ACS Energy Letters* 5(2):663–668.

Bills, A., S. Sripad, W. L. Fredericks, M. Guttenberg, D. Charles, E. Frank, and V. Viswanathan. 2020. Universal battery performance and degradation model for electric aircraft. arXiv preprint arXiv:2008.01527.

Bräuer, T., C. Voigt, D. Sauer, S. Kaufmann, V. Hahn, M. Scheibe, H. Schlager, F. Huber, P. Le Clercq, R. H. Moore, and B. E. Anderson. 2021. Reduced ice number concentrations in contrails from low-aromatic biofuel blends. *Atmospheric Chemistry and Physics* 21(22):16817–16826.

Chester, M.V., and A. Horvath. 2009. Environmental assessment of passenger transportation should include infrastructure and supply chains. *Environmental Research Letters* 4(2):024008.

Cox, K., M. Renouf, A. Dargan, C. Turner, and D. Klein-Marcuschamer. 2014. Environmental life cycle assessment (LCA) of aviation biofuel from microalgae, Pongamia pinnata, and sugarcane molasses. *Biofuels, Bioproducts and Biorefining* 8(4):579–593.

de Jong, S., K. Antonissen, R. Hoefnagels, L. Lonza, M. Wang, A. Faaij, and M. Junginger. 2017. Life-cycle analysis of greenhouse gas emissions from renewable jet fuel production. *Biotechnology for Biofuels* 10(1):64. https://doi.org/10.1186/s13068-017-0739-7.

Fahey, D. W., and D. S. Lee. 2016. *Aviation and Climate Change: A Scientific Perspective*. CCLR, p. 97. OR: T. Fahey, E. N. Wilson, R. O'Loughlin, M. Thomas and S. Klipfel. 2016. A history of weather reporting from aircraft and turbulence forecasting for commercial aviation. In *Aviation Turbulence* (pp. 31–58). Springer, Cham.

Gray, N., McDonagh, S., O'Shea, R., Smyth, B., and Murphy, J.D. 2021. Decarbonising ships, planes and trucks: An analysis of suitable low-carbon fuels for the maritime, aviation and haulage sectors. *Advances in Applied Energy* 1(100008):2666–7924. https://doi.org/10.1016/j.adapen.2021.100008.

Han, J., A. Elgowainy, H. Cai, and M. Q. Wang. 2013. Life-cycle analysis of bio-based aviation fuels. *Bioresource Technology* 150:447–456.

Han, J., L. Tao, and M. Wang. 2017. Well-to-wake analysis of ethanol-to-jet and sugar-to-jet pathways. *Biotechnology for Biofuels* 10(1):21. https://doi.org/10.1186/s13068-017-0698-z.

Hemighaus, G., T. Boval, J. Bacha, F. Barnes, M. Franklin, L. Gibbs, N. Hogue, J. Jones, D. Lesnini, J. Lind, and J. Morris. 2006. *Aviation Fuels Technical Review*. Chevron Products Company.

Huq, N. A., G. R. Hafenstine, X. Huo, H. Nguyen, S. M. Tifft, D. R. Conklin, D. Stück, J. Stunkel, Z. Yang, J. S. Heyne, M. R. Wiatrowski, Y. Zhang, L. Tao, J. Zhu, C. S. McEnally, E. D. Christensen, C. Hays, K. M. Van Allsburg, K. A. Unocic, H. M. Meyer III, Z. Abdullah, and D. R. Vardon. 2021.

Toward net-zero sustainable aviation fuel with wet waste–derived volatile fatty acids. *PNAS* 118(13) https://doi.org/10.1073/pnas.2023008118.

IPCC (Intergovernmental Panel on Climate Change). 1999. *Aviation and the Global Atmosphere.* J. E. Penner, D. H. Lister, D. J. Griggs, D. J. Dokken, and M. McFarland, eds.. Prepared in collaboration with the Scientific Assessment Panel to the Montreal Protocol on Substances that Deplete the Ozone Layer. Cambridge University Press, UK. Pp. 373. https://archive.ipcc.ch/ipccreports/sres/aviation/index.php?idp=0.

Klöwer, M., M. R. Allen, D. S. Lee, S. R. Proud, L. Gallagher, and A. Skowron. 2021. Quantifying aviation's contribution to global warming. *Environmental Research Letters* 16(10):104027.

Lee, D. S., D. W. Fahey, P. M. Forster, P. J. Newton, R. C. N. Wit, L. L. Lim, B. Owen and R. Sausen. 2009. Aviation and global climate change in the 21st century. *Atmospheric Environment* 43(22–23):3520–37. https://doi.org/10.1016/j.atmosenv.2009.04.024.

Lee, D. S., D. W. Fahey, A. Skowron, M. R. Allen, U. Burkhardt, Q. Chen, S. J. Doherty, S. Freeman, P. M. Forster, J. Fuglestvedt, A. Gettelman, R. R. De León, L. L. Lim, M. T. Lund, R. J. Millar, B. Owen, B., J. E. Penner, G. Pitari, M. J. Prather, R. Sausend, and L. J. Wilcox. 2021. The contribution of global aviation to anthropogenic climate forcing for 2000 to 2018. *Atmospheric Environment* 244:117834. https://doi.org/10.1016/j.atmosenv.2020.117834.

Li, J. 2020. Modelling nuclear fuel assembly with thermal-hydraulic feedback and burnup using WIMS-PANTHER-Serpent. *Journal of Physics: Conference Series* 1603(1):012012. IOP Publishing.

Moreira, M., A. C. Gurgel, and J. E. A Seabra. 2014. Life cycle greenhouse gas emissions of sugar cane renewable jet fuel. *Environmental Science & Technology* 48(24):14756–14763. https://doi.org/10.1021/es503217g.

Mukhopadhaya, J., and D. Rutherford. 2022. Performance analysis of evolutionary hydrogen-powered aircraft. https://theicct.org/wp-content/uploads/2022/01/LH2-aircraft-white-paper-A4-v4.pdf (accessed March 8, 2022).

O'Connell, A., M. Kousoulidou, L. Lonza, and W. Weindorf. 2019. Considerations on GHG emissions and energy balances of promising aviation biofuel pathways. *Renewable and Sustainable Energy Reviews* 101:504–515. https://doi.org/10.1016/j.rser.2018.11.033.

Prussi, M., U. Lee, M. Wang, R. Malina, H. Valin, F. Taheripour, C. Velarde, M. D. Staples, L. Lonza, and J. I. Hileman. 2021. CORSIA: The first internationally adopted approach to calculate life-cycle GHG emissions for aviation fuels. *Renewable and Sustainable Energy Reviews* 150:111398. https://doi.org/10.1016/j.rser.2021.111398.

Seber, G., R. Malina, M. N. Pearlson, H. Olcay, J. I. Hileman, and S. R. H. Barrett. 2014. Environmental and economic assessment of producing hydroprocessed jet and diesel fuel from waste oils and tallow. *Biomass and Bioenergy* 67:108–118. https://doi.org/10.1016/j.biombioe.2014.04.024.

Staples, M. D., R. Malina, H. Olcay, M. Pearlson, J. I. Hileman, A. Boies, and S.R.H. Barrett. 2014. Lifecycle greenhouse gas footprint and minimum selling price of renewable diesel and jet fuel from fermentation and advanced fermentation production technologies. *Energy & Environmental Science* 7(5):1545–1554. https://doi.org/10.1039/C3EE43655A.

Stratton, R. W., P. J. Wolfe, and J. I. Hileman. 2011. Impact of aviation non-CO_2 combustion effects on the environmental feasibility of alternative jet fuels. *Environmental Science & Technology* 45(24):10736–10743.

Suresh, P., R. Malina, M. D. Staples, S. Lizin, H. Olcay, D. Blazy, M. N. Pearlson, and S. R. H. Barrett. 2018. Life cycle greenhouse gas emissions and costs of production of diesel and jet fuel from municipal solid waste. *Environmental Science & Technology* 52(21):12055–12065. https://doi.org/10.1021/acs.est.7b04277.

Tan, E. C. D., T. R. Hawkins, U. Lee, L. Tao, P. A. Meyer, M. Wang, and T. Thompson. 2021. Biofuel options for marine applications: Technoeconomic and life-cycle analyses. *Environmental Science & Technology* 55(11):7561–7570.

Ushakov, S., D. Stenersen, and P. M. Einang. 2019. Methane slip from gas fueled ships: A comprehensive summary based on measurement data. *Journal of Marine Science and Technology* 24(4):1308–1325. https://doi.org/10.1007/s00773-018-00622-z.

Van Dyk, S., and J. Saddler. 2021. *Progress in Commercialization of Biojet/Sustainable Aviation Fuels (SAF): Technologies, Potential and Challenges.* https://www.ieabioenergy.com/wp-content/uploads/2021/06/IEA-Bioenergy-Task-39-Progress-in-the-commercialisation-of-biojet-fuels-May-2021-1.pdf (accessed March 8, 2022).

Van, T. C., J. Ramirez, T. Rainey, Z. Ristovski, and R. J. Brown. 2019. Global impacts of recent IMO regulations on marine fuel oil refining processes and ship emissions. *Transportation Research Part D: Transport and Environment* 70:123–134.

Viswanathan, V., A. H. Epstein, Y.-M. Chiang, E. Takeuchi, M. Bradley, J. Langford, and M. Winter. 2022. The challenges and opportunities of battery-powered flight. *Nature* 601(7894):519–525. https://doi.org/10.1038/s41586-021-04139-1.

Voight, C., J. Kleine, D. Sauer, R. H. Moore, T. Bräuer, P. Le Clercq, S. Kaufmann, M. Scheibe, T. Jurkat-Witschas, M. Aigner, U. Bauder, Y. Boose, S. Borrmann, E. Crosbie, G. S. Diskin, J. DiGangi, V. Hahn, C. Heckl, F. Huber, and B. E. Anderson. 2021. Cleaner burning aviation fuels can reduce contrail cloudiness. *Communications Earth & Environment* 2(1):1–10. https://doi.org/10.1038/s43247-021-00174-y.

9
Biofuels

This chapter focuses on biofuels for transportation, how they are produced, and how to use life-cycle analysis (LCA) to assess their climate impacts. First, this chapter briefly summarizes several different broad categories of biofuels, either currently in production or near commercialization, and then evaluates the types of feedstocks used to produce biofuels. Next the special considerations in the assessment of attributional, supply-chain emissions from biofuel production are discussed. Last, several LCA methodologies used to assess the market-mediated effects attributable to biofuels are summarized.

The primary motivating factors to produce biofuels have been to reduce reliance on petroleum products, to enhance energy security, and to create new markets for agricultural products. Scarcity and price volatility of petroleum fuels provided important early motivation for interest in biofuels, going back for a century and peaking during periods of high oil prices or concerns about continued availability. More recently, concern over climate change caused primarily by the combustion of fossil fuels added a new motivation to seek alternative fuels produced from biogenic feedstocks that can in principle reduce greenhouse gas (GHG) emissions while also reducing exposure of U.S. consumers to oil price volatility. Market development, including price support, for agricultural products has also been a key motivation for biofuel production, and agricultural producers have been actively involved in the development and promotion of biofuels (Chambers et al., 1979; Gasparatos et al., 2013; Hill, 2022; Keeney 2009; Khanna et al., 2008; Lark et al., 2022; Robertson et al., 2008; Tilman et al. 2009; Tyner and Taheripour, 2007, 2014).

There have also been long-standing concerns about biofuels, including their contribution to climate change and their merits compared to mitigation efforts. This historically took the form of debates over net energy balance of different biofuels, and more recently have focused on life-cycle GHG emissions, air quality, and water quality. Other concerns include increased costs to taxpayers and consumers, competition for land, increased food prices, net energy balance, GHG impacts, and negative impacts on air and water quality impacts, biodiversity, and other environmental concerns (Chambers et al., 1979; Gasparatos et al., 2013; Hill, 2022; Keeney 2009; Khanna et al., 2008; Lark et al., 2022; Robertson et al., 2008; Tilman et al. 2009).

Potential benefits of biofuels include:
- Replacing petroleum with a less polluting alternative fuel
- Reducing exposure of U.S. fuel consumers to oil price volatility
- Increased demand for agricultural commodities, which is a benefit from the perspective of the producers of these commodities and a concern from competing users of the same commodities or parties concerned about agricultural expansion.
- Potential for other environmental benefits

Concerns about biofuels have included:
- Replacing petroleum with a more polluting alternative fuel
- Increased costs to U.S. fuel consumers both directly from fuel prices and indirectly through federal and state subsidies or mandates.
- Questions about the full life-cycle impact of biofuels once inputs and emissions from feedstock production and conversion are considered. These historically took the form of arguments over net energy balance of ethanol, and more recently have focused on life-cycle GHG emissions, air quality, and water quality.

- "Food versus fuel" concerns that expanded production of biofuels compete with food consumption, driving up food prices and harming food consumers, especially poor food consumers.
- Land use change (LUC) due to biofuels, including claims that expanded use of biofuels leads to expansion of agricultural acreage at the expense of forest or grassland ecosystems in the U.S. Midwest or in the tropics.
- Potential for other environmental damages

For context in considering these potential benefits and concerns, the policy environment both at home and abroad is discussed in this chapter.

BIOFUEL FEEDSTOCKS AND FINISHED FUELS

Categories of Finished Biofuels

At their most basic level, biofuels are composed of molecules that include carbon that was recently removed from the atmosphere by photosynthesis, in contrast to carbon originating from fossil resources. Though the feedstocks for these fuels are biogenic in origin, their production may use fossil inputs in some parts of their supply chain, which must be accounted for in LCAs. Depending on the specific fuel, biofuels may be produced from sugars, starches, lipids, cellulose, lignin, and other compounds derived from mixed organic waste, or from woody or herbaceous biomass. Liquid biofuels target four markets: gasoline (for spark-ignited internal combustion engines), diesel (for compression-ignited internal combustion engines), jet fuel (for jet engines), and marine fuel (for compression-ignited internal combustion engines on marine vessels). Gaseous biofuels require additional changes to vehicles and engines to accommodate their on-board storage and use; these fuels also require separate fueling infrastructure. Compressed biomethane, derived from biogas produced during anaerobic digestion and sometimes referred to as renewable natural gas, can also be used in vehicles that are equipped to run on compressed natural gas. Additionally, hydrogen can be produced from biomass through gasification or reforming of biomethane for use in fuel cell vehicles.

Ethanol

The most common biofuel in use currently in the United States is ethanol, which is produced through microbial fermentation of sugars and is suitable as a blendstock for spark-ignited internal combustion engines. Ethanol production facilities primarily use yeast to convert sugars to ethanol, namely *Saccharomyces cerevisiae* (also known as brewer's yeast). *S. cerevisiae* and other, less common hosts naturally produce ethanol under anaerobic conditions, meaning in the absence of oxygen. If the starting feedstock is a polysaccharide such as starch, as is the case for the corn ethanol industry, a mixture of enzymes is first required to break it down into glucose ($C_6H_{12}O_6$). Sucrose ($C_{12}H_{22}O_{11}$), which is the sugar present in sugarcane juice, can be utilized directly by yeast to produce ethanol. One of the challenges facing cellulosic biofuel producers is associated with the limited ability of some microbial hosts to efficiently utilize sugars derived from biomass. In these cases, a cocktail of enzymes is also required to depolymerize the polysaccharides present in cellulose and hemicellulose portions of cell walls, such as glucan and xylan, to their corresponding monosaccharides. The second-most abundant sugar liberated from plant cell walls is xylose ($C_5H_{10}O_5$), and common hosts for ethanol production do not naturally utilize five-carbon sugars. Co-utilizing strains (meaning they utilize both five- and six-carbon sugars) have been developed and are being continually improved. More advanced hosts are also being developed to utilize other plant-derived monomers, including aromatic compounds liberated from lignin (Linger et al., 2014).

Ethanol in the United States is produced almost exclusively from corn starch, with 15.8 billion gallons of production, 14.6 billion gallons blended with gasoline, and net exports of 1.2 billion gallons in 2019 (U.S. Energy Information Administration, 2019). The volumetric blending rate was about 10.25 percent in that year. At a 10 percent volumetric blend rate, ethanol can be used in most existing fueling stations and combusted in most spark-ignited internal combustion engines without any adjustments. Because ethanol's lower heating value is 21.1 MJ/liter, as compared to a typical gasoline lower heating value of 32.0 MJ/liter, vehicles deemed "flex-fuel" require modifications to the fuel pump and fuel injection system, as well as the engine control module, to account for this difference in heating value and ethanol's oxygen content. These flex-fuel vehicles are capable of using blends of up to 83 percent ethanol in gasoline.[1] Flex-fuel vehicles can run on a blend known as E85 which, contrary to what its name might suggest, is gasoline blended with anywhere between 51 percent and 83 percent ethanol by volume.

Biodiesel

Biodiesel (fatty acid methyl esters) is the second most common liquid biofuel produced in the United States. Biodiesel is suitable as a neat fuel (100 percent) or for blending with diesel for use in compression-ignited internal combustion engines. It is produced by reacting fats or oils with short-chain alcohols such as ethanol or methanol in the presence of a catalyst; this process is called transesterification. The result is a 10:1 ratio of biodiesel and glycerin (a co-product). Total production in 2020 was 1.8 billion gallons, with 45 percent of that volume being used as a neat fuel and 55 percent used in blends with petroleum-derived diesel fuel. Soybean oil is the largest feedstock for producing biodiesel in the United States, followed by smaller, roughly equal amounts of corn oil, canola oil, and yellow grease (a waste product). Algae are also capable of producing oils suitable for conversion to biofuels, although the cost of production and the effort required to lyse the cell walls to extract the oils have hindered commercialization of the technology (Cruce and Quinn, 2019). Other organisms can produce oils; for example, oleaginous yeast strains can convert sugars to oils, but those cells must also be lysed to recover the product (Sitepu et al., 2014).

Aviation Biofuels

A wide array of biomass feedstocks including grains, sugar crops, oilseeds, lignocellulosic crops, and waste materials can be processed into alternative aviation biofuels via chemical, biological, and hybrid biological-chemical routes. These pathways require additional processing, certification and testing to ensure that they meet the physiochemical characteristics of conventional jet fuel at certain technology-specific blending rates. To date, there are no commercially-approved aviation biofuel pathways suitable for use at a 100 percent blend rate, with the maximum allowed share limited to 50 percent. See Chapter 8 for more information on approved pathways and relevant life-cycle methodologies.

Biomethane

Biomethane refers to biogenic methane produced from a variety of feedstocks and conversion processes, including decomposition in landfills, anaerobic digestion of biogenic wastes and residues, and gasification of biomass in conjunction with a methanation reaction. Depending on the process used to produce the biomethane, its purity (i.e., its methane content) can vary significantly; at lower methane concentrations (approximately 50 percent), it is typically referred to as biogas. This biogas can be combusted onsite for electricity or heat, or alternatively, upgraded to remove impurities such as sulfur and carbon monoxide to make it suitable for distribution on the natural gas grid for use in natural gas vehicles or at natural gas power plants.

Drop-In Fuels

The term "drop-in fuel" is something of a misnomer because most alternative fuels cannot serve as neat replacements for petroleum fuels. However, drop-in fuels do tend to have heating values closer to those

---

[1] See: https://afdc.energy.gov/vehicles/how-do-flexible-fuel-cars-work.

of conventional petroleum fuels and may be suitable at higher blending levels than, for example, ethanol (Taptich et al., 2018). Renewable diesel is an example of a drop-in fuel. Unlike biodiesel, renewable diesel can encompass a variety of production processes, including biological or thermochemical, that result in hydrocarbons suitable for use in compression-ignited internal combustion engines.[2] One of the simpler methods for producing renewable diesel is to hydrotreat oils and fats to remove the oxygen molecules (deoxygenate). Hydroprocessed esters and fatty acids (HEFA) fuel is produced by a similar process but is intended for use as a jet fuel blendstock. Hydroprocessed esters and fatty acids synthetic paraffinic kerosene (HEFA-SPK) is produced by hydrodeoxygenation followed by cracking and isomerization of the paraffinic molecules to produce molecules of the desired chain length for jet fuels.

A commonality among renewable diesel and HEFA production processes is the need for hydrogen, which is used to deoxygenate the feedstock and isomerize the finished fuel. Therefore, the source of hydrogen can influence the emissions attributable to these biofuels, ranging from more emissions-intensive fossil hydrogen to alternative sources such as green hydrogen produced using electrolysis of renewable energy (more discussion of hydrogen LCA considerations is available in Chapter 7).

FEEDSTOCKS FOR BIOFUEL PRODUCTION

Agricultural Feedstocks

Corn and soybeans are the most common feedstocks currently used to produce biofuels in the United States. Corn grain is the primary source of domestic ethanol production and, as mentioned earlier, soybean oil is the primary feedstock for domestic biodiesel production. Agricultural residues, such as corn stover and wheat straw, are also potential biofuel feedstocks. Sugarcane, while not a major crop inside the United States, is an important bioenergy crop globally (mostly in Brazil), and the residues (bagasse and trash) can serve as lignocellulosic feedstocks in addition to the sucrose that is currently fermented to produce fuel. Other types of dedicated so-called energy crops that can be used as biofuel feedstocks include energy grasses like switchgrass and Miscanthus, diverse native prairie biomass, and short rotation woody crops, although these are not yet produced at commercial scale for biofuels (Field et al., 2020; Tilman et al., 2006). Cover crops, which are crops grown primarily for the purpose of maintaining or enhancing the productivity of the land between main crop harvests, also can be used as feedstock for biofuel production.

A key distinction among different agricultural feedstocks is between perennial and annual crops. Unlike annual crops which must be replanted each harvest, perennial bioenergy crops may live for a decade or longer. Rather than requiring steady inputs each year, perennial crops have increased inputs during establishment year(s) and reduced inputs once they are fully established. High-yielding perennial grasses like switchgrass and Miscanthus, and diverse mixtures of native prairie species, are sometimes credited for their potential to restore soil carbon because of the reduced disturbances to soils associated with perennial agronomic practices and because of the substantial accumulation of below-ground carbon in the roots and rhizomes, although this net sequestration does not continue indefinitely. This reduced soil disturbance associated with perennial crops is also why they are sometimes cited as being appropriate to grow on marginal lands that are not suitable for food production (Keeler et al., 2013), although the definition of what constitutes marginal land is hotly debated (Khanna et al., 2021).

Woody Biomass

Woody biomass is one of the most abundant feedstock for bioenergy production in the United States. Wood is used to produce pellets, briquettes, and logs, which are often categorized as densified biomass fuel. The total capacity of densified biomass fuel manufactured in the United States is about 13 million tons/year, and most of them are for wood pellet production. In the United States, wood pellets are made from forest residues from managed forest or wood residues from industrial plants such as sawmills and

---

[2] See https://afdc.energy.gov/fuels/emerging_hydrocarbon.html.

wood product manufacturing facilities. Densified wood pellets in the United States are made from diverse feedstocks, including residues from industrial plants (e.g., sawmills and wood product manufacturing facilities and roundwood (also referred as stemwood) harvested from managed forest (about 20 percent of total feedstock in 2018) (Lan et al., 2021; Liu et al., 2017a). As a waste material from forest management and wood processing activities, forest residues can be a source of biorefineries for biofuel production. LCAs have been conducted for biofuels made from forest residues (e.g., Hsu, 2012; Lan et al., 2019; Nie and Bi, 2018). For roundwood, most of the LCA studies have focused on the solid fuel products instead of liquid biofuels (Liu et al., 2017b; Verkerk et al., 2018).

Biomass Wastes and Residues

In addition to purpose-grown crops, biofuels can be produced from wastes and residues of other primary products. This category is extremely varied, with a great variety of physical and economic differences between different wastes and residues, and very different life-cycle implications depending on the material in question. Though many feedstocks in this category are lignocellulosic, they may also include gaseous feedstocks and lipids. For example, straw and residues typically left on the field or used as animal fodder can be converted into biofuel; alternately, biogenic wastes or by-products from industrial processes, such as inedible beef tallow or used cooking oil from restaurants can also be converted into biofuel. The life-cycle approach to study the impacts of these feedstocks will necessarily vary based on the product in question, as discussed in Chapter 6.

Algae

Biofuel can be produced from algae, which is grown in open ponds or in closed photobioreactors (Tu et al., 2017). These systems can be located on land that is not used for agriculture; moreover the productivity per unit land area of algal systems can be high compared to land plants.

KEY CONSIDERATIONS IN BIOFUEL LIFE-CYCLE ANALYSIS

This section assesses several key aspects of life-cycle methods particular to biofuels. First, the existing methods used to assess the attributional supply chain emissions attributable to purpose-grown biomass feedstocks and the emissions attributable to biorefineries are discussed. Next, is a discussion of several market-mediated effects attributable to biofuel policies, and a discussion of existing approaches and methods to assess the emissions impacts of these effects. Last, the degree to which different market-mediated effects have been assessed and factored into existing biofuel policies is discussed.

Attributional Effects of Biofuels

Agricultural Feedstock Production

LCA methods commonly used to estimate GHG emissions associated with crop production in conventional agricultural systems are largely similar regardless of the specific crop in question. Emission sources include the production and use of agricultural equipment (e.g., tractors) and fertilizers, herbicides, and pesticides. For perennial crops, an average may be taken, or other allocations may be made, across the lifetime of the crop to account for year-to-year variations in yields and inputs. An array of land management practices can affect emissions associated with feedstock production. These practices include tillage, the use of nitrification inhibitors, animal manure application, crop rotation, and cover cropping (Lee et al., 2020). Each of these management practices must be evaluated on a life-cycle basis because there can be tradeoffs between on-farm emissions and upstream or market-mediated effects. For example, while nitrification inhibitors reduce nitrous oxide (N_2O) emissions on-field, GHGs are emitted during their production. Furthermore, the effects of land management practices can vary depending on location given differences in soil quality, crop yield, and climate conditions (Hill et al., 2019; Maestrini and Basso, 2018; Qin et al., 2018). Many of these factors, along with soil properties, also affect N_2O emissions from soils, which can vary widely (Lawrence et al., 2021; Sathaye et al., 2011; Shakoor et al., 2021).

There are strong spatial variations in emissions from land management practices for agriculture (Adler et al., 2012). Chapter 4 discusses the influence of methodological aspects of spatial effects in more detail, including for soil carbon changes that are one of the most important outcomes of land management effects.

Because LCA results are spatially dependent, increasing transparency and availability of public data pertaining to agricultural inputs at finer spatial scales may improve LCA parameterization. Currently, the U.S. Department of Agriculture publishes critical statistics of farm energy use and fertilizer consumption every five years with yield estimates published annually. More frequent data collection may also highlight how key parameters such as these change during years with extreme weather events and would facilitate statistical and machine learning based approaches to estimating fertilizer and energy consumption under different farming conditions (e.g., economic, climate, policy). The finest spatial scale generally available is county-level given concerns around protecting farmers' proprietary information. See Chapter 5 for a discussion of verification strategies that can be used to develop farm-level feedstock information in the absence of publicly available databases.

There have been numerous examples of estimates of supply chain GHG emissions for conventional agriculturally-derived feedstocks using a variety of LCA methods (Hill et al., 2019; Smith et al., 2017). It is important to remember that LCA results for feedstocks can vary spatially and temporally because of a variety of factors including technological advances, uptake of land management practices, soil characteristics, market conditions, policy changes, and climate change among other factors. Understanding these spatial and temporal variations and how they change over time may be used to improve aspects of estimates of biofuel life-cycle GHG emissions and how they evolve with changing climate, market, and policy conditions.

> **Conclusion 9-1:** Improved data on biofuel feedstock production, including energy consumption, yield, and fertilizer application at fine spatial resolutions may be useful for some applications. Data quality improvements may support improved GHG accounting in biofuel feedstock production, especially should a performance-based LCFS be developed that accounts for spatially-explicit fertilizer and energy consumption, and land management practices like cover crop planting, land clearing, overfertilization, manure application, use of nitrification inhibitors, or noncompliance with long-term soil carbon storage incentives.

Woody Biomass Production

The GHG emissions associated with the production of woody biomass come from multiple sources, including the use of energy and materials (e.g., fertilizers and soil amendments) for forest management, harvesting, storage and transportation. Energy-related GHG emissions can be estimated by fuel consumption and the GHG emission factors of each fuel. The fuel consumption depends on wood species and forest operations. For example, thinning is a common practice that generates forest residues and is used in the United States to address issues such as bark beetles and wildfires (Knapp et al., 2021). Diesel fuels are often used for activities involved in commercial thinning, such as felling and bunching, skidding, delimbing, debarking, and loading (Markewitz, 2006; McEwan, Brink, and Spinelli, 2019; Oneil et al., 2010). Similar activities are involved in logging (in other words, final harvesting), but the diesel inputs are often different. The GHG emissions of machinery used for thinning and logging can be estimated by using the diesel consumption data and emission factors of diesel (Lan et al., 2019).

Herbicide, fertilizers, and soil amendments used in the forest site preparation and planting generate GHG emissions from the applications and upstream production. The data of upstream GHG emissions are available in many life-cycle inventory (LCI) databases. These upstream GHG emissions have large variations, depending on the chemicals and materials used and production technologies. GHG emissions may also be generated during and after the application of fertilizers, and may depend on fertilizer type. For example, compared to many other nitrogen-containing fertilizers, urea may have lower carbon dioxide (CO_2) emissions during the production stage but higher CO_2 emissions after field application due to the CO_2 generated after urea hydrolysis (Hasler et al., 2015). The GHG emissions released by fertilizers can be estimated by emission factors and application rates (Brentrup et al., 2000). Therefore, the GHG emissions related to the materials and chemicals used in forest management need to be transparently reported with additional information in production technology, supply chain (e.g., transportation modes), and application methods and rates (e.g., in the unit kg/ha).

The storage of most woody biomass generates GHG emissions from energy use and fugitive emissions. The energy use of woody biomass storage is generally low (Sahoo et al., 2018). Fugitive emissions are contributed by chemical and biological degradation, including CO_2, methane (CH_4), and other volatile hydrocarbons (Alakoski et al., 2016; Geronimo et al., 2022), and may vary substantially. The large variations of fugitive emissions depend on many factors such as the shape and sizes of both biomass and piles, moisture control strategies, storage methods, the type of biomass, and the local environment. The research on these fugitive emissions is more-limited compared to GHG emissions studied in other stages related to woody biomass production.

The GHG emissions of woody biomass transportation are driven by the weight of biomass, distances, and energy use. The weight of biomass and distances depend on the biomass supply chain design. Centralized design requires the transportation of woody biomass directly to a biorefinery while decentralized design often pretreats woody biomass by drying and/or size reduction for easier and more economic transportation (Lan et al., 2021). Main GHG emissions sources are energy that is used for activities such as loading, hauling, and transporting (mostly by truck). Transportation GHG emissions can be estimated using either energy-based approach discussed previously, or weight–distance based emission factors (e.g., data in the USLCI database reported on the basis of 1 t*km).

The source of wood used affects the GHG emissions of biomass production in several aspects. In many cases the feedstocks of biofuels are residues generated from forestry operations or wood manufacturing (e.g., sawdust and barks). As residues are also produced from forestry operations, allocating GHG emissions of forestry operations and other activities between the main products (e.g., logs) and other products is often needed, and the allocation methods affect the LCA results (see discussion below on allocation). It may be challenging to use system expansion to avoid allocation in this case due to the lack of alternative product routes for logs and residues. However, system expansion could be helpful to include counterfactual scenarios of alternative uses and/or fates of residues. For example, Lan et al. (2020) explored the counterfactual scenarios of forest residues. Residues from forest operations are currently either left on site to decay are combusted without energy recovery (Booth, 2018). Mill residues are sent to landfills, burned for energy recovery, or sold to market to produce other wood products like particleboard (Lan et al., 2020). These counterfactual scenarios could be considered for estimating avoided emissions (see section on Negative Emissions), although they are commonly not included in biofuel LCAs using forest residues (Lan et al., 2021) except for several studies focusing on developing global warming potential (GWP) indicators for biogenic CO_2 emissions that consider slow residue decomposition if not collected (Liu et al., 2017a, 2017b). The source of wood also affects the carbon flows associated with effects related to forest management practices. The source of wood determines the forest management and relevant direct GHG emissions given the differences in forestry operations for diverse wood species and regional climate.

The source of wood may have different carbon implications due to changes in land use and forest management practices. For example, the wood from afforestation and reforestation brings carbon benefits of increased carbon stocks on lands (Favero et al., 2020; Kauppi et al., 2020). The carbon implications of using woody biomass for biofuels are complicated, and depends on how the forest and related systems will be affected (Klein et al., 2015); see discussion of biogenic emissions accounting in Chapter 6. The use of forest residues may provide financial incentives for using overstocked, low-quality biomass that reduce forest wildfires and associated GHG emissions (Verkerk et al., 2018). Excessive removal of forest residues affects the forest soil and growth (Achat et al., 2015). The rapid increase of roundwood harvest rate may lead to decrease of carbon stock (Pingoud et al., 2020); however, the increased demand of woody biomass could also stimulate new forests and better forest management for additional carbon storage benefits (Cowie et al., 2020; Favero et al., 2020). The forest carbon stock changes are closely linked to the biogenic carbon issues of forests (see section on biogenic carbon in Chapter 6). Previous studies also show the reduced climate benefits if the policy supporting the use of forest reduces timber supply for material applications (Favero et al., 2020). Thereforeit is important that a practitioner consider the impacts of using woody biomass on forest management in LCAs. It may be challenging to directly assess the future forest carbon stock changes caused by different forest management strategies and under diverse climate change situations. However, it is useful to at least explore the potential impacts of key parameters or assumptions of forest management practices that could be changed in response to increased demand of woody biomass, such as changes in tree species, thinning, residue removal, and forest productivity (Klein et al., 2015).

In sum, estimates of the GHG emissions from woody biomass production depend on the source of wood, forest management (i.e., the application of fertilizers, pesticides, and soil amendments), logistics design, and the accounting methods of biogenic carbon. The use of residues and primary product logs from the forest sector may lead to different impacts and resulting carbon implications depending on the category of woody biomass used and where it is produced, such as carbon emissions or reductions from LUC, changing forest management practices in response to increased woody biomass demand, and counterfactual scenarios of alternative uses and fates of residues. However, data gaps exist in fugitive GHG emissions of woody biomass storage.

> **Conclusion 9-2:** Estimates of the GHG emissions associated with biofuels from woody biomass depend on the source of wood, forest management practices, and the carbon accounting method.
>
> **Recommendation 9-1:** Additional research should be done to assess key parameters and assumptions in forest management practices induced by increased woody biomass demand, including: changes in residue removal rates, stand management and forest productivity, and changes in tree species selection during replanting.
>
> **Recommendation 9-2:** Research and data collection efforts should be carried out for improved data and modeling related to forest feedstock production and storage, including energy use, yield, inputs, fugitive emissions, and changes in forest carbon stock should be supported.

CARBON EMISSIONS AND SEQUESTRATION FROM BIOREFINERIES

GHG emissions associated with biomass conversion come from multiple sources, including on-site combustion of fuels (e.g., fossil fuels, biomass, or byproducts), direct emissions from conversion processes, and upstream emissions associated with the production of chemicals, enzymes, and electricity used by biorefineries. There may be a key distinction between biogenic (sometimes referred to as "contemporary") carbon and fossil carbon for purposes of carbon accounting—the atmosphere is not affected differently by them. Some studies include biogenic carbon emitted in the form of CO_2, then balanced with a biogenic CO_2 sequestration credit during the biomass growth phase (e.g., Spatari et al., 2005), while other studies exclude them by using the carbon neutral assumption (e.g., Cai et al., 2013; Scown et al., 2012). It is important to note that the form of carbon can change in conversion (e.g., carbon as CO_2 fixed by photosynthesis but released as CH_4, with a different GWP. It is critical to clearly identify the individual contributors to GHG emissions and sequestration for biomass conversion. If a CO_2 sequestration credit is applied to account for biomass regrowth, this should be differentiated from any soil organic carbon sequestration credits (or net emissions) included in the analysis. GHG emissions from biorefineries can be measured when emission monitoring systems are available, and organic carbon testing has been considered in regulations related to biomass content. They can also be estimated using the mass and energy balance data collected from biorefineries (primary data), literature (secondary data), or derived from mechanistic process simulations (Huo et al., 2009). The energy balances provide energy demand information that can be used together with GHG emissions factors of fuels to compute the energy-related emissions (Lan et al., 2020). The energy-related emissions can also be estimated by mechanistic modeling for the combustion processes in biorefineries (Cai et al., 2018). Depending on the source and quantity of fuel used, the energy-related GHG emissions vary; for example, using renewable electricity or increasing the efficiency of a biorefinery could reduce its energy-related emissions. The mass balance data can provide estimations of process-related GHG emissions as they are parts of gas outputs of biorefineries.

Process-related GHG emissions can be estimated by theoretical estimations or mechanistic process models, and these emissions are often process- and technology-specific. For example, ethanol biorefineries generate significant biogenic CO_2 emissions in the fermentation process, which was estimated to be 45 Mt/year for 216 existing biorefineries in the United States (Sanchez et al., 2018) using an emission factor of 2853 metric ton of CO_2/million gallons of ethanol produced. Such an emission factor was estimated based on the stoichiometry of converting glucose to ethanol and CO_2 (Hornafius and Hornafius, 2015). Another example is biomass gasification that can produce various biofuel products using different downstream processes, such as ethanol, methanol, hydrogen, Fischer–Tropsch fuels and many others. CO_2 is part of the gas products from gasification and the quantity of CO_2 depends on many factors such as feedstock compositions, gasifiers, operational conditions, and downstream cleaning up processes (Sikarwar et al., 2017). Given the complexity of chemical reactions and downstream processes involved in gasification, theoretical estimations of CO_2 emissions based on chemical reactions could be challenging. Mechanistic models such as process simulations using software AspenPlus have been used to simulate the CO_2 gas stream for gasification to different types of biofuels (Susmozas et al., 2016; Salkuyeh et al., 2018).

CO_2 emissions of biomass conversion could be mitigated by integrating carbon capture into biorefineries, though the impact on biorefinery emissions depends on the CO_2 capture rate, the end use of the captured CO_2, and the fate of stored CO_2. Ethanol plants have been considered near-term candidates for carbon capture (Sanchez et al., 2018), while other biomass conversion pathways such as gasification also attract increasing interest given the significant amount of CO_2 generated as by-products (Sikarwar et al., 2017). The quantity of CO_2 that can be removed by carbon capture depends on many factors such as technology choices, CO_2 concentrations, and end-of-life of CO_2 (e.g., geological storage or utilized for other applications such as enhanced oil recovery that may lead to secondary leakage, as well as increased oil use). As CO_2 capture facilities need additional energy input, they may change the energy generation and supply scheme of biorefineries. The energy demand of CO_2 capture facilities in biorefineries could be met by burning byproducts (e.g., solid residues), expanding the capacity of on-site energy generation facilities, or purchasing fuels externally (Kim et al., 2020). That being said, this requirement is less for the CO_2 by-product of fermentation, which does not require gas-phase separation, only dewatering and compression. Different choices will affect the GHG emissions associated with the energy consumption of CO_2 capture facilities, but mass and energy balance approaches have been used to estimate energy use and energy-related GHG emissions (Yang et al., 2020).

There are GHG emissions generated during the upstream production of chemicals, enzymes, and electricity purchased by biorefineries. These emission data can be obtained from LCI databases and literature (e.g., the life-cycle GHG emissions of electricity generation are available in databases such as the USLCI database [National Renewable Energy Laboratory, 2012] and literature [Sathaye et al., 2011]). As these GHG emissions highly depend on the technology pathways and market mix, the GHG emission factors with technology and market mix information are should be disclosed for biofuel LCA.

In addition to biofuels, biorefineries almost always produce byproducts or co-products. The distinction between what materials qualify as co-products and what outputs are truly waste products is often unclear. For corn ethanol facilities, distiller's dried grains with solubles and corn oil are common co-products that have clear applications in the animal feed market and can be accounted for through allocation procedures (see Chapter 3). From the standpoint of carbon flows, it is a safe assumption that all of the carbon in these products will eventually be oxidized, some as methane from ruminants. Co-products or byproducts from other biorefining processes, particularly if the product(s) do not have a well-established market, can be more difficult to account for.

Biochar, a byproduct from most thermochemical conversion processes, is an example of a product for which a robust market does not yet exist (Baker et al., 2020; Campbell et al., 2018). In general, biochar is made of aromatic compounds and is rich in carbon (> 65 percent by mass), although the composition varies widely depending on the biomass feedstock and process conditions under which it was produced (Ippolito et al., 2020). Char can be combusted as a fuel on-site or exported to be used as a fuel elsewhere, in which case the carbon accounting can be calculated using system expansion, assuming the fuel it is displacing (e.g., coal or natural gas) is known. If biochar is applied to soils with the intention of improving soil quality or sequestering carbon, the impact is highly uncertain and is the topic of heated debate (Dumortier et al., 2020; Lan et al., 2021; Maroušek et al., 2017). In the *Getting to Neutral* report by Lawrence Livermore National Lab (Baker et al., 2020), researchers assumed that 80 percent of carbon in biochar buried underground would remain sequestered for at least 100 years but they chose not to attempt to capture the impacts on net primary productivity (if applied to croplands or other lands). This is similar to the assumption used in Woolf et al. (2012) and Breunig et al. (2019), both of which assume that 85 percent of carbon in biochar is recalcitrant over a 100-year period. The assumptions used in these studies are consistent with biochar research that shows high carbon persistence in biochar, although the biochar decomposition rate depends on many factors, such as pyrolysis temperature and biomass composition (Joseph et al., 2021; Lehmann, 2021; Woolf et al., 2021). One advantage for biochar, in terms of its recalcitrance, is that it contains little in the way of nutrients, meaning that microbial breakdown of biochar in the environment, if buried in large quantities, should be slow (Astals et al., 2012; Moset et al., 2015; Pognani et al., 2009).

Another common byproduct of bioenergy production is solid digestate; this is particularly prevalent for anaerobic digestion facilities, where a typical facility only breaks down 20-40 percent of solids in the digester and the remaining material must be directly land-applied, composted, or sent to landfills (Astals et al., 2012; Moset et al., 2015; Pognani et al., 2009). This material will vary in its composition depending on the original feedstock type, residence time in the digester, and digester conditions, but the average carbon content for digestate produced from manure or mixed organic waste is estimated to be 18-25 percent on a dry mass basis (European Commission, 2018). In contrast to biochar, digestate is rich in nutrients, which means it has value as a fertilizer supplement and any carbon sequestration potential is negligible as remaining carbon-containing compounds will be broken down rapidly by microbes once the digestate is directly land-applied or composted. In addition, its use is a potential source of N_2O emissions.

The role of CH_4 emissions in low-carbon fuel production systems is important in the context of anaerobic digestion facilities (e.g., wastewater treatment facilities and manure digesters). Regardless of whether the carbon is biogenic or fossil, methane emissions must be accounted for. For consistency, peer-reviewed LCA literature and regulatory applications tend to use GWP_{100} (i.e., global warming potential calculated over 100 years) values for all non-CO_2 GHGs. In systems that produce CH_4, fugitive emissions are often most easily and transparently calculated as a fraction of total system throughput, as is common practice in government-sponsored GHG inventories, such as the Environmental Protection Agency (EPA) GHG Inventory[3] and the California Air Resources Board's GHG inventory.[4] This can be applied to different parts of the supply chain, including production, processing/cleanup, and pipeline transport. However, measurement and monitoring is required to ensure these estimates are accurate; Alvarez et al. suggested that prior fugitive CH_4 emissions estimates were likely too low, and the leakage from the oil and gas supply chain was the equivalent of 2.3 percent of all natural gas produced, as of 2015 (Alvarez et al., 2018). Though the authors did not estimate the equivalent effect across biomethane supply chains, fugitive CH_4 leaks may pose similar risks (see Chapter 7). Regular monitoring of fugitive CH_4 emissions from anaerobic digestion and any other methane-producing facilities will be necessary to ensure that accurate leakage values are incorporated into CI values.

---

[3] See: https://www.epa.gov/ghgemissions/inventory-us-greenhouse-gas-emissions-and-sinks.
[4] See: https://ww2.arb.ca.gov/ghg-inventory-data.

Using allocation or system expansion to include the substitution benefits of byproducts is common in biofuel LCA. If the substitution benefits of byproducts are included, the use of byproducts and counterpart products to be replaced has direct impacts on the GHG emissions allocated to biofuel. Therefore, the information of byproduct applications, substitution effects, and the GHG emissions of counterpart products and materials is to be disclosed and documented.

To summarize, mass or energy balance are the most common methods used to estimate GHG emissions of biorefineries. Some of the attributional GHG emissions of upstream production of electricity and chemicals used in biomass conversion are available in many LCI databases but have large variations depending on the production technologies and market mix. Research articles vary in their assumptions about the potential for carbon sequestration using biorefinery co-products as soil amendments; but many assume 80–85 percent of biochar is stable for at least 100 years whereas digestate and compost are not assumed to result in accumulation of stable carbon in soils.

Conclusion 9-3: The impact of biorefinery co-products, particularly biochar, compost, digestate, or other products meant to be applied to soils, is highly dependent on how these materials are produced and handled and on what land they are applied. Assigning any GHG offset credit to a biorefinery for producing and exporting these materials requires extensive verification to ensure they deliver the intended benefits. Nutrient-rich materials such as compost only offer fertilizer offset benefits if they are applied in a manner that results in lowered net GHG emissions.

Recommendation 9-3: Policymakers should exercise caution in crediting biorefineries for GHG emissions sequestration as a result of exporting co-products such as biochar, digestate, and compost, as it risks over-crediting producers for downstream behavior that is not necessarily occurring. The committee recommends that any credits generated from these activities must be contingent on verification that these activities are being practiced.

Recommendation 9-4: Applying credits for carbon sequestration to soil or reduced use of fertilizer should require robust measurement and verification to prove the co-products are applied in a manner that yields net climate benefits.

Market-Mediated Effects of Biofuels

Large-scale production of biofuels has an effect on various markets at regional, national, or global scales and can affect market prices at these levels. Changes in market prices can trigger other changes in production and consumption decisions that may have positive or negative effects on GHG emissions from those markets. These secondary effects on GHG emissions are of concern because they affect the assumed savings in GHG emissions obtained by displacing fossil fuels by biofuels. In this section, several relevant effects attributable to increased biofuel demand are identified, then several methods used to assess their impacts are discussed.

Competition for Land

Biofuels produced from agricultural feedstocks, including food and feed crops, perennial energy crops, or short rotation woody crops, compete for land with crops produced for non-biofuel uses. The extent to which this is the case will depend on the type of feedstock, the yield of feedstock per unit of land, and the types and availability of land on which they can be grown. The use of food or feed crops for biofuels could divert some of the produced crops from food or feed consumption. If land resources are not fully employed (idle land is available) and the supply function for a given crop is relatively elastic, then additional biofuel production would increase production of the food crop with a more limited increase in the price of the food crop. However, the diversion of the food crop to biofuel could increase the price of the food/feed crop. Increases in biofuel production could also lead to changes on the extensive margin by bringing additional land into crop production, by conversion of idle land, pastureland, and forestland to cropland.

It may also lead to reduction in the amount of the food crop exported (if the country is an exporter) for agricultural commodities or an increase in imports of agricultural commodities. If the country is a major exporter (importer) of food/feed crops with a large share of world market trade, the increased demand for biofuel crops could reduce exports (increase imports) of food/feed crops, and this can potentially affect world market prices of those crops, leading to increased returns to producing those crops in other countries and to intensive and extensive margin changes in land use in those countries.

The use of non-food feedstocks, such as dedicated energy crops, could also divert land from food crops; the extent to which this is the case may be less than for food crops, because these crops can potentially be grown on low quality/marginal land not suitable for other crops, and they may be high yielding and thus require less land to produce a given amount of biofuel. However, to the extent that their production increases demand for land and diverts some land from food crops and raises land rents, they could also create similar incentives for LUCs at the intensive and extensive margin as in the case of food crops. Higher prices of food crops and agricultural commodities will lead to similar changes in trade in agricultural commodities as described above.

The extent of LUC and the type of land (idle land, pastureland, forestland) conversion induced by biofuels will depend on various features of the feedstock, such as its yield, requirements for soil fertility and quality, climatic conditions, its co-products, the potential to make changes in crop production at the intensive margin (e.g., yield improvement and expansion in harvest frequency due to cultivation of unused cropland), and availability and ease of conversion of various types of non-cropland to crop production. The extent of LUC associated with a given volume of biofuels is expected to change over time in response to changes in crop varieties, crop productivity, management practices, changes in the climate conditions and so on. As the production of biofuel feedstocks increases, there may be an increase in demand for land and an increase in the returns to land and in food or feed crop prices within the country. This effect could be non-linear if additional land brought into production has a yield lower than the existing cropland. Additionally, increases in crop prices can induce changes in management practices that can increase land productivity. Furthermore, the availability of idle land is also expected to change over time, depending on the rate of growth of demand for food or feed relative to the rate of growth of crop productivity, urban development, government policy support for agriculture, and other factors. Induced LUC is therefore a dynamic phenomenon that cannot be proxied by a fixed estimate that is constant over time.

The resulting changes in land use both at the intensive and extensive margins within the country and globally can affect GHG emissions. Intensive use of existing cropland, due to additional chemical applications, can increase GHG emissions not only on land under biofuel feedstocks but on other cropland as well. While fertilizers used for biofuel crop production enter directly into the LCA of biofuels, an increase in the rate of fertilizer application on other cropland could generate some additional GHG emissions as a market-mediated effect in response to biofuel demand. Similarly, if due to extensification carbon-rich grasslands and forestland are converted to crop production, this could release stored carbon into the atmosphere (Fargione et al., 2008; Gibbs et al., 2008; Hertel et al., 2010).

There are several key factors that can influence estimates for ILUC values for a given biofuel pathway, including: changes in the mix of crops induced by biofuel production, modeling yield change per harvest, modeling change in harvest frequency (including multiple cropping and returning unused land to crop production), the share of additional land expansion that goes on high carbon stock land, elasticities, emission factors and amortization time horizon.

Recommendation 9-5: Additional review and research is recommended on the key factors affecting induced land use change.

Displacement of Fossil Fuels

In principle, large scale biofuel production could displace at least a portion of domestic liquid fossil fuels consumption, ostensibly reducing oil demand. If the production of biofuels does reduce the world market price of oil, then it has the potential to increase the demand for oil in the rest of the world. This has been referred to as the "global rebound effect" which might offset the savings due to biofuel (Hill et al., 2016).

The effect of a reduction in the world oil price on the domestic oil price in the biofuel-producing country is complex and depends on the domestic biofuel and agricultural policies used to promote the production of biofuels in a given country (Taheripour and Tyner, 2014). In general, biofuel blend mandates, such as the U.S. Renewable Fuel Standard (RFS), and low-carbon fuel standards (LCFSs), are designed to implicitly penalize the consumption of fossil fuels and implicitly subsidize the consumption of biofuels with a lower carbon intensity (CI) than an LCFS. The implicit subsidy will depend on the policy, for example, in the case of an LCFS, the implicit subsidy increases as the CI of the biofuel decreases relative to an LCFS; in contrast with a biofuel blend mandate, the subsidy may be the same for all biofuels covered by the mandate.

Other Indirect Effects

While the above two effects have received the greatest attention in the literature, biofuel production affects production and consumption decisions in other markets that can also affect GHG emissions. Specifically, the production of corn ethanol results in the production of co-product distillers dried grain solubles which are used as animal feed and can replace corn and soymeal to some extent. Higher crop prices due to corn ethanol are expected to reduce demand for livestock and this could reduce downstream emissions generated during the production of livestock, which are largely in the form of CH_4 and thus have a higher GWP than carbon emissions. Additionally, feeding distiller's dried grain solubles to cattle affects their digestion and reduces or increases (Masse et al., 2014) enteric CH_4 emissions generated by them (Arora et al., 2010; Bremer et al., 2010; Flugge et al., 2017). Similarly, policies that lead to increased corn ethanol production may also support animal feed production by generating additional distillers dried grain solubles, which incentivizes animal production over plant-based protein sources for human consumption.

Another unintended consequence of demand for biomass for advanced biofuels is the increase in the demand for advanced biofuel feedstocks, such as corn stover. This can create incentives to change crop rotations from corn-soybean rotation to continuous corn rotation. Since corn is more nitrogen fertilizer intensive this can lead to an increase in fertilizer use as corn displaces soybean production. Also, corn stover harvesting practices can be accompanied by increased fertilizer inputs, particularly at higher residue removal rates. It can also lead to a change in tillage methods and use of better varieties of seeds that produce larger agricultural biomass (stover) which can increase carbon sequestration in biomass and soils and reduce atmospheric carbon, although this can depress grain yield leading to the unintended consequence of induced LUC.

Another unintended consequence is dietary change, which may include both food availability and dietary consumption in response to biofuel demand. For example, models used to assess LUC have estimated that increased biofuel production has reduced calories from corn and wheat used for food and feed by 25–50 percent (Searchinger et al., 2015). Also, for example, using more biofuels byproducts such as distillers dried grain solubles and oilseed meals has helped the livestock industry to move towards producing more meat from non-ruminants with lower CI than ruminant meat (Taheripour et al., 2021). The consequences of these changes are complex and can both increase and decrease net GHG emissions.

In addition to induced LUCs and the rebound effect, the existing literature has noted several other indirect effects due to biofuels—such as co-product impacts on livestock markets, food availability and dietary change—that may affect GHG implications positively or negatively.

Recommendation 9-6: Beyond research on induced land use change and rebound effects, research should be done to identify and quantify the impacts of other indirect effects of biofuel production, including but not limited to market-mediated effects on livestock markets, land management practices, and dietary change of food type, quantity, and nutritional content.

QUANTIFYING MARKET-MEDIATED EMISSIONS FROM BIOFUELS

Land Use Change

Assessment of the GHG intensity due to market-mediated effects of biofuels uses a comparison of production and consumption in multiple markets in the economy both domestically and in the rest of the world with biofuels and in a counterfactual scenario without biofuels (holding all else constant). This assessment is commonly conducted using various types of economic models; these models can be classified broadly into partial equilibrium (PE) and computable general equilibrium (CGE) models. Some of these models are global and can analyze both domestic and international LUCs, whereas others only analyze domestic LUCs in the country that produces biofuels.

Partial Equilibrium (PE) models consider only a few sectors of the economy that are most closely affected by biofuel production, such as the agricultural, forestry and fuel sectors. Prices, production, and land allocation within these sectors are determined within the model, and it is assumed that conditions in the rest of the economy remain unchanged with biofuel production. These models could be either static or dynamic and used to analyze the effects of a one-time shock in biofuel demand on equilibrium prices and quantities in the markets included in the model. A widely used example of this is the Food and Agricultural Policy Research Institute (FAPRI) model, developed at Iowa State University's Center for Agricultural and Rural Development (CARD) (e.g., Searchinger et al., 2008; Fabiosa et al., 2010). These models include multiple markets, each of which is represented by a demand and supply relationship that are linked across markets through cross-price elasticities. As a result, changes in demand in one market can affect prices and production in other connected markets. These models estimate the demand for land converted based on assumptions about crop productivity and allocates the expansion in demand for cropland in each region among other types of land cover including forest and grassland categories based on a land allocation process.

These models are suited to analyze the effects of changes in demand for existing crops that are already being produced but not readily adaptable to analyze the effects of introducing new biofuel feedstocks, such as dedicated energy crops that are yet to be commercially produced. Early applications of the FAPRI-CARD model to examine the induced LUC effect of biofuels ignored the potential for conversion of cropland pasture or idle land to crop production before converting forestland. The inclusion of cropland pasture or idle land in later versions of this model was found to substantially reduce the modeled conversion of forestland to crop production by the model (Dumortier et al., 2011).

Other well-known PE models used in assessing induced LUCs due to biofuels are the Forest and Agricultural Sector Optimization Model (FASOM) and the Global Biomass Optimization Model (GLOBIOM). Unlike static multimarket models that are reduced form models of supply and demand, these are dynamic programming models. They are multi-period, structural models that represent the behavior of utility maximizing consumers and profit maximizing producers. They typically include detailed biophysical data to model the dynamics of crop yields, soil carbon changes, and GHG emissions. These models can be regional or global in scope: with examples including FASOM and the Biofuel and Environmental Policy Analysis Model (BEPAM), which are national scale models for the United States and GLOBIOM, which is at the global scale. These models examine the effects of biofuel production on land allocation, equilibrium production, and prices subject to constraints on land, technology, and various material balances, and solve for endogenous prices, production and consumption decisions in multiple markets within a limited number of sectors of the economy. In these models, land is allocated among alternatives based on returns to land and LUCs can occur at the intensive and extensive margin based on relative returns to alternative uses. These "bottom-up" models can incorporate detailed representation of the technology, introduce new technologies and crops and substantial spatial heterogeneity in land availability, land suitability, and costs of production across fine spatial scales (Khanna and Crago, 2012; Khanna et al., 2014). The national scale models are not suited to calculate the global induced LUC but can calculate domestic induced LUC. They can also be linked to other global models to calculate the international portion of the induced LUC as in

Beach and McCarl (2010). EPA (2010) has used FASOM in combination with FAPRI-CARD to assess induced LUC values for corn ethanol, soy biodiesel, sugarcane ethanol, and switchgrass ethanol.

CGE models simulate economy-wide effects of a biofuels shock, include intersectoral linkages and constraints on labor and capital and determine all prices and incomes in the economy simultaneously. These models are typically global in scope, represent multiple economic sectors in each region, and include factor markets for labor and capital. While these models are broad in geographic and sectoral scope, many CGE models have limited spatial resolution and usually partition the world into several large homogenous geographical units called agro-ecological zones. In these models each region or country has a representative household that allocates resources across uses domestically and consumes goods and services produced domestically or imported. Some CGE models may have more than one representative household. On the production side, multiple producers demand primary and intermediate inputs (produced domestically or imported) and produce goods and services. Primary inputs are typically land, labor, capital, and resources. In these models the government collects taxes and pays subsidies or makes transfer payments. Each region interacts with other regions through trade. These models assume that consumers maximize utility and producers maximize profits in a perfectly competitive market setting, leading to endogenously determined prices and quantities of goods and services and also factors of production. These models typically limit the number of agricultural products considered by categorizing individual commodities into large groups (e.g., all coarse grains, oilseeds) and imposing the same behavioral and market assumptions on the individual components. Among the CGE models being used are the Global Trade Analysis Project (GTAP-BIO), the Modeling International Relationships in Applied General Equilibrium (MIRAGE) and the Emissions Prediction and Policy Analysis (EPPA) models (Gurgel et al., 2007; Hertel et al., 2010; Laborde and Valin 2012; Taheripour et al., 2017). In determining induced LUCs due to biofuels, these models do not divide induced LUCs into direct and indirect. However, CGE models usually provide induced LUCs by geographical regions. These models distinguish between different types of land and the ease of conversion of land from one type to another is determined by an elasticity of transformation in GTAP-BIO and by a cost of conversion in EPPA. For more details on modeling LUCs in CGE models see Taheripour et al. (2020).

In addition to structural differences across the PE and CGE models, they also differ in key assumptions that affect LUC in response to biofuel production. Some of the key assumptions are described in the rest of this section.

Land Productivity in Intensive and Extensive Margins. The effect of a biofuel policy shock on land use depends on some key assumptions such as (1) the productivity of new cropland which is brought into crop production, (2) the crop yields in the base data, (3) the potential for intensive margin changes in response to biofuel induced increase in crop price, and (4) cultivation of unused cropland for cropland. These intensive margin changes can take two forms, a change in crop yields per unit of harvest area and a change in the potential to increase double cropping on existing land (or conversion of unused cropland) which can reduce the amount of expansion in cropland. In the life-cycle accounting of a fuel, some portion of GHG benefits from intensification may be counterbalanced by increased emissions from other effects such as increased fertilizer application.

Inclusion of By-Products: Biofuels from some types of food crops, such as corn, produce by-products such as distillers dried grains with solubles that can substitute for corn and soymeal used for animal feed and substitute for products that would otherwise require land. The inclusion of these feed byproducts in models affects the land requirements for biofuels. The modeling approach of inclusion of byproducts significantly affects the land use effects. Some co-products or byproducts such as distillers dried grain solubles and soymeal are used for livestock feed and can displace corn and other feed crops, within certain dietary constraints which may positively or negatively affect livestock emissions. Models differ in the mechanisms by which this displacement occurs and in the modeling of the consequences of this displacement. For example, GLOBIOM maintains complementarity relationships among feed rations in livestock; as a result, for example, it generates a large rebound effect since more soymeal leads to more production of other types of feed and need for more land to produce that feed and more demand for meat products. In contrast, GTAP-BIO allows greater substitution among feed rations at the macro level and allows for the mix of livestock to change in response to availability of byproducts, leading to a shift towards poultry and pork from cattle.

Availability of Various Types of Land and Ease of Substitution of Land from One Use to Another: Models differ in the extent to which they include idle or marginal land and the availability of this land as well as the mechanism by which land can change from one use to another. In CGE models that use the constant elasticity transformation approach, LUC is specified by a constant elasticity of transformation parameter that governs the ease with which land is converted from forests to cropland and marginal land to cropland. In PE models, LUC is dependent on the assumptions of the cost of converting land from one use to another and limits on conversion based on historically observed crop mixes and other constraints on the extent of conversion. Some models use other approaches such as extreme distribution functions to model cost of land conversion. These assumptions affect the ease/costs of changes in land use at the extensive margin as well as which land (forestland or marginal cropland) is the first to convert to cropland.

Estimates of induced LUC are sensitive to assumptions about which land is likely to convert in response to higher crop prices and the productivity of this land for producing crops. Replacing forests with cropland may release on average about four to five times more carbon emissions per unit land than converting pasture to cropland (Plevin et al., 2010), although there is wide variation in estimates. Melillo et al. (2009) show that allowing conversion of unmanaged natural areas to cropland leads to a seven-fold higher induced effect of biofuels compared to a scenario that only allows more intensive use of existing managed lands. The larger the potential for intensive management of land for biofuel feedstocks to displace other crops, the lower the induced land use effect of biofuels because it reduces extensive margin effects. A recent analysis of how models for assessing LUC treat availability of land showed how this parameter affects estimated CIs. Figure 9-1 provides a comparison of the different quantities of land categories available for conversion across a selection of models used for estimating biofuel-driven land use.

Price Responsiveness of Consumer Demand for Agricultural Commodities

The greater the elasticity of demand for agricultural commodities, the smaller the increase in crop prices due to the demand for biofuels and the smaller the resulting increase in land conversion to meet the additional demand for food crop-based biofuels.

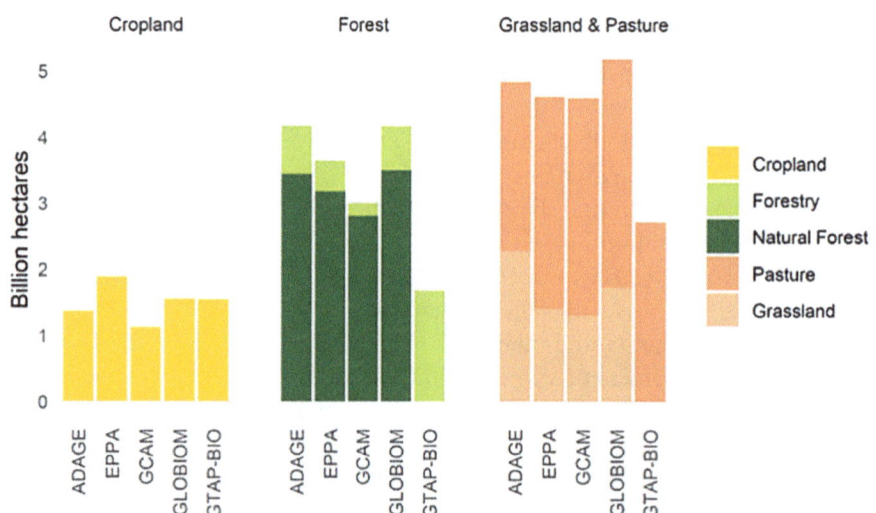

FIGURE 9-1 Land available for conversion in commonly-used models for assessing land use change in biofuel production. NOTE: ADAGE = Applied Dynamic Analysis of the Global Economy Model; EPPA = European Paper Packaging Alliance model; GCAM = Global Change Assessment Model; GLOBIOM = Global Biosphere Management Model; GTAP-BIO = Global Trade Analysis Project-BIO model. SOURCE: Plevin et al. (2022). Creative Commons CC-BY Attribution 4.0 International.

Changes in Food Availability

Along with LUC, one of the consequences of increased competition for land due to biofuels could be a decrease in available food for consumption by humans. In a review of major models used to estimate market-mediated LUC, including GTAP-BIO (used by the California Air Resources Board), FAPRI-CARD (used by U.S. EPA), and MIRAGE (used by the European Union [EU]), Searchinger et al. (2015) found that the models resulted in decreased food availability for humans, the long-term GHG emissions consequences of which have gone largely unexplored, which is particularly important as global food demand is expected to increase substantially over the coming decades (Clark et al., 2020; Tilman et al., 2011). Further research is needed to assess the effect of biofuel production on food availability.

Ease of Transmission of Price Shocks in World Markets

The impact of increased biofuel production on LUCs in the rest of the world depends on the ease with which price shocks are transmitted from domestic markets to the rest of the world. This, in turn, depends on assumptions about the ease with which goods can be traded across countries. Some CGE models typically use the Armington approach[5] which differentiates otherwise homogeneous goods by country of origin (Armington, 1969). Other CGE and all PE models assume that there is one world price for homogenous goods (an integrated world market) and goods will be produced where it is least costly to do so. These models allow for an easier transmission of a shock throughout the world economy which could result in "unrealistic" trade patterns. The Armington approach leads to results that follow observed trade patterns. A potential pitfall of this approach is that it allows price differentials for homogenous goods, such as imported ethanol and domestic ethanol, to persist. Estimates of the induced LUC effect are sensitive to the use of the Armington assumption as compared with the integrated world market assumption. With the Armington assumption in the GTAP-BIO model, land conversions are primarily concentrated in the United States and EU while with the integrated world market assumption they are more evenly distributed across the world and the share of global forest land converted to cropland is higher (Golub and Hertel, 2011).

Types of Biofuels

The mix and quantity of biofuels produced will affect the magnitude of the LUC caused by a low-carbon fuel policy. Low-carbon fuel policies can affect the mix of biofuels depending on the incentives they provide for biofuels from different feedstocks that differ in their CI, cost of production and other factors. This mix of biofuels and overall quantity will depend on the policy inducing the production of biofuels. The use of crop residues and high yielding dedicated energy crops may reduce the need for land for a given volume of biofuels compared to the use of food crops that have relatively lower yield. The overall ILUC intensity of a given target for biofuel consumption will depend on the mix of biofuels and the policy design, although it may not affect the feedstock specific LUC.

Size of the Biofuel Policy Shock

The induced LUC effect of a given target for biofuel consumption is likely to be scale and policy dependent. The magnitude of biofuel produced can affect the magnitude of the induced LUC due to non-linearities in models (Laborde, 2011; Tyner et al., 2010). In CGE models, the concave shape of the constant elasticity of transformation function and the constant elasticity of substitution functions used to represent production possibilities and substitutability among consumption choices causes nonlinearity in effects. Additionally, non-linearity can also arise because the size of the policy shock and the type of policy can affect

---

[5] The Armington assumption is that each country produces a distinct variety of a good. Some CGE models use an Armington elasticity to represent the elasticity of substitution for products from different countries or regions.

the mix of biofuels. The mix of policies can affect the indirect effect of a given volume of biofuel because they can change the mix and levels of different biofuels produced (Chen and Khanna, 2012; Laborde, 2011).

Time Horizon for Assessing the Land Use Change Effect

Induced LUC leads to the release of stored carbon in soils and vegetation if land is converted to crop production. To attribute these GHG emissions to each unit of biofuel produced and compare the induced emissions to the direct flow of carbon savings from using biofuels to displace fossil fuels, studies have amortized the induced LUC emissions over the time horizon that the land is expected to remain in crop production.

Conversion of Induced Land Use Changes to GHG Emissions

Assessing the effects of induced LUC on carbon emission requires two components: (1) An economic model to assess land use and land cover changes induced by biofuel production, or in the case of the approach of Searchinger et al. (2018) the land area used in production, and (2) a set of emissions factors to evaluate the potential emissions associated with each of land conversion or land type in production. Different methods for estimating effects of LUC attributable to biofuel production have been used. In the ALCA context, the opportunity cost of using land for biofuels has been explored with the development of a carbon benefits index that measures the relative output of land use of different types (Searchinger et al., 2018). While the existing literature has intensively discussed the land use modeling component, less attention has been paid to the implications of the choice of emission factors on the estimated ILUC values.

Conversion of Induced Land Use Changes to GHG Emissions

Assessing the effects of induced LUC on carbon emission requires two components: (1) An assessment of land use and land cover changes induced by biofuel production and (2) a set of emissions factors to evaluate the potential emissions associated with each of land conversion or land type in production. While the existing literature has intensively discussed the issue of assessing land use changes, less attention has been paid to the implications of the choice of emission factors on the estimated ILUC values.

The suitability, accuracy, and validity of the implemented emissions factors across the existing modeling efforts have not been reviewed extensively. The early papers in this field applied a set of land use emissions factors provided by the Woods Hole Research Center, Winrock International, or Intergovernmental Plan on Climate Change to estimate these emissions. Then a set of emissions factors developed by Gibbs et al. (2014) and Plevin et al. (2014) have been used in some studies. In support of its GREET model, Argonne National Laboratory developed a separate set of emissions factors (Kwon et al., 2020).[6] Some modelers have developed their own emissions factors using the existing publications following the Intergovernmental Panel on Climate Change guidelines and its reference tables. The implemented emissions factors differ significantly across studies and are a major source of variation in land use emission estimates provided by various studies (Leland et al., 2018).

As models used to assess induced LUC have been updated over the last decade, they have incorporated new elements to reflect agricultural practices in finer detail, including multi-cropping, new land categories such as idled or marginal cropland, and new forms of market mediated responses to biofuel demand. It is worth noting at the beginning of this section that challenges with estimating the emissions consequences of market mediated LUC have led some researchers to call into question whether the modeling approaches described here can adequately quantify these emissions, even as such emissions are expected to occur (Daioglou et al., 2020; Malins et al., 2020). However, counter views have been presented by Taheripour et al. (2021).

---

[6] This sentence was altered after the release of the pre-publication version of the report to the sponsor to acknowledge that Argonne National Laboratory developed a separate set of emissions factors.

Recommendation 9-7: Though the study of induced land use changes from biofuels has been the topic of intense study over the last decade, substantial uncertainties remain on many key components of economic models used to assess these impacts. Further work is warranted to update these estimates of market-mediated land use change and the models so as to inform the development and implementation of an LCFS.

Recommendation 9-8: Assessment of the consequential effects from a future proposed policy, such as induced land use change, should be further developed in order to assess the risk of market-mediated effects and emissions attributable to the policy. Consequential assessment can inform the implementation of safeguards within policies such as limits on high-risk feedstocks, can inform the development of supplementary policies, identify hotspots, and reduce the likelihood of unintended consequences.

Recommendation 9-9: To improve understanding of market-mediated effects of biofuels, research should be supported on different modeling approaches, including their treatment of baselines and opportunity costs, and to investigate key parameters used in national and international modeling based on measured data, including various elasticity parameters, soil carbon sequestration, land cover, and emission factors and others.

Recommendation 9-10: Because other market-mediated effects of biofuel production, such as livestock market impacts, land management practices, and changes in diets and food availability may be linked to land use and biofuel demand assessed using induced land use change models, additional research should be done and model improvements undertaken to include these effects.

Recommendation 9-11: Current and future low-carbon fuel policies should strive for transparency in their modeling efforts.

Rebound Effect from Fossil Fuel Displacement

The rebound effect is also measured using economic models that include the domestic transportation sector, the global fuel sector and global trade in fossil fuels. These models include supply and demand relationships for fossil fuels and biofuels (or in some case vehicle miles travelled that are used to derive demand for alternative types of fuels) in domestic and global markets that are influenced by the biofuel policies. The extent to which these models cover the energy sector varies across the modeling practices. Some models only cover markets for ethanol and gasoline, while others cover a broader perspective including markets for fossil fuels across uses including but not limited to transportation and the link between the crude oil and refinery products. Estimates of the global rebound effect due to U.S. biofuels have varied substantially across studies depending on assumptions about the elasticity of supply of oil in the world market, market power in the oil market, and the quantity of biofuel produced. There are various determinants of the magnitude of the rebound effect, these include:

Market structure: Models differ in their assumption about the structure of the oil market. Some studies assume perfectly competitive markets with free trade in oil among countries. An alternative assumption is that there is sufficient market power in the oil market and that oil prices can be set by large oil producers in response to different market factors.

Elasticity of the world oil supply: Studies differ in their assumptions about the elasticity of world oil supply; this may depend on whether they are analyzing short term or long term effects. The more inelastic the supply curve the larger the reduction in world oil price in response to a reduction in biofuel induced demand for oil.

Elasticity of demand for fuel: The greater the elasticity of demand, the larger the reduction in demand for fuel due to a small policy induced increase in the price of fuel.

Comprehensiveness of fuel and energy markets: Studies differ in modeling the details of energy and fuel markets. A more comprehensive model provides a lower rebound effect.

Extent of trade in fuel with rest of the world: If the biofuel producing country is a small trader in oil then it is commonly assumed that, with a small open economy and a small amount of oil use, the effect on the global market will be negligible. If the biofuel producing country is a large trader in the world market, then it has greater potential to affect world price of oil assuming a competitive oil market (Chen et al., 2021).

Stringency of biofuel policy: The larger the biofuel mandate, the greater the need for displacing fossil fuel with biofuel and the larger the implicit penalty on fossil fuel use and implicit subsidy on biofuel use. This could have a larger negative effect on fossil fuel consumption and can lead to a larger rebound effect, depending on the various factors explained above.

Validation of Models Quantifying Indirect Land Use Change

As with any life-cycle model, since emissions due to LUC or fuel market effects that can be attributed to biofuels are not directly observable, the credibility of the estimates obtained depends on the quality of the model being used to quantify it. Validation of model outcomes by comparing the simulated outcomes with observed data is one approach for assessing the ability of models to predict realistic outcomes. Another approach to checking the credibility of the model is the extent to which the parametric assumptions of the model are supported by empirical evidence. Both of these approaches are however extremely difficult to implement. This is because the global models described above rely on a very large number of parameters and the empirical evidence for these parameters is often thin or not available at the spatial scale or for the time period most relevant for the model. Similarly, these models generate a very large number of numerical outcomes. While some of these may be close to observed data, others may not be; see Arndt et al. (2002) and Liu et al. (2004). The validation methodology offers no guidance on the acceptable deviation between observed and simulated outcomes and what constitutes a model as being valid or invalid. This is a subjective judgment similar to determining what constitutes a "reasonable" degree of uncertainty. Moreover, the time frame for model calibration or validation is typically shorter than the time frame over which predictions are to be made. Although the model may accurately reproduce observed data at a point in time, there is no assurance that it can perform equally well over a longer time frame. The effect of modeling uncertainties and errors on estimates of induced land use change are likely to be lower when one is comparing a difference (change) between two scenarios that hold all assumptions the same and only differ in the amount of biofuel produced. Despite the limitations of using models to make precise predictions, models serve a useful purpose of providing scientific understanding of the intended and unintended system-wide effects of a policy or technology change. Having a range of estimates from multiple models can create uncertainty but can also point to areas where there is greater consistency in findings and areas of large differences. This can be useful for policy design by showing the order of magnitude of effects.

As outlined above, validation of model results for future projections is not an easy task to accomplish. However, *ex post* validation is a more practical option. Models' projections can be validated against new evidence and observation as time moves forward. Over the past 15 years various models have been used to evaluate the extent to which biofuel production could affect land use and LUCs. The results of these modeling practices vary widely. As discussed above, the existing literature shows a few key factors that explain most of the differences across models' results: modeling yield improvement, the extent to which biofuel production occurs on the existing idled land or causes extensification, and the extent to which biofuel produced in one country affect land use changes in other countries due to trade. It is practical to assess the models' assumptions on these key drivers and their land use projections and verify them according to the existing recent evidence. It may not be practical to assess all models' parameters. However, after producing large quantities of biofuel in practice and having various observations, comparing hypothetical modeling projections with observations is a practical and valid practice and can help policy decision-making.

Inclusion of Market-Mediated Effects in Biofuel Policies

Policies to promote low-carbon fuels differ in the way in which they include induced effects in the CI of biofuels. The U.S. RFS specifies minimum thresholds for the CI of various types of biofuels, conventional, advanced and cellulosic, relative to conventional gasoline. For the RFS, the CI of a particular type of biofuel is the sum of a set of fuel supply chain emissions, emissions from induced agricultural practices, and induced LUC emissions intensity; together this sum needs to be lower than the threshold for that type of biofuel as assessed in 2022 to ensure compliance. The RFS regulatory impact assessment also estimated the 2012 and 2017 CI of biofuels, though the 2022 estimate was used to determine the categorization of biofuels within the policy. Uncertainty in estimates of induced LUC is dealt with by taking the mean value of the distribution of estimates of CIs.

In contrast with the RFS, the California Low Carbon Fuel Standard (CA-LCFS) requires specific estimates of the CI of a fuel pathway that can be compared with the goal of the policy (the desired CI of transportation fuel) in order to determine its compliance. The CA-LCFS determines the total CI of a biofuel by adding together the supply chain emissions intensity and induced LUC intensity of the biofuel.

Each of these policies uses an estimated ILUC factor for each type of biofuel that is treated to be invariant to scale of biofuel produced, time period or location of feedstock produced for that biofuel, within a country. To the extent that the ILUC effect is non-linearly related to the scale of production of a biofuel, this approach may under-estimate or over-estimate the magnitude of the effect. It also ignores potential circularity in the magnitude of the effect, since the scale of production depends on the deterrent effects of including the ILUC factor in determining compliance with the policy which in turn will influence the magnitude of the ILUC effect.

Precision in the specific estimate of CI of a biofuel can have meaningful implications for the fuel mix supplied under both an RFS and LCFS-like policy. For the purposes of the RFS, the CI can dictate which subcategory a fuel is eligible within, or its eligibility altogether, whereas in an LCFS a fuel's estimated CI determines its compliance value within the policy. These policies ignore other types of indirect effects, such as those due to fuel market or livestock market effects. However, in the case of the CA-LCFS, the objective of including market-mediated effects in the CI of the biofuel is to deter (implicitly penalize) consumption of biofuels with large emissions from market-mediated effects and incentivize consumption of biofuels with lower total effects. However, the policy can still incentivize biofuels with higher relative ILUC emissions estimates if their total CI estimate is comparable to other fuels. A performance standard such as the CA-LCFS creates incentives to continuously reduce the CI of the low-carbon fuel over time and to substitute lower-CI biofuels for higher-CI biofuels, even within a type or across types, as compared to the RFS.

Since an LCFS is a CI standard, full compliance with it may achieve a reduction in CI (carbon emissions per unit fuel), but there is no guarantee that it will reduce the aggregate GHG emissions from the transportation sector. This will depend on the total volume of fuel consumed, which may offset more than the reduction in the CI of the fuel, and the degree to which the CI reflects the actual emissions attributable to the fuels. The increase in the volume of fuel use may occur for two reasons: a direct effect and an indirect effect. The direct effect arises because an LCFS implicitly penalizes the fuel with a CI higher than the standard and implicitly subsidizes the fuel with a CI lower than the standard. Depending on the magnitudes of these implicit taxes and subsidies, an LCFS can lower the overall cost of the blended transportation fuel and lead to greater demand for fuel. The indirect effect arises because the reduction in the demand for the high CI (fossil) fuels can lower market price of the fuel and lead to a rebound effect in the fuel market, as discussed above (see Holland et al., 2008; Khanna et al., 2014).

REFERENCES

Achat, D. L., C. Deleuze, G. Landmann, N. Pousse, J. Ranger, and L. Augusto. 2015. Quantifying consequences of removing harvesting residues on forest soils and tree growth – A meta-analysis. *Forest Ecology and Management* 348:124–141. https://doi.org/10.1016/j.foreco.2015.03.042.

Adler, P. R., S. J. Del Grosso, D. Inman, R. E. Jenkins, S. Spatari, and Y. Zhang. 2012. Mitigation opportunities for life-cycle greenhouse gas emissions during feedstock production across heterogeneous landscapes. In *Managing Agricultural Greenhouse Gases* 203–219, Elsevier.

Alakoski, E., M. Jämsén, D. Agar, E. Tampio, and M. Wihersaari, M. 2016. From wood pellets to wood chips, risks of degradation and emissions from the storage of woody biomass: A short review. *Renewable and Sustainable Energy Reviews* 54:376–383. https://doi.org/10.1016/j.rser.2015.10.021.

Alvarez, R. A., D. Z.- Araiza, D. R. Lyon, D. T. Allen, Z. R. Barkley, A. R. Brandt, K. J. Davis, S. P. Hamburg, S. C. Herndon, D. J. Jacob, A. Karion, E. A. Kort, B. K. Lamb, T. Lauvaux, J. D. Maasakkers, A. J. Marchese, M. Omara, S. W. Pacala, J. Peisch, A. L. Robinson, P. B. Shepson, C. Sweeney, A. Townsend-Small, and S. C. Wofsy. 2018. Assessment of methane emissions from the US oil and gas supply chain. *Science* 361(6398):186–188. https://www.science.org/doi/full/10.1126/science.aar7204.

Armington, P. S. 1969. *A Theory of Demand for Products Distinguished by Place of Production (Une théorie de la demande de produits différenciés d'après leur origine) (Una teoría de la demanda de productos distinguiéndolos según el lugar de producción)*. Staff Papers (*International Monetary Fund*) 16(1):159–178. https://doi.org/10.2307/3866403.

Arndt, C., S. Robinson, and F. Tarp. 2002. Parameter estimation for a computable general equilibrium model: A maximum entropy approach. *Economic Modelling* 19:375–398.

Arora, S., M. Wu, and M. Wang. 2008. Update of Distillers Grains Displacement Ratios for Corn Ethanol Life-Cycle Analysis. ANL/ESD/11-1. Center for Transportation Research Energy Systems Division, Argonne National Laboratory. https://publications.anl.gov/anlpubs/2011/01/69222.pdf.

Astals, S., V. Nolla-Ardevol, and J. Mata-Alvarez. 2012. Anaerobic co-digestion of pig manure and crude glycerol at mesophilic conditions: Biogas and digestate. *Bioresource Technology* 110:63–70.

Baker, S. E., J. K. Stolaroff, G. Peridas, S. H. Pang, H. M. Goldstein, F. R. Lucci, W. Li, E. W. Slessarev, J. Pett-Ridge, F. J. Ryerson, and J. L. Wagoner. 2020. *Getting to Neutral: Options for Negative Carbon Emissions in California*. No. LLNL-TR-796100. Lawrence Livermore National Lab. (LLNL), Livermore, CA (United States); University of California, Berkeley, CA (United States); Negative Carbon Consulting, Half Moon Bay, CA (United States); University of Calgary, AB (Canada); University of Queensland, Brisbane, QLD (Australia); University of California, Davis, CA (United States); Worcester Polytechnic Institute, MA (United States); Georgetown University, Washington, DC (United States); Valence Strategic, Washington, DC (United States). https://www-gs.llnl.gov/content/assets/docs/energy/Getting_to_Neutral.pdf.

Beach, R. H. and B. A. McCarl. 2010. *US Agricultural and Forestry Impacts of the Energy Independence and Security Act: FASOM Results and Model Description.* Research Triangle Park, NC: RTI International. http://yosemite.epa.gov/Sab/Sabproduct.nsf/962FFB6750050099852577820072DFDE/$File/FASOM+Report_EISA_FR.pdf.

Booth, M.S. 2018. Not carbon neutral: Assessing the net emissions impact of residues burned for bioenergy. *Environmental Research Letters* 13(3):035001.

Bremer, V.R., A. J. Liska, T. J. Klopfenstein, G. E. Erickson, H. S. Yang, D. T. Walters, and K. G. Cassman. 2010. Emissions savings in the corn-ethanol life cycle from feeding coproducts to livestock. *Journal of Environmental Quality*, 39(2):472-482.

Brentrup, F., J. Küsters J. Lammel and H. Kuhlmann. 2000. Methods to estimate on-field nitrogen emissions from crop production as an input to LCA studies in the agricultural sector. *The International Journal of Life Cycle Assessment* 5(6):349. https://doi.org/10.1007/BF02978670.

Breunig, H. M., J. Amirebrahimi, S. Smith, and C. D. Scown. 2019. Role of digestate and biochar in carbon-negative bioenergy. *Environmental Science & Technology* 53(22):12989–12998. https://pubs.acs.org/doi/abs/10.1021/acs.est.9b03763.

Cai, H., J. B. Dunn, Z. Wang, J. Han, and M. Q Wang. 2013. Life-cycle energy use and greenhouse gas emissions of production of bioethanol from sorghum in the United States. *Biotechnology Biofuels* 6:141 https://doi.org/10.1186/1754-6834-6-141.

Cai, H., J. Han, M. Wang, R. Davis, M. Biddy, and E. Tan. 2018. Life-cycle analysis of integrated biorefineries with co-production of biofuels and bio-based chemicals: Co-product handling methods and implications. *Biofuels, Bioproducts and Biorefining* 12(5):815–833. https://doi.org/10.1002/bbb.1893.

Campbell, R. M., N. M. Anderson, D. E. Daugaard, and H. T. Naughton. 2018. https://www.sciencedirect.com/science/article/pii/S0306261918312558.

CARB. 2017. *Co-Processing of Biogenic Feedstocks in Petroleum Refineries: Draft Staff Discussion Paper.* https://ww2.arb.ca.gov/sites/default/files/classic/fuels/lcfs/lcfs_meetings/020717_staffdiscussionpaper.pdf.

CARB. 2022. *GHG Emissions and Sinks.* https://www.epa.gov/ghgemissions/inventory-us-greenhouse-gas-emissions-and-sinks.

CARB. 2021. *Inventory Data.* https://ww2.arb.ca.gov/ghg-inventory-data.

Chambers, R.S., R. A. Herendeen, J. J. Joyce, and P. S. Penner. 1979. Gasohol: Does it or doesn't it produce positive net energy? *Science* 206(4420):789–795. https://www.science.org/doi/10.1126/science.206.4420.789.

Chen, L., D. Debnath, J. Zhong, K. Ferin, A. VanLoocke, and M. Khanna, M. 2021. The economic and environmental costs and benefits of the renewable fuel standard. *Environmental Research Letters* 16(3):034021. https://doi.org/10.1088/1748-9326/abd7af.

Chen, X., Huang, H., Khanna, M., and Onal, H. 2014. Alternative Fuel Standards: Welfare Effects and Climate Benefits. *Journal of Environmental Economics and Management* 67:241–257.

Clark, M.A., N. G. Domingo, K. Colgan, S.K. Thakrar, D. Tilman, J. Lynch, I. L. Azevedo, and J. D. Hill. 2020. Global food system emissions could preclude achieving the 1.5 and 2 C climate change targets. *Science*, 370(6517):705-708.

Cowie, A. L., G. Berndes, N. S. Bentsen, M. Brandão, F. Cherubini, G. Egnell, B. George, L. Gustavsson, M. Hanewinkel, Z. M. Harris, and F. Johnsson. 2021. Applying a science-based systems perspective to dispel misconceptions about climate effects of forest bioenergy. *GCB Bioenergy* 13(8):1210–1231. https://onlinelibrary.wiley.com/doi/epdf/10.1111/gcbb.12844.

Cruce, J. R. and J. C. Quinn. 2019. Economic viability of multiple algal biorefining pathways and the impact of public policies. *Applied Energy* 233:735–746.

Daioglou, V., G. Woltjer, B. Strengers, B. Elbersen, G. B. Ibañez, D. S. Gonzalez, J. G. Barno, and D. P. van Vuuren. 2020. Progress and barriers in understanding and preventing indirect land-use change. *Biofuels, Bioproducts and Biorefining* 14(5):924–934.

Dumortier, J., H. Dokoohaki, A. Elobeid, D. J. Hayes, D. Laird, and F. E. Miguez. 2020. Global land-use and carbon emission implications from biochar application to cropland in the United States. *Journal of Cleaner Production* 258:120684. https://www.sciencedirect.com/science/article/pii/S0959652620307319#bib31.

Dumortier, J., D. J. Hayes, M. Carriquiry, F. Dong, X. Du, A. Elobeid, J. F. Fabiosa, and S. Tokgoz. 2011. Sensitivity of carbon emission estimates from indirect land-use change. *Applied Economic Perspectives and Policy* 33(3):428–48. https://doi.org/10.1093/aepp/ppr015.

European Commission. 2018. *Phyllis: ECN Database of Biomass and Waste Properties.*

Fabiosa, J. F., J. C. Beghin, F. Dong, A. Elobeid, S. Tokgoz, and T.-H. T.-H. Yu. 2010. Land Allocation Effects of the Global Ethanol Surge: Predictions from the International FAPRI Model. *Land Economics* 86(4):687–706.

Fargione, J., J. Hill, D. Tilman, S. Polasky, and P. Hawthorne. 2008. Land clearing and the biofuel carbon debt. *Science* 319:1235–8.

Favero, A., A. Daigneault and B. Sohngen. 2020. Forests: Carbon sequestration, biomass energy, or both? *Science Advances* 6(13):6792. https://www.science.org/doi/10.1126/sciadv.aay6792.

Flugge, M., J. Lewandrowski, J. Rosenfeld, C. Boland, T. Hendrickson, K. Jaglo, S. Kolansky, K. Moffroid, M. Riley-Gilbert, and D. Pape. 2017. A Life-Cycle Analysis of the Greenhouse gas emissions of corn-based ethanol.

Gasparatos, A., P. Stromberg, and K. Takeuchi. 2013. Sustainability impacts of first-generation biofuels, *Animal Frontiers* 3(2):12–26. https://doi.org/10.2527/af.2013-0011.

Geronimo, C., S. E. Vergara, C. Chamberlin, and K. Fingerman. 2022. Overlooked emissions: Influence of environmental variables on greenhouse gas generation from woody biomass storage. *Fuel* 319(123839):0016–2361, https://doi.org/10.1016/j.fuel.2022.123839.

Gibbs, H. K., M. Johnston, J. A. Foley, T. Holloway, C. Monfreda, N. Ramankutty, and D. Zaks. 2008. Carbon payback times for crop-based biofuel expansion in the tropics: The effects of changing yield and technology. *Environmental Research Letters* 3(3):034001.

Gibbs, H., S. Yui, and R. J. Plevin. 2014. *New Estimates of Soil and Biomass Carbon Stocks for Global Economic Models.* GTAP Technical Paper No. 33. West Lafayette, Indiana: Purdue University.

Golub A. A., T. W. Hertel, and S. K. Rose. 2017. Global Land Use Impacts of U.S. Ethanol: Revised Analysis Using GDyn-BIO Framework. In *Handbook of Bioenergy Economics and Policy: Volume II*, M. Khanna, and D. Zilberman, eds. *Natural Resource Management and Policy* 40. Springer, New York, NY. https://doi.org/10.1007/978-1-4939-6906-7_8.

Golub, A. and T. W. Hertel. 2011. "Modeling Land Use Change Impacts of Biofuels in the GTAP-BIO Framework." https://web.ics.purdue.edu/~hertel/data/uploads/publications/golub-hertel-climate-change-economics.pdf.

Gurgel, A., J. M. Reilly, and S. Paltsev. 2007. *Potential Land Use Implications of a Global Biofuels Industry* 36.

Hasler, K., S. Bröring, S. W. F. Omta, and H. W. Olfs. 2015. Life cycle assessment (LCA) of different fertilizer product types. *European Journal of Agronomy* 69:41–51. https://doi.org/10.1016/j.eja.2015.06.001. https://www.sciencedirect.com/science/article/pii/S1161030115000714.

Heijungs, R., K. Allacker, E. Benetto, M. Brandão, J. Guinée, S. Schaubroeck, T. Schaubroeck, and A. Zamagni. 2021. System Expansion and Substitution in LCA: A Lost Opportunity of ISO 14044 Amendment 2. *Frontiers in Sustainability* 2(40). https://doi.org/10.3389/frsus.2021.692055.

Hertel, T. W., W. E. Tyner, and D. K. Birur. 2010. The global impacts of biofuel mandates. *Energy Journal* 31(1):75–100.

Hill, J. 2022. "The sobering truth about corn ethanol." *PNAS* 119:11. https://www.pnas.org/doi/10.1073/pnas.2200997119.

Hill, J., A. Goodkind, C. Tessum, S. Thakrar, D. Tilman, S. Polasky, T. Smith, N. Hunt, K. Mullins, M. Clark, and J. Marshall. 2019. Air-quality-related health damages of maize. *Nature Sustainability* 2:397–403. https://doi.org/10.1038/s41893-019-0261-y.

Hill, J., L. Tajibaeva and S. Polasky. 2016. Climate consequences of low-carbon fuels: The United States renewable fuel standard. *Energy Policy* 97:351–353. https://www.sciencedirect.com/science/article/pii/S0301421516303962?via%3Dihub.

Holland, S. P., J. E. Hughes, and C. R. Knittel. 2009. Greenhouse gas reductions under low carbon fuel standards? *American Economic Journal: Economic Policy* 1(1):106–46.

Hornafius, K. Y. and J. S. Hornafius. 2015. Carbon negative oil: A pathway for CO_2 emission reduction goals. *International Journal of Greenhouse Gas Control* 37:492–503. https://doi.org/10.1016/j.ijggc.2015.04.007.

Hsu, D. D. 2012. Life cycle assessment of gasoline and diesel produced via fast pyrolysis and hydroprocessing. *Biomass and Bioenergy* 45:41–47. https://doi.org/10.1016/j.biombioe.2012.05.019.

Huo, H., M. Wang, C. Bloyd, and V. Putsche. 2009. Life-cycle assessment of energy use and greenhouse gas emissions of soybean-derived biodiesel and renewable fuels. *Environmental Science & Technology* 43(3):750756. https://doi.org/10.1021/es8011436.

International Organization for Standardization. 2006. *Environmental Management: Life Cycle Assessment; Requirements and Guidelines.* ISO, Geneva.

International Standard Organization. 2020. *ISO 14044:2006/AMD 2:2020 Environmental management — Life cycle assessment — Requirements and guidelines — Amendment 2.* ISO.76, Geneva, Switzerland.

Ippolito, J. A., L. Cui, C. Kammann, N. Wrage-Mönnig, J. M. Estavillo, T. Fuertes-Mendizabal, M. L. Cayuela, G. Sigua, J. Novak, K. Spokas, and N. Borchard. 2020. Feedstock choice, pyrolysis temperature and type influence biochar characteristics: A comprehensive meta-data analysis review. *Biochar* 2:421–438. https://link.springer.com/article/10.1007/s42773-020-00067-x.

Joseph, S., Cowie, A. L., Van Zwieten, L., Bolan, N., Budai, A., Buss, W., Cayuela, M. L., Graber, E. R., Ippolito, J. A., Kuzyakov, Y., and Luo, Y. 2021. How biochar works, and when it doesn't: A review of mechanisms controlling soil and plant responses to biochar. *GCB Bioenergy* 13(11):1731–1764.

Kauppi, P. E., P. Ciais, P. Högberg, P., A. Nordin, J. Lappi, T. Lundmark, and I. K. Wernick. 2020. Carbon benefits from Forest Transitions promoting biomass expansions and thickening. *Global Change Biology* 26(10):5365–5370. https://doi.org/10.1111/gcb.15292.

Keeler, B. L., B. J. Krohn, T. A. Nickerson, and J. D. Hill. 2013. US federal agency models offer different visions for achieving renewable fuel standard (RFS2) biofuel volumes. https://pubs.acs.org/doi/10.1021/es402181y.

Keeney, D. 2009. Ethanol USA. *Environmental Science & Technology* 43:8–11.

Khanna, M. and C. L. Crago. 2012. Measuring indirect land use change with biofuels: Implications for policy. *Annual Review of Resource Economics* 4(1):161–84. https://doi.org/10.1146/annurev-resource-110811-114523.

Khanna M., A. Ando and F. Taheripour. 2008. Welfare effects and unintended consequences of ethanol subsidies. *Review of Agricultural Economics* 30(3):411–421.

Khanna, M., L. Chen, B. Basso, X. Cai, J. L. Field, K. Guan, C. Jiang, T. J. Lark, T. L. Richard, S. A. Spawn-Lee, and P. Yang. 2021. Redefining marginal land for bioenergy crop production. *GCB Bioenergy* 13(10):1590–1609. https://onlinelibrary.wiley.com/doi/10.1111/gcbb.12877.

Khanna, M., P. Dwivedi and R. Abt. 2017. Is Forest. Bioenergy Carbon Neutral or Worse than Coal? Implications of Carbon Accounting Methods. *International Review of Environmental and Resource Economics* 10(3–4):299–346.

Khanna, M., D. Rajagopal and D. Zilberman. 2021. Lessons learned from US experience with biofuels: Comparing the hype with the evidence. *Review of Environmental Economics and Policy* 15(1):67–861.

Khanna, M., W. Wang, T. Hudiburg, and E. DeLucia. 2017. The social inefficiency of regulating indirect land use change due to biofuels. *Nature Communications* 8:15513.

Khanna, M., D. Zilberman, and C. Crago. 2014. Modeling Land Use Change with Biofuels. In *The Oxford Handbook of Land Economics*. Oxford University Press, Oxford, UK.

Kim, S., X. Zhang, A. D. Reddy, B. E. Dale, K. D. Thelen, C. D. Jones, R. C. Izaurralde, T. Runge, and C. Maravelias. 2020. Carbon-negative biofuel production. *Environmental Science & Technology* 54(17):10797–10807. https://doi.org/10.1021/acs.est.0c01097.

Klein, D., C. Wolf, C. Schulz, and G. Weber-Blaschke. 2015. 20 years of life cycle assessment (LCA) in the forestry sector: State of the art and a methodical proposal for the LCA of forest production. *International Journal of Life Cycle Assessment* 20:556–575. https://doi.org/10.1007/s11367-015-0847-1.

Knapp, E. E., A. A. Bernal, J. M. Kane, C. J. Fettig, and M. P. North. 2021. Variable thinning and prescribed fire influence tree mortality and growth during and after a severe drought. *Forest Ecology and Management* 479:118595. https://doi.org/10.1016/j.foreco.2020.118595.

Kwon, H., X. Liu, J. B. Dunn, S. Mueller, M. M. Wander, and M. Wang. 2020. *Carbon Calculator for Land Use and Land Management Change from Biofuels Production (CCLUB)*. Argonne National Laboratory, ANL/ESD/12-5, Rev.6.

Laborde, D. and H. Valin. 2012. Modeling land-use changes in a global CGE: Assessing the EU biofuel mandates with the MIRAGE-BioF model. *Climate Change Economics* 3(3):1250017. https://doi.org/10.1142/S2010007812500170.

Lammens, T. M. 2020. Methodologies for biogenic carbon determination when co-processing fast pyrolysis bio-oil. *Energy Conversion and Management: X*, 10, 100069. https://doi.org/10.1016/j.ecmx.2020.100069.

Lan, K., L. Ou, S. Park, S. S. Kelley, P. Nepal, H. Kwon, H. Cai, and Y. Yao. 2021. Dynamic life-cycle carbon analysis for fast pyrolysis biofuel produced from pine residues: implications of carbon temporal effects. *Biotechnology for Biofuels and Bioproducts* 14:191. https://doi.org/10.1186/s13068-021-02027-4.

Lan, K., L. Ou, S. Park, S. S. Kelley, and Y. Yao. 2020. Life cycle analysis of decentralized preprocessing systems for fast pyrolysis biorefineries with blended feedstocks in the southeastern United States. *Energy Technology* 8(11):1900850. https://doi.org/10.1002/ente.201900850.

Lan, K., L. Ou, S. Park, S. S. Kelley, B. C. English, E. T. Yu, J. Larson, and Y. Yao. 2021. Techno-economic analysis of decentralized preprocessing systems for fast pyrolysis biorefineries with blended feedstocks in the southeastern United States. *Renewable and Sustainable Energy Reviews* 143:110881. https://doi.org/https://doi.org/10.1016/j.rser.2021.110881.

Lan, K., S. S. Kelley, P. Nepal, and Y. Yao. 2020. Dynamic life cycle carbon and energy analysis for cross-laminated timber in the Southeastern United States. *Environmental Research Letters*.

Lark, T. J., N. P. Hendricks, A. Smith, N. Pates, S. A. Spawn-Lee, M. Bougie, E. G. Booth, C. J. Kucharik, and H. K. Gibbs. 2022. Environmental outcomes of the US Renewable Fuel Standard. *Proceedings of the National Academy of Sciences* 119(9):e2101084119.

Lawrence, N. C., C. G. Tenesaca, A. VanLoocke, and S. J. Hall. 2021. Nitrous oxide emissions from agricultural soils challenge climate sustainability in the US Corn Belt. *Proceedings of the National Academy of Sciences*, 118(46), e2112108118. https://doi.org/10.1073/pnas.2112108118.

Lee, U., Z. Lu, P. Su, M. Wang, V. DaVita, and D. Collings. 2022. *Carbon Intensities of Refining Products in Petroleum Refineries with Co-processed Biofeedstocks*. Argonne National Laboratory. ANL/ESD-21/20 https://publications.anl.gov/anlpubs/2022/02/173380.pdf.

Lee, E. K., X. Zhang, P. R. Adler, G. S. Kleppel, and X. X. Romeiko. 2020. Spatially and temporally explicit life cycle global warming, eutrophication, and acidification impacts from corn production in the US Midwest. *Journal of Cleaner Production* 242:118465. https://www.sciencedirect.com/science/article/abs/pii/S0959652619333359.

Lehmann, J., A. Cowie, C. A. Masiello, C. Kammann, D. Woolf, J. E. Amonette, M. L. Cayuela, M. Camps-Arbestain, and T. Whitman. 2021. Biochar in climate change mitigation. *Nature Geoscience* 14(12):883–892.

Leland, S., S. K. Hoekman, and X. Liu. 2018. "Review of modifications to indirect land use change modeling and resulting carbon intensity values within the California Low Carbon Fuel Standard regulations." *Journal of LCeaner Production* 180:698-707. https://doi.org/10.1016/j.clepro.2018.01.077.

Lewandrowski, J., J. Rosenfeld, D. Pape, T. Hendrickson, K. Jaglo, and K. Moffroid, K. 2020. The greenhouse gas benefits of corn ethanol: Assessing recent evidence. *Biofuels* 11(3):361–375. https://doi.org/10.1080/17597269.2018.1546488.

Linger, J. G., D. R. Vardon, M. T. Guarnieri, E. M. Karp, G. B. Hunsinger, M. A. Franden, C. W. Johnson, G. Chupka, T. J. Strathmann, P. T. Pienkos, and G. T. Beckham. 2014. Lignin valorization through integrated biological funneling and chemical catalysis. *Proceedings of the National Academy of Sciences* 111.33:12013–12018. https://doi.org/10.1073/pnas.1410657111.

Liu, J., C. Arndt and T. W. Hertel. 2004. Parameter estimation and measures of fit in a global, general equilibrium model. *Journal of Economic Integration* 19(3):626–649.

Liu, W., Z. Zhang, X. Xie, Z. Yu, K. von Gadow, J. Xu, S. Zhao, and Y. Yang. 2017a. Analysis of the global warming potential of biogenic CO_2 emission in life cycle assessments. *Scientific Reports* 7:39857. https://doi.org/10.1038/srep39857.

Liu, W., Z. Yu, X. Xie, K. V. Gadow, and C. Peng. 2017b. A critical analysis of the carbon neutrality assumption in life cycle assessment of forest bioenergy systems. *Environmental Reviews* 26(1):93–101. https://doi.org/10.1139/er-2017-0060.

Liu, X., H. Kwon and M. Wang. 2021. Varied farm-level carbon intensities of corn feedstock help reduce corn ethanol greenhouse gas emissions. *Environmental Research Letters* 16(6):064055. https://iopscience.iop.org/article/10.1088/1748-9326/ac018f.

Maestrini, B. and B. Basso. 2018. Drivers of within-field spatial and temporal variability of crop yield across the US Midwest. *Scientific Reports* 8:14833. https://doi.org/10.1038/s41598-018-32779-3.

Malins, C., R. Plevin and R. Edwards. "How robust are reductions in modeled estimates from GTAP-BIO of the indirect land use change induced by conventional biofuels?" *Journal of Cleaner Production* 258:120716. doi:10.1016/j.clepro.2020.120716.

Markewitz, D. 2006. Fossil fuel carbon emissions from silviculture: Impacts on net carbon sequestration in forests. *Forest Ecology and Management* 236(2):153–161. https://doi.org/10.1016/j.foreco.2006.08.343.

Maroušek, J., M. Vochozka, J. Plachý and J. Žák. 2017 Glory and misery of biochar. *Clean Technologies and Environmental Policy*, 19, 311–317. https://link.springer.com/article/10.1007/s10098-016-1284-y.

McEwan, A., Brink, M., and Spinelli, R. 2019. Efficiency of Different Machine Layouts for Chain Flail Delimbing, Debarking and Chipping. *Forests* 10(2):126. https://www.mdpi.com/1999-4907/10/2/126.

Melilio, J. M., P. A. Steudler, J. D. Aber, K. Newkirk, H. Lux, F. P. Bowles, C. Catricala, A. Magill, T. Ahrens, and S. Morrisseau. 2002. "Soil Warming and Carbon-Cycle Feedbacks to the Climate System." *Science* 298(2173). https://doi.org/10.1126/science.1074153.

Moset, V., M. Poulsen, R. Wahid, O. Højberg, and H. B. Møller. 2015. Mesophilic versus thermophilic anaerobic digestion of cattle manure: Methane productivity and microbial ecology. *Microbial Biotechnology* 8(5):787–800.

National Renewable Energy Laboratory (NREL). 2012. *U.S. Life Cycle Inventory Database.*

Nie, Y. and X. Bi. 2018. Life-cycle assessment of transportation biofuels from hydrothermal liquefaction of forest residues in British Columbia. *Biotechnology for Biofuels* 11(1):23. https://doi.org/10.1186/s13068-018-1019-x.

Lan, K., S. Park, and Y. Yao. 2019. "Key issue, challenges, and status quo of models for biofuel supply chain design." In: Biofuels for a More Sustainable Future: Life Cycle Sustainability Assessment and Multi-Criteria Decision Making. Biofuels for a More Sustainable Future: Life Cycle Sustainability Assessment and Multi-Criteria Decision Making. pp. 273-315. https://doi.org/10.1016/B978-0-12-815581-3.00010-5.

Oneil, E. E., L. R. Johnson, B. R. Lippke, J. B. McCarter, M. E. McDill, P. A. Roth, and J. C. Finley. 2010. Life-cycle impacts of inland northwest and northeast/north central forest resources. *Wood and Fiber Science* 42:29–51.

Pingoud, K., T. Ekholm, R. Sievänen, S. Huuskonen, and J. Hynynen. 2018. Trade-offs between forest carbon stocks and harvests in a steady state: A multi-criteria analysis. *Journal of Environmental Management* 210:96–103. https://www.sciencedirect.com/science/article/pii/S0301479717312641?via%3Dihub.

Plevin, R. J., J. Jones, P. Kyle, A. W. Levy, M. J. Shell ,and D. J. Tanner. 2022. Choices in Land Representation Materially Affect Modeled Biofuel Carbon Intensity Estimates. *Journal of Cleaner Production* 349:131477. https://doi.org/10.1016/j.jclepro.2022.131477.

Plevin, R., H. Gibbs, J. Duffy, S. Yui, and S. Yeh. 2014. *Agro-ecological Zone Emission Factor (AEZEF) Model (V47)*, GTAP Technical Paper No. 34. West Lafayette, Indiana: Purdue University.

Pognani, M., G. D'Imporzano, B. Scaglia, and F. Adani. 2009. Substituting energy crops with organic fraction of municipal solid waste for biogas production at farm level: A full-scale plant study. *Process Biochemistry* 44(8):817–821.

Powlson, D. S., M. J. Glendining, K. Coleman, and A. P. Whitmore. 2011. Implications for Soil Properties of Removing Cereal Straw: Results from Long-Term Studies 1. *Agronomy Journal*. https://doi.org/10.2134/agronj2010.0146s.

Qin, Z., C. E. Canter, J. B. Dunn, S. Mueller, H. Kwon, J. Han, M. M. Wander, and M. Wang. 2018. Land management change greatly impacts biofuels' greenhouse gas emissions. *GCB Bioenergy* 10(6):370–381.

Robertson, G. P., V. H. Dale, O. C. Doering, S. P. Hamburg, J. M. Melillo, M. M. Wander, W. J. Parton, P. R. Adler, J. N. Barney, R. M. Cruse, and C. S. Duke. 2008. Sustainable biofuels redux. *Science* 322(5898):49–50.

Sahoo, K., E. M. Bilek, and S. Mani. 2018. Techno-economic and environmental assessments of storing woodchips and pellets for bioenergy applications. *Renewable and Sustainable Energy Reviews* 98:27–39. https://doi.org/10.1016/j.rser.2018.08.055.

Salkuyeh, Y. K., B. A. Saville, and H. L. MacLean. 2018. Techno-economic analysis and life cycle assessment of hydrogen production from different biomass gasification processes. *International Journal of Hydrogen Energy* 43(20):9514–9528. https://doi.org/10.1016/j.ijhydene.2018.04.024.

Sanchez, D. L., N. Johnson, S. T. McCoy, P. A. Turner, and K. J. Mach. 2018. Near-term deployment of carbon capture and sequestration from biorefineries in the United States. *Proceedings of the National Academy of Sciences* 115(19):4875–4880. https://doi.org/10.1073/pnas.1719695115.

Sathaye, J., O. Lucon, A. Rahman, J. Christensen, F. Denton, J. Fujino, G. Heath, M. Mirza, H. Rudnick, and A. Schlaepfer. 2011. *Renewable Energy in the Context of Sustainable Development. IPCC Special Report on Renewable Energy Sources and Climate Change Mitigation.*

Schmidt, J. H., B. P. Weidema, and M. Brandão. 2015. A framework for modelling indirect land use changes in life cycle assessment. *Journal of Cleaner Production* 99(0959–6526):230–238.

Scown, C. D., W. W. Nazaroff, U. Mishra, B. Strogen, A. B. Lobscheid, E. Masanet, N. J. Santero, A. Horvath, and T. E. McKone. 2012. Lifecycle greenhouse gas implications of US national scenarios for cellulosic ethanol production. *Environmental Research Letters* 7(1):014011. https://doi.org/10.1088/1748-9326/7/1/014011.

Searchinger, T., R. Edwards, D. Mulligan, R. Heimlich, and R. Plevin. 2015. Do biofuel policies seek to cut emissions by cutting food? *Science* 347(6229):1420–1422.

Searchinger, T., R. Heimlich, R. A. Houghton, F. Dong, A. Elobeid, J. Fabiosa, S. Tokgoz, D. Hayes and T.-H Yu. 2008. Use of U.S. croplands for biofuels increases greenhouse gases through emissions from land-use change. *Science* 319(5867):1238–1240. https://doi.org/10.1126/science.1151861.

Shakoor, A., S. Shakoor, A. Rehman, F. Ashraf, M. Abdullah, S. M. Shahzad, T. H. Farooq, M. Ashraf, M. A. Manzoor, M. M. Altaf, and M. A. Altaf. 2021. Effect of animal manure, crop type, climate zone, and soil attributes on greenhouse gas emissions from agricultural soils—A global meta-analysis, *Journal of Cleaner Production* 278(124019):0959–6526. https://doi.org/10.1016/j.jclepro.2020.124019.

Sikarwar, V. S., M. Zhao, P. S. Fennell, N. Shah, and E. J. Anthony. 2017. Progress in biofuel production from gasification. *Progress in Energy and Combustion Science* 61:189–248. https://doi.org/10.1016/j.pecs.2017.04.001.

Sitepu, I. R., L. A. Garay, R. Sestric, D. Levin, D. E. Block, J. B. German, and K. L. Boundy-Mills. 2014. Oleaginous yeasts for biodiesel: Current and future trends in biology and production." *Biotechnology Advances* 32(7):1336–1360.

Smith, T. M., A. L. Goodkind, T. Kim, R. E. O. Pelton, K. Suh, and J. Schmitt. 2017. *Subnational Mobility and Consumption-Based Environmental Accounting of US Corn in Animal Protein and Ethanol Supply Chains.* PNAS.

Spatari, S., Y. Zhang, and H. L. MacLean. 2005. Life cycle assessment of switchgrass-and corn stover-derived ethanol-fueled automobiles. *Environmental Science & Technology* 39(24):9750–9758. https://doi.org/10.1021/es048293.

Susmozas, A., D. Iribarren, P. Zapp, J. Linßen, and J. Dufour. 2016. Life-cycle performance of hydrogen production via indirect biomass gasification with CO_2 capture. *International Journal of Hydrogen Energy* 41(42):19484–19491. https://doi.org/10.1016/j.ijhydene.2016.02.053.

Taheripour, F. and W. Tyner. 2014. Welfare Assessment of the Renewable Fuel Standard: Economic Efficiency, Rebound Effect, and Policy Interactions in a General Equilibrium Framework. In *Modeling, Optimization and Bioeconomy*, A. Pinto and D. Zilberman, eds.

Taheripour, F., X. Zhao, and W. E. Tyner. 2017. The Impact of Considering Land Intensification and Updated Data on Biofuels Land Use Change and Emissions Estimates. *Biotechnology for Biofuels* 10(1):191. https://doi.org/10.1186/s13068-017-0877-y.

Taptich, M. N., C. D. Scown, K. Piscopo, and A. Horvath. 2018. Drop-in biofuels offer strategies for meeting California's 2030 climate mandate. *Environmental Research Letters*, 13(9):094018. https://doi.org/10.1088/1748-9326/aadcb2.

Tilman, D., J. Hill and C. Lehman. 2006. Carbon-negative biofuels from low-input high-diversity grassland biomass. *Science* 314(5805):1598–1600.

Tilman, D., R. Socolow, J. Foley, J. Hill, E. Larson, L. Lynd, S. Pacala, J. Reilly, T. Searchinger, C. Somerville, and R. Williams. 2009. Beneficial biofuels – The food, energy, and environment trilemma. *Science* 32:270–271.

Tu, Q., M. Eckelman, and J. Zimmerman. 2017. Meta-analysis and harmonization of life cycle assessment studies for algae biofuels. *Environmental Science & Technology* 51(17):9419–9432.

Tyner, W. and F. Taheripour. 2007. Renewable Energy Policy Alternatives for the Future. *American Journal of Agricultural Economics* 89(5):1303–1310.

Tyner, W. and F. Taheripour. 2014. Advanced Biofuels: Economic Uncertainties, Policy Options, and Land Use Impacts. In *Plants and Bioenergy*, B. McCann, and Carpita, eds. Advances in Plant Biology, Vol. 4, Springer Science+Business Media, New York.

Verkerk, P. J., I. M. De Arano, and M. Palahí. 2018. The bio-economy as an opportunity to tackle wildfires in Mediterranean forest ecosystems. *Forest policy and economics* 86:1–3.

Wang, C., B. Amon, K. Schulz, and B. Mehdi. 2021. Factors that influence nitrous oxide emissions from agricultural soils as well as their representation in simulation models: A review. *Agronomy* 11:770. https://doi.org/10.3390/agronomy11040770.

Weidema, B. P., Ch. Bauer, R. Hischier, Ch. Mutel, T. Nemecek, J. Reinhard, C. O. Vadenbo, and G. Wernet. 2013. *The Ecoinvent Database: Overview and Methodology, Data Quality Guideline for the Ecoinvent Database* Version 3.

Whittaker, C., N. E. Yates, S. J. Powers, N. Donovan, T. Misselbrook, and I. Shield. 2017. Testing the use of static chamber boxes to monitor greenhouse gas emissions from wood chip storage heaps. *BioEnergy Research* 10(2):353–362. https://doi.org/10.1007/s12155-016-9800-9.

Wihersaari, M. 2005. Evaluation of greenhouse gas emission risks from storage of wood residue. *Biomass and Bioenergy* 28(5):444–453. https://doi.org/10.1016/j.biombioe.2004.11.011.

Woolf, D. and J. Lehmann. 2012. Modelling the long-term response to positive and negative priming of soil organic carbon by black carbon. *Biogeochemistry* 111(1–3):83–95.

Yang, M. N. R. Baral, A. Anastasopoulou, H. M. Breunig, and C. D. Scown. 2020. Cost and life-cycle greenhouse gas implications of integrating biogas upgrading and carbon capture technologies in cellulosic biorefineries. Environmental Science & Technology 54 (20): 12810-12819. https://doi.org/10.1021/acs.est.0c02816.

10
Electricity as a Vehicle Fuel

Plug-in electric vehicles (PEVs) use energy stored in an onboard battery for propulsion and charge the battery using electricity from the power grid. PEVs include battery electric vehicles (BEVs), which rely entirely on energy from the power grid, and plug-in hybrid electric vehicles (PHEVs), which can power the vehicle from a mix of energy from the battery and from another fuel source, typically gasoline.

Because PEVs require substantial battery capacity, vehicle production emissions can differ from those of liquid fuel vehicles. Vehicle production emissions are discussed in Chapter 6. Here the focus is on the fuel: electricity.

When PEVs are charged, they add load to the power grid. This demand is satisfied by increased output from power plants, which has implications for air emissions and other power system impacts. Additionally, broad adoption of PEVs could add a large enough load to the power grid to trigger construction of new power plants in some regions (known as capacity expansion). The potential for flexible PEV charging profiles acting as demand response units could plausibly change the economic and logistical factors associated with increasing penetration of intermittent, non-dispatchable generators, like wind and solar (Weis et al., 2014). The focus here is primarily on grid emissions from operation of existing or projected future generators.

In this chapter the focus is on the consequential effects of PEV charging for power grid emissions. There are also other consequential effects of electric vehicle (EV) adoption and PEV use beyond the power sector, similar to biofuels, such as rebound effects and fuel market effects as well as potential land use changes. The focus here is on power sector effects because they are specific to PEVs.

This chapter begins by (1) comparing attributional life-cycle assessment (ALCA) and consequential life-cycle assessment (CLCA) of air emissions from electricity consumption; (2) providing an overview of approaches to estimating consequential emissions of PEV charging, including regression, simulation, proxies and real time data; and (3) summarizing several key issues, including upstream emissions, uncertainty and dynamics, energy efficiency, effects of public policy, and data sources.

COMPARING ATTRIBUTIONAL AND CONSEQUENTIAL LIFE-CYCLE ASSESSMENT FOR ELECTRICITY

Past life-cycle studies of EVs have accounted for power sector emissions using attributional or consequential methods. As discussed in Chapter 2, the two approaches are intended to answer different questions. In the context of PEVs:

- ALCA seeks to answer what emissions a PEV is associated with or responsible for, given some judgments about how to assign power grid emissions to demand sources.
- CLCA seeks to answer how emissions will change if a technology or policy is adopted given some judgments about how to predict future counterfactual scenarios.

Figure 10-1 provides a conceptual illustration of the difference between these questions and approaches for PEV charging. In attributional approaches, a portion of total power grid greenhouse gas (GHG) emissions are assigned to PEV charging. In contrast, in consequential approaches power grid emissions are estimated in two scenarios—one without PEV charging and one with PEV charging— and the difference between emissions in the two scenarios is the consequential effect of PEV charging.

LCA GHG estimates of PEVs can vary substantially as a result of the type of LCA estimate used and regional boundaries used, so it is important to understand the differences in methods and regional boundary choices to identify which questions each approach can answer (Ryan et al., 2016).

Electricity as a Vehicle Fuel

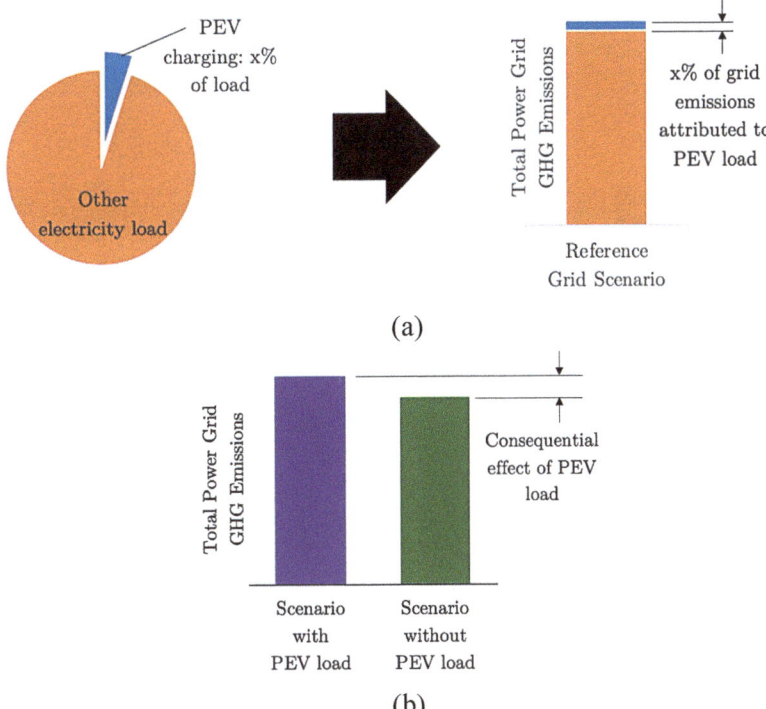

FIGURE 10-1 Conceptual illustration of the difference in approaches for assessing power sector emissions from PEV charging. NOTES: (a) in ALCA (top), a portion of power grid emissions are assigned to PEV load (usually proportional to PEV charging load); (b) in CLCA (bottom), total emissions are compared across two scenarios: with and without PEV load.

The most common attributional approach for estimating emissions from PEV charging is the simplest approach: compute the average emissions per unit of energy produced in the power system and assign this rate of emissions to PEV charging. Results from this approach can vary widely depending on boundary definitions (Weber et al., 2010). Some past U.S. studies assigned to PEV charging the average emission rate from power plants located in the country (Miotti et al., 2016) or the state (Yawitz et al., 2013) where the vehicle is charged. These boundaries are easy to identify, but because the grid is not generally organized around political boundaries, in most cases political boundaries have little to do with the impact of load at a given location. Most attributional studies use the average emissions from generators in regions defined by the power grid (Yuksel et al., 2016), rather than political boundaries, with the idea that load within one of these grid regions is more likely to affect generation within the region than across the boundary. However, electricity is constantly being traded across boundaries, and attributional assessments can vary widely, depending on how boundaries for analysis are chosen (Ryan et al., 2016; Weber et al., 2010).

Figure 10-3 provides a simplified conceptual illustration of why consequential emissions from PEV charging can differ from average grid emission rates. In this example, Region 1 has both coal and nuclear power plants, and Region 2 has only nuclear plants. In Region 1 the nuclear plants are fully utilized to meet existing load, and the coal plants are partly utilized (represented as height in red). If consumption is increased in Region 1 to charge a PEV, the nuclear plants cannot increase generation to supply that load, so the coal plants will increase generation. Although the average emissions in Region 1 are those associated with a mix of nuclear and coal generation, the effect of adding load to that region is the emissions associated with increasing coal generation. Although Region 2 contains only nuclear generators, they are already fully utilized and cannot increase power generation. If new PEV charging load is added to Region 2, it will need to increase trade with Region 1 in order to meet its demand. So, even though power generation in Region 2 is entirely from a zero-carbon source, the effect of charging a PEV in Region 2 may be to increase emissions from coal generators.

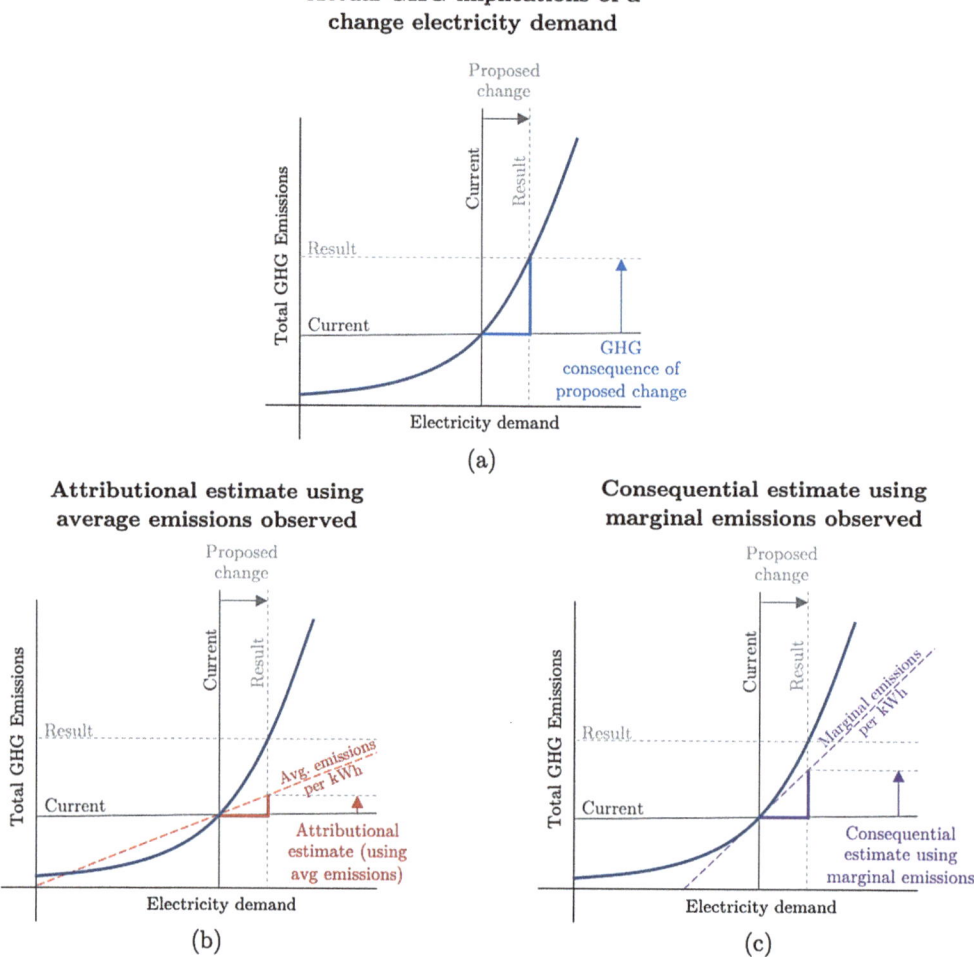

FIGURE 10-2 Relationship of average and marginal power grid emission factors to attributional and consequential LCA. NOTE: The emissions consequences of a change in electricity load from current levels to new levels are the difference between emissions levels with and without the change (a). In contrast, average emission factors, used in ALCA, assign average emissions per kWh to PEV load (b). Marginal emission factors are often used to estimate consequential emissions when the change in load is small (c).

A related example raised frequently is PEV owners who have rooftop solar generation at their homes. Rooftop solar generation reduces emissions by displacing fossil fuel generation. However, whether or not a household has rooftop solar generation, adding a PEV does not (usually) increase the amount of solar power generated. Instead, adding PEV load will increase the amount of energy that the household demands from the power grid (or reduce the amount of rooftop solar sold to the grid), triggering increased generation from plants on the grid. Like Region 2 in the example above, even in a household with rooftop solar the effect of charging a PEV may be to increase generation at a fossil fuel plant.[1]

These examples show why the emissions associated with a technology change (such as a household purchasing a PEV instead of a gasoline vehicle) or a policy change (e.g., a policy encouraging or mandating PEV adoption) can look quite different from the average power generation emissions in a region.

---

[1] It is worth noting that when decisions are coupled, such as comparing the adoption of PEV with rooftop solar to adoption of a gasoline vehicle and no solar, the difference in consequential emissions between the two scenarios involves both the effects of adding rooftop solar and the effects of replacing gasoline demand with electricity demand.

TABLE 10-1 Sample of Published Studies Assessing PEV Emissions in the United States

Study	Vehicle Type	Regional Resolution	Life-Cycle Scope	LCA Approach for Estimating PEV Emissions
EPRI-NRDC (2007)	PHEV	NERC regions	Use phase	Consequential
Hadley and Tsvetkova (2009	PHEV	13 NERC subregions	Portion of use phase	Consequential
Anair and Mahmassani (2012)	ICV, HEV, PHEV, BEV	eGRID subregion	Use phase	Attributional
MacPherson et al, (2012)	PHEV	NERC regions, NERC	Life cycle	Attributional
Thomas (2012)	HEV, PHEV, BEV	13 NERC subregions	Use phase	Consequential
Yawitz et al. (2013)	HEV, PHEV, BEV	State	Life cycle	Attributional
Graff Zivin et al. (2014)	ICV, HEV, PHEV, BEV	eGRID subregion	Portion of use phase	Consequential
Onat et al. (2015)	ICV, HEV, PHEV, BEV	13 NERC subregions	Life cycle	Consequential
Tamayao et al. (2015)	ICV, HEV, PHEV, BEV	NERC region	Life cycle	Consequential
Yuksel and Michalek (2015)	BEV	NERC region	Portion of use phase	Consequential
Nealer et al. (2015)	BEV	eGRID subregions	Life cycle	Attributional
Archsmith et al. (2015)	ICV, BEV	NERC regions	Life cycle	Consequential
Yuksel et al. (2016)	ICV, HEV, PHEV, BEV	County-level estimates	Life cycle	Consequential
Miotti et al. (2016)	ICV, HEV, PHEV, BEV	United States – average	<Life cycle	Attributional
Holland et al. (2016)	ICEV, BEV	NERC region	Part use phase	Consequential
Hoehne and Chester (2016)	PHEV, BEV	NERC region	Portion of use phase	Consequential
Nopmongcol et al. (2017)	CV, HEV, PHEV, BEV	US-REGEN region	Use ohase	Consequential
Elgowainy et al. (2018)	CV, HEV, PHEV, BEV	National	Life cycle	Attributional
Holland et al. (2019a)	CV, BEV	NERC region	Portion of use phase	Consequential
Holland et al. (2019b)	CV, BEV	NERC region	Portion of use phase	Consequential
Kawamoto et al. (2019)	CV, BEV	National	Life cycle	Attributional
Desai et al. (2020)	CV, HEV, PHEV, BEV	State	Life cycle	Consequential
Jenn et al. (2020)	PHEV, BEV	RTO	Portion of use phase	Consequential
Tong et al. (2020)	CV, HEV, BEV	NERC regions	Life cycle	Attributional and consequential
Sheppard et al. (2021)	CV, BEV	National	Life cycle	Consequential

NOTE: Vehicle types: BEV = battery electric vehicle; CV= commercial vehicle; HEV = hybrid electric vehicle; ICEV= internal combustion engine vehicle; ICV = internal combustion vehicle; PHEV = plug-in-hybrid-electric vehicle. Regional resolution: NERC = North American Electric Reliability Corporation; eGRID = EPA's Emissions & Generation Resource Integrated Database; US-REGEN = U.S. Regional Economy, Greenhouse Gas, and Energy; RTO = regional transmission organization.

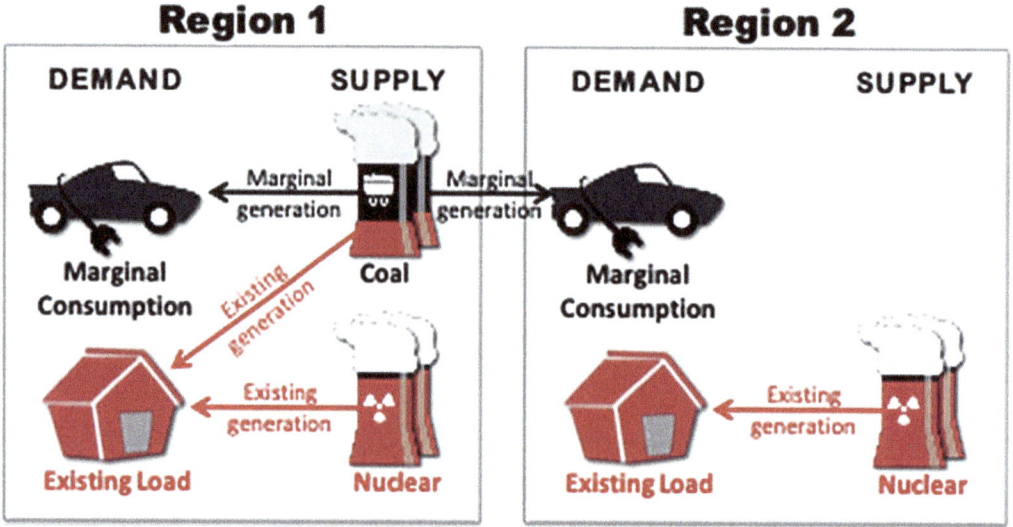

FIGURE 10-3 Illustration of why the emissions implications of charging a PEV in a region can differ from the emissions of the average grid mix in that region. SOURCE: Tamayao et al. (2015). Reprinted with permission from Environmental Science & Technology. Copyright 2015 American Chemical Society.

Because the power grid is highly interconnected in many regions, with many generators adding to the system and many demand sources drawing from the system, it is generally not possible to know precisely which power plants increase generation in response to increased load, especially when projecting into the future, such as over the life of a PEV. When assigning emissions to a PEV, many researchers and advocates with an attributional view find that it seems unfair to assign certain grid emissions to existing loads and different grid emissions to new loads, so it is common in ALCA to assign equal emission rates to all loads. Such allocations do not estimate how a technology or policy will change net emissions. To answer these questions, CLCA is needed.

With CLCA no assignment of emissions is made. Rather, the question of interest is how emissions will change if more PEVs are adopted (and charged). The effect of adding PEV charging load to the power grid depends on when and where the load is added. Figure 10-4 shows a hypothetical dispatch curve, which orders power plants available for dispatch based on their marginal generation cost. In the early morning hours in this example, the existing load is 67 GW. Adding new load at this time will increase generation from the specific generators on the margin (located just above the 67 GW load line): the natural gas combined cycle (red). In the afternoon on a hot day, the existing load is 114 GW, so adding new load at this time will increase generation from plants next in the dispatch order: natural gas and other (yellow). If fossil generation assets were to be replaced by dispatchable low-carbon energy sources, the addition of PEV charging demand would not create additional fossil fuel emissions from the electricity system.

APPROACHES TO CONSEQUENTIAL LIFE-CYCLE ASSESSMENT FOR PLUG-IN VEHICLE CHARGING EMISSIONS

CLCA approaches attempt to estimate emissions from the plants that will change generation in response to a change in load. A dispatch curve like the one in Figure 10-4 helps to visualize why this answer may change with time and location. In practice there are a number of factors that make operations more complicated than a simple dispatch curve, including transmission constraints, regulations, ramp rate limits, and other factors.

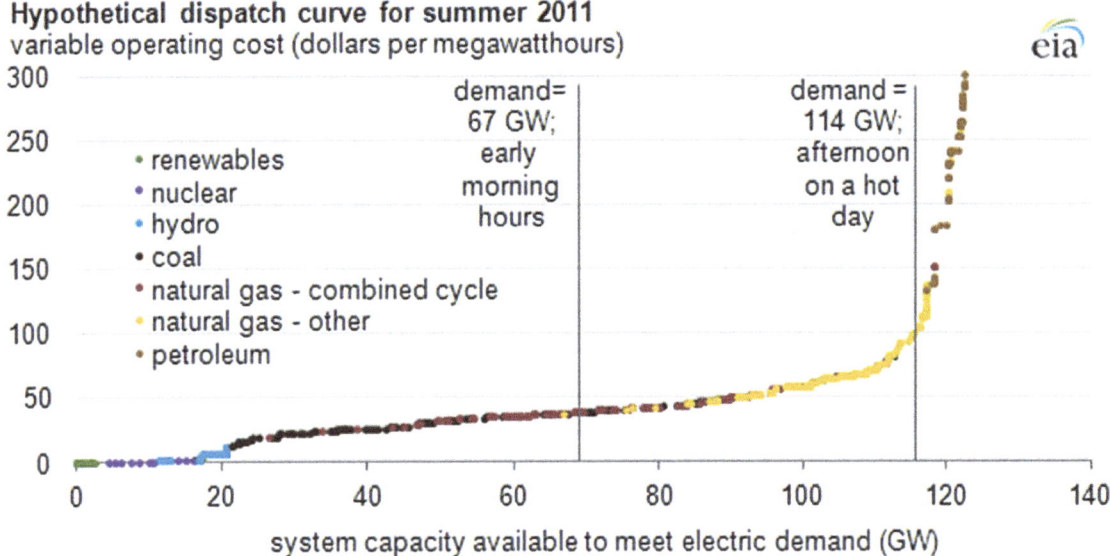

FIGURE 10-4 Hypothetical dispatch curve. NOTE: See text for discussion; GW = gigawatt. SOURCE: Energy Information Administration (2012, August)..

In broad terms, CLCA approaches to estimating PEV emissions fall into four main categories: regression, simulation, proxies, and real-time data.

Regression

Regression approaches use past data on power grid operations and emissions to statistically estimate how a marginal change in load affects emissions output across different operating conditions (Ryan et al., 2016). Figure 10-5 provides an example—for the Midwest Reliability Organization (MRO), which covers the Midwest region[2] in which marginal generation is estimated to be primarily from coal plants during low demand hours but with more from gas plants at high demand hours, with implications for emissions.

The main advantage of regression-based approaches is that they are based on real data about how the power system has operated. The main limitations are that they typically can only model the effect of marginal changes in load and they only look backward at how the power grid worked when the data were collected, making it difficult to predict how future technologies or policies might affect future loads and emissions. Regression approaches vary, and each has its advantages and limitations. There are two main approaches, one that uses total generation in a region and one that uses total consumption.

The approach originally proposed by Siler-Evans et al. (2012) uses total generation in a region as the independent variable and considers only dispatchable plants (fossil fuel plants that can change generation on demand). An advantage of this approach is that it avoids a potential correlation/causality confusion of counting temporary changes in generation timing from hydroelectric plants as consequential emissions. (Unlike fossil plants, a hydroelectric plant has finite supply limited by lake levels, so increasing generation at one time to satisfy demand reduces the supply available to generate electricity at a future time). A disadvantage is that it ignores marginal trade across regions by focusing on regional generation rather than regional consumption, and assumptions are needed to translate marginal consumption to marginal generation.

---

[2] The region covers the provinces of Saskatchewan and Manitoba, and all or parts of the states of Arkansas, Illinois, Iowa, Kansas, Louisiana, Michigan, Minnesota, Missouri, Montana, Nebraska, New Mexico, North Dakota, Oklahoma, South Dakota, Texas, and Wisconsin.

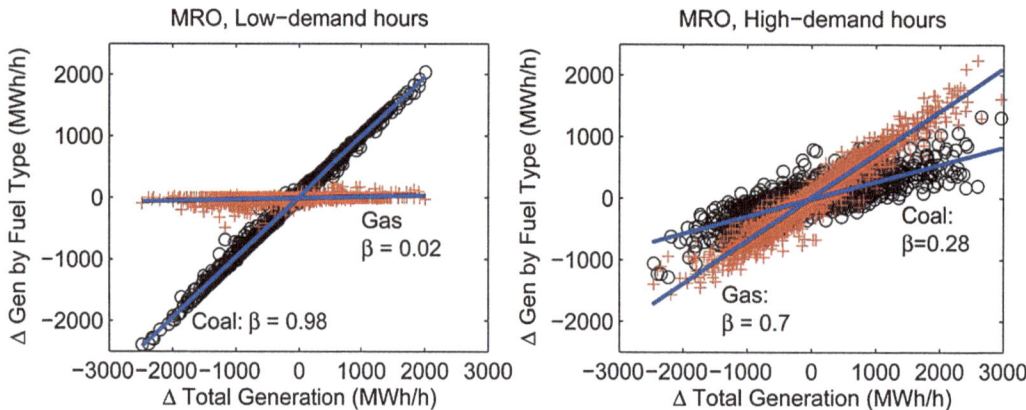

FIGURE 10-5 Example of regression results identifying that during low-demand hours in the MRO grid region, when total generation changes, the change comes overwhelmingly from changes in coal generation and not from changes in gas generation. During high-demand hours, changes in generation come more from changes in gas generation than from coal generation. NOTE: MRO = Midwest Reliability Organization; MWh = Megawatt hour. SOURCE: Siler-Evans (2012, p. 4744). Reprinted with permission from Environmental Science & Technology, Copyright 2012 American Chemical Society.

In contrast, the approach originally proposed by Graff Zivin et al. (2014) uses total consumption in a region as the independent variable and considers all plants in the broader interconnect. An advantage of this approach is that it measures the relationship between consumption and generation directly and allows for trade across regions at the margin within an interconnected area. A disadvantage is that it can conflate increased generation with shifted generation timing for hydroelectric plants.

Simulation

Simulation approaches model the power grid mathematically. Figure 10-6 show simulations of operations under scenarios that include and exclude PEV load, observing the difference in emissions across the two scenarios. These approaches typically model the power system as optimally satisfying load at minimum cost subject to practical constraints, such as transmission constraints, ramp rate limits, and capacity constraints. The main advantage of simulation-based approaches is that they can be used to study future scenarios or large changes in load. The main limitation is that it is difficult for a model to capture all of the factors that might affect grid operations in practice, and therefore there is generally some expected deviation between what an idealized model predicts and what would happen in practice. In particular, models that make more simplifying assumptions, such as using simple dispatch ordering without constraints on transmission or generation, can typically model larger systems at some expense of fidelity, while models that include detailed operational constraints typically limit scope to a particular region and may therefore miss effects of PEV load on marginal trade with other regions.

Proxies

Some analyses use proxies to approximate marginal emissions. For example, the U.S. Environmental Protection Agency's (EPA) Emissions & Generation Resource Integrated Database (eGRID) data[3] provide estimates of non-baseload generation by region, and non-baseload generation is sometimes used as a proxy for marginal generation.

---

[3] See https://www.epa.gov/egrid.

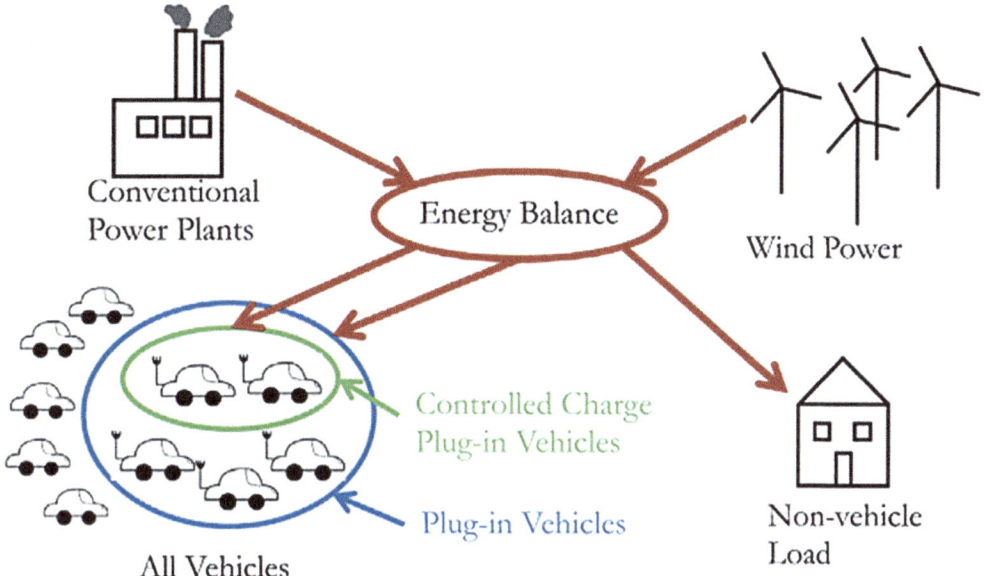

FIGURE 10-6 Conceptual illustration of energy balance maintained at every time step of a simulated dispatch model. SOURCE: Weis et al. (2014). Reprinted from Applied Energy, Elsevier.

Real-Time Data

Regression and simulation models as well as non-baseload eGRID data are ultimately based on actual generation resources dispatch and interchange data. This is the function that the regional transmission operators (RTOs) and independent system operators (ISOs) are performing (Greer, 2012):

"The operation of a wholesale power system requires one centralized power system operator to integrate the generation and transmission of electricity in order to ensure reliability. These system operator functions include determining which generation units to start up and shut down, dispatching of operating units, ensuring that the system is being operated reliably, and responding to changing system conditions."

Each of the RTOs/ISOs has its own control area where they are responsible for operating the electric grid reliably. There are currently seven RTOs or ISOs in the United States. Their names imply but are not limited to their general regional coverage:

- Pennsylvania New Jersey Maryland Interconnection (PJM).
- Midcontinent Independent System Operator (formerly, Midwest ISO) (MISO).
- Electric Reliability Council of Texas (ERCOT).
- California ISO (CAISO).
- New York ISO (NYISO).
- Southwest Power Pool (SPP).
- ISO New England (ISO-NE).

Several RTOs and ISOs publish real-time market and dispatch data as well as real-time electricity interchanges with other control areas. If not already publicly available, RTOs and ISOs can provide real-time marginal data, which can potentially be combined with data on EV charging patterns to estimate consequential emissions of EV charging.

Table 10-2 summarizes the advantages and disadvantages of regression-based, simulation-based, proxy, and real-time CLCA approaches and compares them with ALCA approaches that use average grid emissions. While regression-based approaches have advantages for shorter term analyses with incremental changes to operations, simulation-based approaches are likely needed to understand the impacts of PEVs over typical vehicle lifetimes, given the substantial changes to the grid that would be required to achieve climate stabilization targets and the potential for future feedstock prices and policy to change dispatch order. Proxies can be easy to use, though how well they estimate marginal emissions can vary, and real-time marginal emission estimates can provide observed dispatch information but may not, on their own, provide a basis for projecting future scenarios. Ryan et al. (2016) summarize additional recommendations for appropriate models estimating power-sector emissions.

TABLE 10-2 Comparison of Approaches to Estimate Grid Emissions from PEV Charging

Approach	Advantage	Disadvantage
ALCA Average Grid Emissions	Easy to find data and implement	Does not answer the question of how emissions will change if a technology or policy is adopted
CLCA Regression-Based Marginal Emissions	Based on real-world data of how changes in load have affected power sector emissions in the past	Limited to modeling small changes in load (marginal emissions only); examines only past grid behavior; does not predict how future technology or policy will affect a future power grid
CLCA Simulation-Based Marginal or Non-marginal Emissions	Can model effects of large load changes; can model future power grid scenarios	Difficult to model all factors that affect power grid operations in practice so idealized model predictions may differ from practice
CLCA Marginal Emission Proxies	Easy to find data and implement	Accuracy for estimating marginal emissions can vary
CLCA Real-Time Marginal Emissions from RTOs and ISOs	Captures actual dispatch implications of changes in load real time	Not known in advance; does not provide a basis for modeling future scenarios; addresses only marginal changes; may not account for marginal trade across RTOs and ISOs

NOTES: ALCA = attributional life-cycle assessment; CLCA = consequential life-cycle assessment; ISOs = independent system operators; RTOs = regional transmission operators.

In the context of a low-carbon fuel standard (LCFS) policy, for regulatory impact assessment (see Chapter 3), consequential grid emissions in future grid scenarios are needed, so simulation-based CLCA approaches may be most appropriate. For verification (see Chapter 5), real-time marginal emission factors from RTOs and ISOs have the potential to provide useful data. For carbon intensities assigned to fuels, there is no single agreed-upon estimate for PEVs. The choice depends on a policymaker's goals, and there are different views, both in the research community broadly and among the members of this committee.

Conclusion 10-1: ALCA is sometimes used to estimate emissions from electricity consumption because it is easy or because the modeler is interested in an attributional, rather than consequential, question. However, using average emission factors does not answer the question of how emissions will change if PEVs or a PEV policy is adopted. CLCA aims to answer how PEV or PEV policy adoption would change emissions from the power sector.

Conclusion 10-2: For CLCA, regression-based approaches are useful for grounding in data, but simulation-based approaches are needed to project consequential effects of large changes in PEV charging or PEV charging on future grids.

Conclusion 10-3: CLCA for future PEV loads is inherently uncertain, as is any term related to the future, given unknown future conditions that affect consequential emissions, including feedstock prices, regulations, non-vehicle load, and other factors.

Recommendation 10-1: Regulatory impact assessment or other analyses estimating the emissions implications of a change in PEV charging load should use a CLCA approach to estimate the implications of power grid emissions and clearly characterize uncertainty of estimates due to assumptions, especially for future scenarios.

UPSTREAM EMISSIONS

The emissions consequences of increasing electricity demand are not limited to emissions from combustion at the power plants that increase generation to serve that load. Power plants also have upstream emissions from feedstock production, processing, and transport. For example, increasing generation from coal-fired power plants implies increased coal consumption, which triggers additional emissions from coal mining, coal transportation, and other supply chain activities. Attributional estimates of these upstream emissions have been used in ALCA studies (Anair and Mahmassani, 2012; Elgowainy et al., 2018; Kawamoto et al., 2019; MacPherson et al., 2012; Miotti et al., 2016; Nealer et al., 2015; Tong et al., 2020; Yawitz et al., 2013) and in policy.[4] However, consequential or marginal emissions from changes in these supply chain activities are not well characterized in the literature, and these emissions are typically ignored or average attributional emissions estimates are used as a proxy. Upstream emissions can vary, but some studies have estimated upstream GHG emissions as 5–10 percent of electricity GHG emissions (Michalek et al., 2011). In addition, land use implications have not been well characterized.

Emissions from sources upstream of power plants are omitted in some electricity LCA studies and included with ALCA estimates in other LCA studies. CLCA estimates of how emissions upstream of power plants change with generation are generally not available in the literature.

Recommendation 10-2: Research should be done to estimate how upstream emissions in the power sector change in response to changes in generation.

UNCERTAINTY AND DYNAMICS

Consistent with the general findings about uncertainty in LCA (see Chapter 4), all approaches to estimating the power grid emissions consequences of PEV charging involve some degree of uncertainty. Furthermore, because the power grid will change over time in ways that cannot be fully predicted, including during the life of a vehicle, consequential emissions have important dynamic sources of uncertainty. For example, future marginal emissions may look different if the price of coal drops and the price of natural gas increases than they do if the price of natural gas drops and the price of coal increases (Weis et al., 2016). Many such factors can affect the emissions consequences of future PEV charging.

In sum, the emissions consequences of PEV charging are inherently uncertain, and effects of future PEV charging depend on future factors that cannot now be known.

Recommendation 10-3: Analyses that estimate the emissions implications of changing PEV adoption or PEV policy should provide a transparent assessment of how sensitive or robust the results of the analyses are to reasonable variations in modeling assumptions and future scenarios.

One key potential benefit of PEVs is the potential to make electricity from low emissions sources, such as wind, solar, hydro, or nuclear power. Such sources typically do not operate on the margin in the United States today, but if the capacity of these sources increases in the future such that renewable sources are routinely curtailed, renewable generators could be on the margin in a future power grid, implying low

---

[4] See https://ww2.arb.ca.gov/our-work/programs/low-carbon-fuel-standard.

consequential emissions from increasing the EV charging load. Several governments have announced targets to substantially increase renewable generation in the coming decades. The California LCFS policy provides incentives and alternative crediting for smart charging timed to coincide with low-emission grid composition, though low emission average composition does not necessarily imply low marginal emissions. Changes in power grid emissions caused by PEV charging could be low if PEV charging coincides with times when renewables would otherwise be curtailed.

> **Recommendation 10-4:** Analyses estimating the emissions implications of PEV adoption in future power grid scenarios should consider changes in power grid emissions caused by PEV charging in each power grid scenario.

The committee notes that studies that examine GHG emissions in isolation may miss co-benefits or tradeoffs with other externalities that can be larger in magnitude. For example, studies have found that the change in external costs of health effects from conventional air pollution can be larger than the change in external costs from GHG emissions when gasoline vehicles are replaced by PEVs (Michalek et al., 2011; Tessum et al., 2014; Weiss et al., 2016).

ENERGY EFFICIENCY

As discussed in Chapter 6, the life-cycle emissions of transportation fuels cannot be fully understood in isolation from the vehicles that use them. In particular, vehicle efficiency affects how much fuel must be consumed to serve a given travel need. PEV efficiency can be substantially affected by a number of factors including:

1. Which specific PEV design is being studied: BEVs in 2021 ranged in efficiency from 24 kWh/100 mi for the Tesla Model 3 to 50 kWh/100 mi for the Porsche Taycan Turbo S.[5]
2. Driving conditions: In city driving conditions with frequent stops, PEVs have substantial efficiency benefits over gasoline vehicles, but for highway cruising, PEVs are more comparable to gasoline vehicles (Karabasoglu and Michalek, 2013; Lee et al., 2017).
3. Climate: Both BEVs and gasoline vehicles are less efficient in cold weather, but BEVs typically experience a greater efficiency loss, in part because, unlike gasoline vehicles that use waste heat from the engine to heat the cabin, BEVs must use energy from the battery to heat the cabin instead of propelling the vehicle. BEVs can lose half of their range in extreme hot and cold weather climates (Lee and Thomas, 2017; Yuksel et al., 2016; Yuksel and Michalek, 2015). Other climate-related factors, such as humidity and precipitation, can also have substantial effects on vehicle efficiency.

The effects of these sources of heterogeneity, in addition to the effects of location and charge timing on grid emissions, can be larger than the differences in emissions among vehicle technologies. Therefore, studies that select a single vehicle design to represent each technology, a single estimate of grid emissions, and a single set of assumptions about charging, driving, and climate conditions have significant limitations. Variation in these factors can qualitatively affect the outcome of an LCA, as illustrated in Figure 10-7.

DATA SOURCES FOR RESEARCH

There are various data sources that researchers can use to estimate marginal emissions and externalities of electricity consumption (to estimate consequential emissions from PEV charging): two key ones come from EPA and the Center for Climate and Energy Decision Making at Carnegie Mellon University.

---

[5] See https://www.fueleconomy.gov/.

- EPA maintains estimates from two models: eGRID[6] and AVERT.[7] AVERT, in particular, supports calculation of marginal emissions factors on a regional, state, and county-levels. The user manual of AVERT (EPA, 2020) states: "within each region across the country, system operators decide when, how, and in what order to dispatch generation from each power plant in response to customer demand for electricity in each moment and the variable cost of production at each plant." AVERT analyzes how hourly changes in demand change the output of fossil generators and, with that, their hourly generation, heat input, and emissions of $PM_{2.5}$, SO_2, NO_x, and CO_2." The carbon intensity returned from marginal models such as AVERT can differ significantly from average eGRID data. Table 10-3 shows that, for selected states, this difference can be greater than 60 percent (Mueller and Unnasch, 2021).

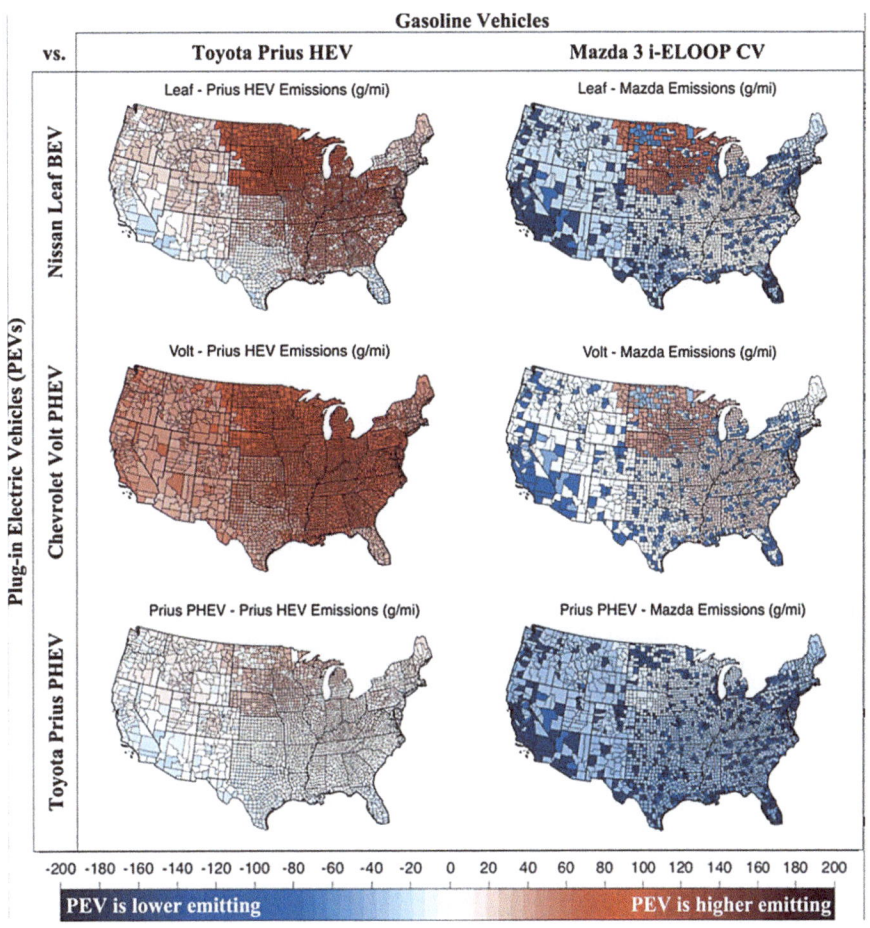

FIGURE 10-7 Illustration of how the relative life-cycle GHG emissions of a particular PEV compare with a gasoline vehicle can depend on many factors, including vehicle design, and regional factors such as the power grid, driving conditions, and climate. SOURCE: Yuksel et al. (2016). NOTES: Areas colored in blue are regions where the PEV has lower estimated life-cycle GHG emissions than the gasoline vehicle; areas colored in red are regions where the PEV has higher life-cycle GHG emissions than the gasoline vehicle. CV = conventional vehicle; HEV = hybrid electric vehicle; BEV = battery electric vehicle; g/mi = gallons per mile. Reprinted with permission from Environmental Research Letters, p.044007. © 2016 IOP Publishing Ltd.

---

[6] See https://www.epa.gov/egrid.
[7] See https://www.epa.gov/avert.

- The Center for Climate and Energy Decision Making at Carnegie Mellon University maintains a database of marginal emission factors that include GHG emissions and air pollutants:[8] These estimates are summarized in Table 10-4: they include regression-based marginal emission factors, simulation-based marginal emission factors, and average emissions from fossil generators (a potential proxy for marginal emissions).

TABLE 10-3 Comparison of Marginal AVERT Factors with eGrid, by Selected States and Regions

State/Region	AVERT Region	AVERT 2019 (lbs/MWh)[a]	eGRID Region[b]	eGRID 2018 (lbs/MWh)[c]	eGRID Transmission Loss (%)	eGRID with Transmission Loss (lbs/MWh)	% Diff Marginal to eGRID Average
Colorado	Rocky Mountain	1,904	RMPA	1,171	4.88%	1,231	55%
Illinois – Chicago	Mid-Atlantic	1,540	RFCW	1,174	4.88%	1,234	25%
Illinois – Rural	Midwest	1,860	SRMW	1,677	4.88%	1,763	6%
Indiana	Midwest	1,860	RFCW	1,174	4.88%	1,234	51%
Iowa	Midwest	1,860	MROW	1,249	4.88%	1,313	42%
Kansas	Central	1,800	SPNO	1,172	4.88%	1,232	46%
Kentucky	Midwest	1,800	SRTV	1,038	4.88%	1,091	65%
Michigan	Midwest	1,860	RFCM	1,321	4.88%	1,389	34%
Minnesota	Midwest	1,860	MROW	1,249	4.88%	1,313	42%
Missouri	Midwest	1,860	SRMW	1,677	4.88%	1,763	6%
Nebraska	Central	1,800	MROW	1,249	4.88%	1,313	37%
North Dakota	Midwest	1,860	MROW	1,249	4.88%	1,313	42%
Ohio	Mid-Atlantic	1,540	RFCW	1,174	4.88%	1,234	25%
South Dakota	Midwest	1,800	MROW	1,249	4.88%	1,313	37%
Wisconsin	Midwest	1,860	RFCW, MROE/MROW	1,420	4.88%	1,493	25%

[a] Values already adjusted for transmission loss.
[b] eGRID output factors not adjusted for transmission loss.
[c] eGRID Regions: RMPA = Western Electricity Coordinating Council (WECC) Rockies; RFCW = Reliability First Corporation (RFC) West; SRMW = SERC Reliability Corporation (SERC) Midwest; MROW = Midwest Reliability Organization (MRO) West; SPNO = Southwest Power Pool (SPP) North; SRTV = SERC Tennessee Valley; MROE = MRO East.

TABLE 10-4 Comparison of Estimated Emission Factors for Changing Electricity Load, by North American Electric Reliability Corporation (NERC) Region of the U.S. Power Grid (2017) averaged over seasons and time of day.

Approach	U.S. Grid Region							
	FRCC	MRO	NPCC	RFC	SERC	SPP	TRE	WECC
Regression-based marginal (kg/MWh)	483	789	441	671	640	665	583	552
Simulation-based marginal (kg/MWh)	586	881	459	773	700	706	606	549
Fossil fuel average emission factors (proxy) (kg/MWh)	534	907	473	743	677	772	688	686

NOTES: The data are averaged over seasons and time of day in kg/MWh; actual EV charging load may occur at different times with different marginal emission implications. U.S. grid regions: SOURCE: Data from Center for Climate and Energy Decision Making at Carnegie Mellon University. NOTES: CEDM = Center for Climate and Energy Decision Making at Carnegie Mellon University; U.S. Grid Regions: FRCC = Florida Reliability Coordinating Council, Inc.; MRO = Midwest Reliability Organization; NPCC = Northeast Power Coordinating Council, Inc.; RFC = Reliability First Corporation; SERC = SERC Reliability Corporation; SPP = Southwest Power Pool; TRE = Texas Reliability Entity; WECC = Western Electricity Coordinating Council.

---

[8] See https://cedm.shinyapps.io/MarginalFactors/.

EFFECTS OF PUBLIC POLICY ON CONSEQUENTIAL PLUG-IN ELECTRIC VEHICLE EMISSIONS

The consequential GHG emissions implications of vehicle electrification can be substantially affected by policy. In the United States, emission rates of new light-duty vehicles are capped by corporate average fuel economy (CAFE) standards (regulated by the National Highway Traffic Safety Administration) and light-duty vehicle GHG emissions standards (regulated by EPA).[9] Both agencies treat some alternative fuels, including electricity, favorably in compliance calculations; effectively counting some alternative fuel vehicles as lower emitting than they actually are and therefore permitting higher emissions from other vehicles to meet the same standard, resulting in increased overall permitted fleet emissions when alternative fuel vehicles are sold (Gan et al., 2021; Jenn et al., 2016, 2019). For example, in 2021 each BEV sold counts in compliance calculations as though 1.5 BEVs were sold, and BEV charging emissions are multiplied by zero in compliance calculations (Jenn et al., 2016) For this reason, the consequential implications of policies that increase BEV market share, such as some LCFS policies or California's "zero emission vehicle" policy, is to increase permitted fleet emissions, at least as long as these incentives are in place in national fleet standards (Choi et al., 2013).

Every PEV sold in the United States increases permitted fleet emissions in federal standards, and because the auto industry is constrained by fleet GHG standards, they tend to sell fleets that emit as much as permitted by the standards. Studies find that each time an EV is sold, permitted fleet GHG emissions increase by up to 60 tons, depending on the year and vehicle type (Jenn et al., 2016), and policies like California's zero emission vehicle mandate result in increased emissions due to these fleet standards (Jenn et al., 2019). It is possible that such policies may also induce innovation, trigger increased adoption, or enable stricter future standards that may reduce long-run emissions, though these factors are more difficult to quantify.

The emissions implications of policies like an LCFS that encourage fuel and vehicle technology switching can be substantially affected by interactions with other policies, including fleet vehicle emission standards.

Recommendation 10-5: LCA to estimate the change in GHG emissions induced by a policy or a change in technology adoption should consider how interaction with existing policies may affect outcomes. For cars and trucks, national fleet standards are key to understanding the net GHG outcomes of technology or policy actions.

Conclusion 10-3: Transportation fuel policies can have co-benefits and tradeoffs in terms of near-term human health effects, climate impacts and other factors.

Recommendation 10-6: Methods for LCA of low-carbon transportation fuels can evaluate co-benefits and tradeoffs of transportation policies in terms of climate impact, human health, and other factors.

Recommendation 10-7: Continuing and improved data are needed to support evaluation of the GHG emissions of electricity used as a transportation fuel.

REFERENCES

Anair, D., and A. Mahmassani. 2012. *State of Charge. Union of Concerned Scientists* 10.
Archsmith, J., A. Kendall, and D. Rapson. 2015. From cradle to junkyard: Assessing the life cycle greenhouse gas benefits of electric vehicles. *Research in Transportation Economics* 52:72–90.

---

[9] See https://www.govinfo.gov/content/pkg/FR-2021-12-30/pdf/2021-27854.pdf.

Choi, D.-G., F. Kreikebaum, V. M. Thomas, and D. Divan. 2013. Coordinated EV adoption: Double digit reductions in emissions and fuel use for $40/vehicle-year. *Environmental Science & Technology* 47(18):10703–10707. http://dx.doi.org/10.1021/es4016926.

Desai, R. R., R. B. Chen, E. Hittinger, and E. Williams. 2019. Heterogeneity in economic and carbon benefits of electric technology vehicles in the US. *Environmental Science & Technology* 54(2):1136–1146.

Elgowainy, A., J. Han, L. Poch, M. Wang, A. Vyas, M. Mahalik, and A. Rousseau, A. 2010. *Well-to-Wheels Analysis of Energy Use and Greenhouse Gas Emissions of Plug-in Hybrid Electric Vehicles.* ANL/ESD/10-1. Argonne National Lab. (ANL), Argonne, IL (United States). https://doi.org/10.2172/982352.

Elgowainy, A., J. Han, J. Ward, F. Joseck, D. Gohlke, A. Lindauer, T. Ramsden, M. Biddy, M. Alexander, M., S. Barnhart, and I. Sutherland. 2018. Current and future United States light-duty vehicle pathways: Cradle-to-grave lifecycle greenhouse gas emissions and economic assessment. *Environmental Science & Technology* 52(4):2392–2399.

EPA (US Environmental Protection Agency). 2020. *AVoided Emissions and geneRation Tool AVERT).* User Manual Version 3.0 https://www.epa.gov/sites/default/files/2020-09/documents/avert_user_manual_09-12-20_508.pdf. (accessed February 4, 2022).

EPA. *40 CFR Parts 86 and 600.* Federal Register / Vol. 86, No. 248 / Thursday, December 30, 2021. https://www.govinfo.gov/content/pkg/FR-2021-12-30/pdf/2021-27854.pdf.

EPA. n.d. Emissions & Generation Resource Integrated Database (eGRID). https://www.epa.gov/egrid (accessed February 4, 2022).

Gan, Y., M. Wang, Z. Lu, and J. Kelly, J. 2021. Taking into account greenhouse gas emissions of electric vehicles for transportation de-carbonization. *Energy Policy* 155(2021):112353. https://doi.org/10.1016/j.enpol.2021.112353.

Graff, J. S. G. Z., M. J. Kotchen, and E. T. Mansur. 2014. Spatial and temporal heterogeneity of marginal emissions: Implications for electric cars and other electricity-shifting policies. *Journal of Economic Behavior & Organization* 107:248–268.

Greer, M., ed. 2012. Chapter 3 – U.S. Electric Markets, Structure, and Regulations. Pages 39–100 in *Electricity Marginal Cost Pricing.* Butterworth-Heinemann, ISBN 9780123851345. https://doi.org/10.1016/B978-0-12-385134-5.00003-X.

Hadley, S. W., and A. A. Tsvetkova. 2009. Potential impacts of plug-in hybrid electric vehicles on regional power generation. *The Electricity Journal* 22(10):56–68.

Hoehne, C. G., and M. V. Chester. 2016. Optimizing plug-in electric vehicle and vehicle-to-grid charge scheduling to minimize carbon emissions. *Energy* 115:646–657.

Holland, S. P., E. T. Mansur, N. Z. Muller, and A. J. Yates. 2016. Are there environmental benefits from driving electric vehicles? The importance of local factors. *American Economic Review* 106(12):3700–3729.

Holland, S. P., E. T. Mansur, N. Z. Muller, and A. J. Yates. 2019a. Distributional effects of air pollution from electric vehicle adoption. *Journal of the Association of Environmental and Resource Economists* 6(S1):S65–S94.

Holland, R.A., K. Scott, P. Agnolucci, C. Rapti, F. Eigenbrod, and G. Taylor. 2019b. The influence of the global electric power system on terrestrial biodiversity. *Proceedings of the National Academy of Sciences* 116(51):26078–26084.

Jenn, A., I. L. Azevedo, and J. J. Michalek. 2016. Alternative fuel vehicle adoption increases fleet gasoline consumption and greenhouse gas emissions under United States corporate average fuel economy policy and greenhouse gas emissions standards. *Environmental Science & Technology* 50(5):2165–2174.

Jenn, A., I. L. Azevedo, and J. J. Michalek. 2019. Alternative-fuel-vehicle policy interactions increase U.S. greenhouse gas emissions. *Transportation Research Part A: Policy and Practice* 124:397–407.

Jenn, A., J. H. Lee, S. Hardman, and G. Tal. 2020. An in-depth examination of electric vehicle incentives: Consumer heterogeneity and changing response over time. *Transportation Research Part A: Policy and Practice* 132:97–109.

Karabasoglu, O., and J. J. Michalek. 2013. Influence of driving patterns on lifetime cost and life cycle emissions of hybrid and plug-in electric vehicle powertrains. *Energy Policy* 60:445–461.

Kawamoto, R., H. Mochizuki, Y. Moriguchi, T. Nakano, M. Motohashi, Y. Sakai, and A. Inaba. 2019. Estimation of CO_2 emissions of internal combustion engine vehicle and battery electric vehicle using LCA. *Sustainability* 11(9):2690.

Lee, D.-Y., and V.M. Thomas. 2017. Parametric modeling approach for economic and environmental life cycle assessment of medium-duty trucks. *Journal of Cleaner Production* 142(4):3300–3321. http://dx.doi.org/10.1016/j.jclepro.2016.10.139.

MacPherson, N. D., G. A. Keoleian, and J. C. Kelly. 2012. Fuel economy and greenhouse gas emissions labeling for plug-in hybrid vehicles from a life cycle perspective. *Journal of Industrial Ecology* 16(5):761–773.

Michalek, J. J., M. Chester, P. Jaramillo, C. Samaras, C. N. Shiau, and L. B. Lave. 2011. "Valuation of plug-in vehicle life-cycle air emissions and oil displacement benefits." *PNAS* 108(40):16554-16558. https://www.pnas.org/content/108/40/16554.

Michalek, J. J., M. Chester., P. Jaramillo, C. Samaras, C.-S.N. Shiau, and L. B. Lave. 2011. Valuation of plug-in vehicle life-cycle air emissions and oil displacement benefits. *PNAS* 108(40):16554–16558. https://doi.org/10.1073/pnas.1104473108.

Miotti, M., G. J. Supran, E. J. Kim, and J. E. Trancik. 2016. Personal vehicles evaluated against climate change mitigation targets. *Environmental Science & Technology* 50(20):10795–10804. https://doi.org/10.1021/acs.est.6b00177.

MISO (Midcontinent Independent System Operator). n.d. https://www.misoenergy.org/markets-and-operations/real-time–market-data/real-time-displays/.

Mueller, S., and S. Unnasch. 2021. High octane low carbon fuels: the bridge to improve both gasoline and electric vehicles. University of Illinois at Chicago. https://erc.uic.edu/wp-content/uploads/sites/633/2021/03/UIC-Marginal-EV-HOF-Analysis-DRAFT-3_22_2021_UPDATE.pdf (accessed February 4, 2022).

Nealer, R., D. Reichmuth, and D. Anair. 2015. *Cleaner Cars from Cradle to Grave: How Electric Cars Beat Gasoline Cars on Lifetime Global Warming Emissions*. Union of Concerned Scientists.

Nopmongcol, U., J. Grant, E. Knipping, M. Alexander, R. Schurhoff, D. Young, J. Jung, T. Shah, and G. Yarwood. 2017. Air quality impacts of electrifying vehicles and equipment across the United States. *Environmental Science & Technology* 51(5):2830–2837.

Onat, N. C., M. Kucukvar, and O. Tatari, O. 2015. Conventional, hybrid, plug-in hybrid or electric vehicles? State-based comparative carbon and energy footprint analysis in the United States. *Applied Energy* 150:36–49.

Ryan, N. A., J. X. Johnson, and G. A. Keoleian. 2016. Comparative assessment of models and methods to calculate grid electricity emissions. *Environmental Science & Technology* 50:8937–8953. https://doi.org/10.1021/acs.est.5b05216.

Sheppard, C. J., A. T. Jenn, J. B. Greenblatt, G. S. Bauer, and B. F. Gerke. 2021. Private versus shared, automated electric vehicles for US personal mobility: Energy use, greenhouse gas emissions, grid integration, and cost impacts. *Environmental Science & Technology* 55(5):3229–3239.

Siler-Evans, K., I. L. Azevedo, and M. G. Morgan. 2012. Marginal emissions factors for the U.S. electricity system. *Environmental Science & Technology* 46(9):4742–4748.

Tamayao, M. A. M., J. J. Michalek, C. Hendrickson, and I. M. Azevedo. 2015. Regional variability and uncertainty of electric vehicle life cycle CO_2 emissions across the United States. *Environmental Science & Technology* 49(14):8844–8855.

Tamayao, M-A. M., J. J. Michalek, C. Hendrickson, and I. M. L. Azevedo. 2015. Regional Variability and Uncertainty of Electric Vehicle Life Cycle CO_2 Emissions across the United States. *Environmental Science & Technology* 49(14):8844–8855. https://doi.org/10.1021/acs.est.5b00815.

Tessum, C. W., J. D. Hill, and J. D. Marshall. 2014. Life cycle air quality impacts of conventional and alternative light-duty transportation in the United States. *Proceedings of the National Academy of Sciences* 111(52):18490–18495.

Thomas, C. S. 2012. How green are electric vehicles? *International Journal of Hydrogen Energy* 37(7):6053–6062.

Tong, X., S. Ovtar, K. Brodersen, P. V. Hendriksen, and M. Chen. 2020. Large-area solid oxide cells with La0. 6Sr0. 4CoO3-δ infiltrated oxygen electrodes for electricity generation and hydrogen production. *Journal of Power Sources* 451:227742.

Weber, C. L., P. Jaramillo, J. Marriott, and C. Samaras. 2010. Life cycle assessment and grid electricity: what do we know and what can we know? *Environmental Science & Technology* 44(6):1895–1901. https://doi.org/10.1021/es9017909.

Weis, A., P. Jaramillo, and J. Michalek. 2014. Estimating the potential of controlled plug-in hybrid electric vehicle charging to reduce operational and capacity expansion costs for electric power systems with high wind penetration. *Applied Energy* 115(190–204):0306–2619. https://doi.org/10.1016/j.apenergy.2013.10.017.

Weis, A., P. Jaramillo, and J. Michalek. 2016. Consequential life cycle air emissions externalities for plug-in electric vehicles in the PJM interconnection. *Environmental Research Letters* 11(2). https://iopscience.iop.org/article/10.1088/1748-9326/11/2/024009.

Weis, A., J. J. Michalek, P. Jaramillo, and R. Lueken. 2015. Emissions and cost implications of controlled electric vehicle charging in the U.S. PJM interconnection. *Environmental Science & Technology* 49(9):5813–5819. https://doi.org/10.1021/es505822f.

Yawitz, D., A. Kenward, and E. D. Larson. 2013. *A Roadmap to Climate-Friendly Cars*. http://assets.climatecentral.org/pdfs/ClimateFriendlyCarsReport_Final.pdf.

Yuksel, T., and J. J. Michalek. 2015. Effects of regional temperature on electric vehicle efficiency, range, and emissions in the United States. *Environmental Science & Technology* 49(6):3974–3980.

Yuksel, T., M. Tamayao, C. Hendrickson, I. Azevedo, and J. J. Michalek. 2016. Effect of regional grid mix, driving patterns and climate on the comparative carbon footprint of electric and gasoline vehicles. *Environmental Research Letters* 11(4):044007.

Appendix A
Conclusions and Recommendations

Conclusions

FUNDAMENTALS OF LIFE-CYCLE ASSESSMENT

Conclusion 2-1: The approach to LCA needs to be guided on the basis of the question the analysis is trying to answer. Different types of LCA are better suited for answering different questions or achieving different objectives, from fine tuning a well-defined supply chain to reduce emissions, to understanding the global, economy-level effect of a technology or policy change.

Conclusion 2-2: Process-based ALCAs entail bottom-up accounting where emissions are assigned to products or processes based on modeling approach of a static world. Process-based ALCA can identify major sources of emissions in well-defined supply chains and identify opportunities to reduce supply chain carbon intensity, especially when case-specific process-data can be used instead of generic data. Economic input-output life cycle assessment (EIO LCA) identifies implications of interactions across broad sectors of the economy. It can capture emissions that may not be immediately apparent if only a well-defined supply chain is evaluated. It also is helpful in flagging emissions sources that are far-removed from the foreground system but are major contributors to total environmental effects. Hybrid Process/EIO ALCA identifies major sources of emissions beyond well-defined supply chains to include economy-wide effects. CLCA assesses the net effect of a decision or action, such as a change in fuel use or a change in policy, on total GHG emissions.

Conclusion 2-3: LCA results can vary depending on which methods are used, which data are used, which assumptions are made, what scope is defined, and what question is asked.

LIFE-CYCLE ASSESSMENT IN A LOW-CARBON FUEL STANDARD POLICY

Conclusion 3-1: The carbon intensities of fuels used in an LCFS are not necessarily equivalent to the full climate consequences of their adoption. Increased use of a fuel with a low carbon intensity, as defined in an LCFS, could potentially decrease or increase carbon emissions relative to the baseline, depending on policy design and other factors. Regulatory impact assessments that use CLCA to project the consequences of policy can help assess the extent to which a given policy design with particular carbon intensity estimates will result in reduced GHG emissions.

Conclusion 3-2: More research is needed to evaluate effective methods to collectively leverage the strengths of CLCA, ALCA, and verification methods in achieving LCFS objectives.

KEY CONSIDERATIONS: DIRECT AND INDIRECT EFFECTS, UNCERTAINTY AND VARIABILITY, AND SCALE OF PRODUCTION

Conclusion 4-1: Dividing emissions into direct and indirect can be used when identifying and classifying sources of emissions, but it can cause confusion, even if carefully defined and transparently presented in an LCA.

Conclusion 4-2: Direct and indirect emissions are concepts distinct from the concepts of attributional and consequential LCA.

Conclusion 4-3: Explicitly considering parametric, scenario, and model uncertainty can help to represent the degree of confidence in model results.

Conclusion 4-4: Up-to-date LCA studies are needed to inform policy.

Conclusion 4-5: LCA studies can produce different estimates depending on regional scope or assumptions

Conclusion 4-6: ALCA studies may produce substantially different results depending on modeling choices about how emissions are assigned to co-products.

Conclusion 4-7: LCA of commercial-scale production for processes that have not been commercialized involve assumptions that can introduce substantial uncertainty, including effects of interactions among multiple uncertain data or parameters, and so may be particularly sensitive to uncertainty.

Conclusion 4-8: Variability in methods and circumstances under which fuels are produced may be associated with differential economic returns. When this is the case, a techno-economic analysis may be helpful to understand the conditions under which market actors will produce the fuel.

Conclusion 4-9: Research is warranted on how the carbon intensity and economics of fuel production may change over time.

Conclusion 4-10: The scale of production can affect life-cycle GHG emissions, and current LCA methods often do not explicitly incorporate changes in production scale into their calculations.

Conclusion 4-11: More research is needed to develop LCA methodologies for incorporating scale dependence.

VERIFICATION

Conclusion 5-1: In verification to evaluate land use change at a national level, specifying the approach used to evaluate the extent, location, and type of agricultural expansion and the degree of uncertainty aids in transparency and clarity.

Conclusion 5-2: Insight into the degree of agricultural expansion domestically into ecologically important, but potentially small, land parcels requires more frequent data with higher spatial resolution and ideally high producer and user accuracy.

Conclusion 5-3: In verification to evaluate electricity load shifts from national electric vehicle policies, specifying the approach used to evaluate the extent, location, and type of load expansion to be verified and the degree of uncertainty will aid transparency and clarity.

Conclusion 5-4: While smart charging has potential to provide information about the carbon intensity of retail electric vehicle load, the assignment of specific generators to specific loads relies on assumptions from either an attributional frame (e.g.: under what conditions renewable generation should be assigned to electric vehicle load or to another load) or from a consequential perspective (e.g.: what emissions would look like in a counterfactual scenario without electric vehicle load).

Conclusion 5-5: Since satellite data allow for monitoring of international land use change, it would be possible to use satellite data to monitor international land use change, support calculations of LUC impacts, and support results from economic models used to estimate international land use change GHG emissions.

Conclusion 5-6: Certification and verification approaches have been implemented in contemporary LCFSs to inform values for many parameters that influence emissions.

Conclusion 5-7: Certification through protocols and methods that are consistent or compatible across regions and countries may mitigate global trade barriers.

Conclusion 5-8: There are a number of issues relating to the choice of certification protocols that use verification, including the cost to fuel providers, the benefits of reciprocity among protocols, and whether protocols act as trade barriers. These should be weighed against the net costs or benefits that verification provides to society including the carbon footprint of the certification process itself.

Conclusion 5-9: Certification protocols that use verification strategies can complement initial fuel pathway modeling with LCA and associated models (e.g., economic models used to estimate land use changes) to lessen the impacts of uncertainty in LCA results and to inform policymakers of the effects of an LCFS as they unfold. This insight can aid in policy adjustments if undesirable effects arise over the course of the policy.

SPECIFIC METHODOLOGICAL ISSUES RELEVANT TO A LOW CARBON FUEL STANDARD

Conclusion 6-1: The carbon intensity of fuels derived from methane that would otherwise be released (e.g., methane from manure or landfill) is strongly influenced by assumptions in the LCA of the alternative fate of methane pollution and is subject to dramatic change if relevant regulations or practices change.

Conclusion 6-2: Different biogenic carbon accounting methods exist and the choice of method affects the carbon intensity of fuels.

Conclusion 6-3: Given the importance of soil organic carbon changes in influencing life-cycle GHG emissions of biofuels, investments are needed to enhance data availability and modeling capability to estimate soil organic carbon change. Capabilities to evaluate permanence of soil organic carbon changes should also be developed.

Conclusion 6-4: Several metrics in addition to global warming potential for 100 years are now available with differing emphases such as short-term, long-term, or cumulative impacts.

Conclusion 6-5: To make a meaningful comparison of the LCA of transportation fuels, the vehicles that use those fuels should be considered.

Conclusion 6-6: If an LCA uses a single point estimate for efficiency of each vehicle type, its conclusions may vary substantially depending on which vehicle design (make-model-trim) is used to represent each fuel type.

Conclusion 6-7: If an LCA uses a single point estimate for efficiency of each vehicle type, its conclusions may vary substantially depending on which use conditions are assumed.

Conclusion 6-8: Specifically formulated high-octane fuels in combination with dedicated fuel engine technologies can provide efficiency improvements in fuel combustion that affect LCA results.

Conclusion 6-9: Ignoring vs. including vehicle production emissions in an LCA could affect its conclusion about which transportation fuels have the lowest carbon emission implications per unit of transportation services delivered.

Conclusion 6-10: A per-vehicle-mile functional unit is, on its own, not fully informative for comparing transportation fuels for weight-constrained or space-constrained applications, such as Class 8 trucks.

FOSSIL AND GASEOUS FUELS FOR ROAD TRANSPORTATION

Conclusion 7-1: Additional data, reporting, and transparency are needed for petroleum sector operations, including improved information on venting and flaring of methane.

Conclusion 7-2: More emissions inventory data from natural gas systems are needed, particularly regarding emissions from storage tanks.

Conclusion 7-3: The share of natural gas extracted from shale as a share of overall domestic consumption in the United States has increased rapidly and additional research and data collection will be necessary to better understand its production process and climate implications.

Conclusion 7-4: Assumptions on co-product handling methods have broad implications on natural gas LCAs. Additional research and data collection on industry practices can assist in the understanding of choice of co-product handling for natural gas production.

Conclusion 7-5: The selection of methane emissions leakage rate within an LCA has profound impacts on the overall estimated climate impact of natural gas production. Additional research and data collection is necessary to identify representative leakage rates for the natural gas industry and is essential to enable comparison of natural gas LCA across studies.

Conclusion 7-6: The life-cycle emissions attributable to green hydrogen are sensitive to assumptions on the upstream source of electricity used for electrolysis, as the difference in emissions between hydrogen produced from renewable electricity and even grid-average electricity is substantial.

AVIATION AND MARITIME FUELS

Conclusion 8-1: Non-CO_2 effects from aviation fuels may be high but remain uncertain. The largest non-CO_2 impact from aviation fuel combustion may be aviation-induced cloudiness, with the remaining contributions being much smaller.

Conclusion 8-2: Reduced aviation fuel aromatic content, whether through processing of fossil fuels or blending of alternative aviation fuels may have beneficial climate effects on non-CO_2 emissions. However, additional research is necessary to more accurately assess the contribution of non-CO_2 effects from the fuel cycle on a consistent basis with existing LCA of alternative aviation fuels. Due to the non-linearity of these effects, additional testing is necessary to evaluate the effect of alternative aviation fuel blending on non-CO_2 emissions.

Conclusion 8-3: The overall addition of CO_2 and NO_x to the atmosphere from aviation fuel combustion is well-characterized; however, there is substantial uncertainty on the emissions of sulfur, soot, and aviation-induced cloudiness.

Conclusion 8-4: The combustion emissions from aviation fuel are proportional to the quantity of fuel consumed. However, non-CO_2 impacts are non-linear and do not necessarily correspond proportionally to fuel switching. Furthermore, changes in airplane routing, such as location, altitude, and time of day may also influence non-CO_2 impacts of aviation.

Conclusion 8-5: Though there is evidence that fuel blending can mitigate the impact of some non-CO_2 climate forcing; attributing these impacts to fuel switching in policies may result in inaccurate crediting of these fuels.

Conclusion 8-6: The blending of alternative aviation fuels with different energy densities, as well as the introduction of alternative technologies such as electric-drive or hydrogen-powered airframes, may change the operating weight and efficiency of aircraft. In order to compare the emissions of these alternative fuels to conventional jet fuel on a consistent basis, these changes in efficiency must be taken into consideration.

Conclusion 8-7: There are some variations in the life-cycle emissions attributable to alternative aviation fuels at facilities for some fuel pathways, depending on whether they are configured to maximize alternative aviation fuel output or to maximize yields of other co-products, such as middle distillates.

Conclusion 8-8: Estimating the life-cycle GHG emissions of very low sulfur fuel oil will depend upon information about individual refinery choices in meeting marine fuel sulfur requirements.

BIOFUELS

Conclusion 9-1: Improved data on biofuel feedstock production, including energy consumption, yield, and fertilizer application at fine spatial resolutions may be useful for some applications. Data quality improvements may support improved GHG accounting in biofuel feedstock production, especially should a performance-based LCFS be developed that accounts for spatially-explicit fertilizer and energy consumption, and land management practices like cover crop planting, land clearing, overfertilization, manure application, use of nitrification inhibitors, or noncompliance with long-term soil carbon storage incentives.

Conclusion 9-2: Estimates of the GHG emissions associated with biofuels from woody biomass depend on the source of wood, forest management practices, and the carbon accounting method.

Conclusion 9-3: The impact of biorefinery co-products, particularly biochar, compost, digestate, or other products meant to be applied to soils, is highly dependent on how these materials are produced and handled and on what land they are applied. Assigning any gGHG offset credit to a biorefinery for producing and exporting these materials requires extensive verification to ensure they deliver the intended benefits. Nutrient-rich materials such as compost only offer fertilizer offset benefits if they are applied in a manner that results in lowered net GHG emissions.

ELECTRICITY AS A VEHICLE FUEL

Conclusion 10-1: ALCA is sometimes used to estimate emissions from electricity consumption because it is easy or because the modeler is interested in an attributional, rather than consequential, question. However, using average emission factors does not answer the question of how emissions will change if PEVs or a PEV policy is adopted. CLCA aims to answer how PEV or PEV policy adoption would change emissions from the power sector.

Conclusion 10-2: For CLCA, regression-based approaches are useful for grounding in data, but simulation-based approaches are needed to project consequential effects of large changes in PEV charging or PEV charging on future grids.

Conclusion 10-3: CLCA for future PEV loads is inherently uncertain, as is any term related to the future, given unknown future conditions that affect consequential emissions, including feedstock prices, regulations, non-vehicle load, and other factors.

Conclusion 10-5: Transportation fuel policies can have co-benefits and tradeoffs in terms of near-term human health effects, climate impacts and other factors.

Recommendations

FUNDAMENTALS OF LIFE-CYCLE ASSESSMENT

Recommendation 2-1: When emissions are to be assigned to products or processes based on modeling choices including functional unit, method of allocating emissions among co-products, and system boundary, ALCA is appropriate. Modelers should provide transparency, justification, and sensitivity or robustness analysis for modeling choices.

Recommendation 2-2: When a decision-maker wishes to understand the consequences of a proposed decision or action on net GHG emissions, CLCA is appropriate. Modelers should provide transparency, justification, and sensitivity/robustness analysis for modeling choices for the scenarios modeled with and without the proposed decision or action.

LIFE-CYCLE ASSESSMENT IN A LOW-CARBON FUEL STANDARD POLICY

Recommendation 3-1: When some emissions consequences of fuel use are excluded from carbon intensity values in an LCFS, the rationale, justification, and implications for these exclusions should be documented.

Recommendation 3-2: Public policy design based on LCA should ensure through regulatory impact assessment that, at a minimum, the consequential life-cycle impact of the proposed policy is likely to reduce net GHG emissions and increase net benefits to society. Regulatory impact assessments should consider changes in production and use of multiple fuel types (e.g., gasoline, electricity, biofuels, hydrogen).

Recommendation 3-3: LCA practitioners who choose to combine attributional and consequential LCA estimates should transparently document these choices and clearly identify the implications of combining these different types of estimates for the given application, scope and research question.

Recommendation 3-4: Research programs should be created to advance key theoretical, computational, and modeling needs in LCA, especially as it pertains to the evaluation of transportation fuels. Research needs include:
- Further development of robust methods to evaluate the GHG emissions from development and adoption of low-carbon transportation fuels, and development or integration of process-based, economic input-output, hybrid, and CLCA methodologies
- Products could include the following:
  o development of national, open-source, transparent CLCA models for use in LCFS development and assessment
  o continued development of national, open-source ALCA models from new or existing models
  o evaluation of different approaches to creating, using, or combining ALCA, CLCA, and verification for evaluation of policy outcomes
  o quantification of variation between marginal and average GHG emissions for various feedstock-to-fuel pathways; and
  o quantification and characterization of the implications of approximations and proxies in LCA, such as comparisons of marginal and average emissions.

KEY CONSIDERATIONS: DIRECT AND INDIRECT EFFECTS, UNCERTAINTY AND VARIABILITY, AND SCALE OF PRODUCTION

Recommendation 4-1: Because the terms "direct" and "indirect" are used differently in different contexts, these terms should be carefully defined and transparently presented when used in LCA studies or policy. Another option is to avoid using the terms "direct" and "indirect" altogether, as they are not considered necessary elements of LCA and may lead to greater confusion.

Recommendation 4-2: Current and future LCFS policies should strive to reduce model uncertainties and compare results across multiple economic modeling approaches and transparently communicate uncertainties.

Recommendation 4-3: LCA studies used to inform policy should explicitly consider parameter uncertainty, scenario uncertainty, and model uncertainty.

Recommendation 4-4: When LCA results are used in policy design or policy analysis, the implications of parameter uncertainty, scenario uncertainty, and model uncertainty for policy outcomes should be explicitly considered, including an assessment of the degree of confidence that a proposed policy will result in reduced GHG emissions and increased social welfare.

Recommendation 4-5: Regulatory agencies should formulate a strategy to keep LCAs up to date, which may involve periodic reviews of key inputs to assess whether sufficient changes have taken place to warrant a re-analysis, and agencies should be aware that substantial changes to LCAs on timescales of less than a decade can occur.

Recommendation 4-6: LCA studies used to inform transportation fuel policy should be explicit about the feedstock and regions to which the study applies and to the extent possible should explicitly report the sensitivity of the results to variation in these assumptions.

Recommendation 4-7: ALCA studies used to inform fuel policy should justify the approach used to handle co-products, and as necessary report sensitivity of results to variation in approaches to assigning emissions to co-products.

Recommendation 4-8: LCA studies used to inform transportation policy regarding processes that do not yet exist at scale should explicitly report sensitivity of findings to uncertainty, in order to produce bounding estimates.

Recommendation 4-9: Modelers should conduct sensitivity analysis to understand implications of variation.

Recommendation 4-10: To effectively inform policymaking, LCA studies should document results for a range of input values.

Recommendation 4-11: Researchers and regulatory agencies should identify additional information to assess impacts of large changes in fuel systems.

Recommendation 4-12: Because LCA-based carbon intensities in current LCFS policy are often not structured to capture nonlinear and non-life cycle implications of large changes in fuel and fuel pathway production volume, policymakers should consider potential complementary policy mechanisms.

VERIFICATION

Recommendation 5-1: Estimates of historical land use change—which may be used to inform economic models that evaluate market-mediated land use change —based on survey or remote sensing data should rely on more than one data source and should include estimates of uncertainty. Higher resolution, higher accuracy, and more frequently collected data sources should be made accessible to the public.

Recommendation 5-2: The research and policy communities should develop frameworks and methodologies for use of satellite data to characterize national and international land use change that may be in part attributable to an LCFS. Examples of framing questions include:
- Should an LCFS include measures to mitigate undesirable international land use change, or is it sufficient to monitor international land use change that may be due to the LCFS and these GHG emissions to the associated fuel?
- What are the guardrails (e.g., amount and type of land converted to agriculture in a certain region) that a monitoring approach would put in place and, if approached or exceeded, what action would be undertaken as a result?
- How can satellite data and economic modeling be most effectively used synergistically to limit GHG emissions from international land use change?
- What public data sources will be used to track land use change?
- How should uncertainty in land use change estimates be reported?

Recommendation 5-3: If applied, verification requirements should be used consistently and comparably across pathways to encourage technology development and deployment.

Recommendation 5-4: Baselines, if used, should consider (1) the state of technology, (2) inputs from multiple stakeholders, (3) implications for cost of implementation, and (4) incentives that the baselines create for innovation to reduce emissions and for data collection to demonstrate emissions reductions.

Recommendation 5-5: Combinations of newly developed sensor (including satellite) and supply chain technologies (e.g., database systems, blockchain) could be considered to improve land use change assessments. Policies need to be consistent with verification technology and set realistic expectations for verified LCA values. Data should be made publicly available for external verification. The GHG footprint of verification technologies should be included in the LCA as well.

Recommendation 5-6: An LCFS should consider inclusion of a certification protocol with verification. The protocol and its implementation should be overseen by an agency or group of agencies with the complementary expertise sets needed for success. These expertise sets include insights into multiple energy systems and new technologies, economics, environmental effects of fuels and their production routes, agriculture, fossil fuel production, and electricity generation.

Recommendation 5-7: Certification protocols should be revisited periodically to adapt to the emergence of new verification technology, national and global trends in the energy, transportation, and agriculture sectors, and to update baselines as needed based on evolving common practice.

Recommendation 5-8: Economic modeling and verification processes are complementary to each other and should both be used. Verification processes to assess international- and national-level land use change should use state-of-the art remote sensing technologies, when appropriate, which are evolving toward increased frequency and spatial resolution.

SPECIFIC METHODOLOGICAL ISSUES RELEVANT TO A LOW CARBON FUEL STANDARD

Recommendation 6-1: LCA for LCFS policies should provide as much transparency as possible on the different carbon removal elements of fuel life cycles allowed under the policy, as well as insight into how these may change over time, to inform policymakers and stakeholders. Specifically, LCA pathway analyses used to determine carbon intensity scores should separately indicate the contributions from negative elements (if any) and the counterfactual scenarios, such as avoided CO_2 emissions, avoided methane emissions, carbon capture and sequestration in geologic reservoirs or soil, and use in enhanced oil recovery.

Recommendation 6-2: All biogenic carbon emissions and carbon sequestration generated during the lifecycle of a low-carbon fuel should be accounted for in LCA estimates.

Recommendation 6-3: Research should be conducted to improve the methods for accounting and reporting biogenic carbon emissions.

Recommendation 6-4: Research should be conducted to collect existing soil organic carbon data from public and private partners in an open source database, standardize methods of data reporting, and identify highest priority areas for soil organic carbon monitoring. These efforts could align with the recommendations made in the 2019 National Academies report on negative emissions technologies t to study soil carbon dynamics at depth, to develop a national on-farm monitoring system, to develop a model-data platform for soil organic carbon modeling, and to develop an agricultural systems field experiment network. These efforts should also be extended internationally.

Recommendation 6-5: Research should be conducted to explore remote-sensing and in situ sensor-based methods of measuring soil carbon that can generate more data quickly.

Recommendation 6-6: Use of more than one climate change metric should be considered in the analysis of low-carbon fuel policies.

Recommendation 6-7: Further research should be conducted to better understand the suitability of different GHG metrics for LCA.

Recommendation 6-8: Further research should be conducted to develop a framework to include albedo effects from land cover change, and near-term climate forcers, in LCA of low-carbon fuels.

Recommendation 6-9: Further research should be conducted to better understand the climate implications of increased GHG emissions on the short-term (carbon debt) to support the selection of an appropriate approach to account for the timing of GHG emissions and uptakes in LCA.

Recommendation 6-10: LCA of transportation fuels may include analysis using functional units based on the transportation service provided, such as passenger-mile or ton-mile, or otherwise be based on comparison of comparable transportation services. This may be reported in addition to an energy-based functional unit. LCAs should clearly describe their assumptions for the energy- and service-based functional units, such as through vehicle efficiency, market share, or other factors.

Recommendation 6-11: When comparing life-cycle emissions of different transportation fuels, LCA studies that assess or inform policy should consider the range of vehicle efficiencies within each fuel type to ensure that the comparisons are made on comparable transportation services, such as passenger capacity, payload capacity, and performance.

Recommendation 6-12: When comparing life-cycle emissions of different transportation fuels, LCA studies should avoid relying on a single point estimate for efficiency of each vehicle fuel type and instead consider the range of vehicle efficiencies within each fuel type across vehicles and common or likely operating conditions.

Recommendation 6-13: LCAs of high-octane fuels should consider the impact of fuel octane on vehicle efficiency, but for the purpose of broad policy assessment LCA should be based on the actual and anticipated vehicle fleet, and following common practice for fuel vehicle assessments include only combinations that reflect reality.

Recommendation 6-14: For regulatory impact assessment, LCA of transportation fuels and transportation fuel policy should consider a range of estimates for possible changes in the emissions of vehicle production required to convert transportation fuels into transportation services, and the resulting changes in vehicle fleet composition.

Recommendation 6-15: LCA comparing transportation fuels for weight-constrained applications should present a per-ton-mile functional unit and/or explicitly model the logistical implications of payload effects by fuel type.

FOSSIL AND GASEOUS FUELS FOR ROAD TRANSPORTATION

Recommendation 7-1: Policymakers may consider recognizing the variation in GHG emissions across different petroleum fuel pathways, and include mechanisms to reduce these emissions in fuel policies.

Recommendation 7-2: Further research should be done on the key parameters used to assess the climate impacts of natural gas production, such as methane leakage rates. These parameters will evolve as technology advances, data availability increases, and statistical methods may be used to translate the additional data into improved emissions estimates.

Recommendation 7-3: Further research on the climate impacts of natural gas production should draw upon real world activity data in part supplied by the natural gas industry and in part from independent studies using satellite and remote sensing technology to improve methane emissions rate estimates; these should be revisited frequently— at least every five years.

Recommendation 7-4: To ensure renewable electricity is supplied via the grid to produce green hydrogen in the context of an LCFS, certification is necessary to ensure that the source of the electricity and its additionality.

Recommendation 7-5: In the context of an LCFS, LCAs of hydrogen should be well documented with choices of key parameters supported with facility-measured data or well-supported citations from the literature. These key parameters include the choice of energy source for steam-methane reforming or authothermal reforming, the carbon capture level from the waste gaseous stream, source of upstream electricity, and the rate of methane or CO_2 leakage. Where relevant, the approach to quantifying emissions of upstream natural gas production should align with those used elsewhere in an LCFS for other fuels produced from natural gas.

AVIATION AND MARITIME FUELS

Recommendation 8-1: Because the non-CO_2 effects from aviation fuels remain uncertain, research should be done to clarify the magnitude and direction of these effects.

Recommendation 8-2: Alternative fuels and airframe combinations, particularly those with large density differences such as battery electric technology and hydrogen, may impact airplane efficiency and thus influence overall emissions. The comparative LCA of these technologies should use functional units based on the transportation service provided or otherwise be based on comparison of consistent transportation services.

Recommendation 8-3: Alternative aviation fuel LCA estimates developed for fuel policy should reflect existing practices at facilities or the expected behavior in response to future policies.

Recommendation 8-4: LCA of oil-derived marine fuels should use new data as available for the feedstock conversion life-cycle stage. A body such as the International Maritime Organization should strive to collect data that will enable reliable marine fuel LCAs.

Recommendation 8-5: The baseline life-cycle GHG emissions for marine fuels should reflect current industry trends stemming from MARPOL Annex VI and potentially be updated after several years' time once the industry adjusts more fully to the new regulations through, for example, deployment of more liquefied natural gas-fueled vessels.

Recommendation 8-6: Marine fuel pathways should be evaluated with methods that are consistent with on-road and aviation fuels while considering unique factors in the oil refining and use phase aspects of a marine fuel's life cycle.

BIOFUELS

Recommendation 9-1: Additional research should be done to assess key parameters and assumptions in forest management practices induced by increased woody biomass demand, including: changes in residue removal rates, stand management and forest productivity, and changes in tree species selection during replanting.

Recommendation 9-2: Research and data collection efforts should be carried out for improved data and modeling related to forest feedstock production and storage, including energy use, yield, inputs, fugitive emissions, and changes in forest carbon stock should be supported.

Recommendation 9-3: Policymakers should exercise caution in crediting biorefineries for GHG emissions sequestration as a result of exporting co-products such as biochar, digestate, and compost, as it risks over-crediting producers for downstream behavior that is not necessarily occurring. The committee recommends that any credits generated from these activities must be contingent on verification that these activities are being practiced.

Recommendation 9-4: Applying credits for carbon sequestration to soil or reduced use of fertilizer should require robust measurement and verification to prove the co-products are applied in a manner that yields net climate benefits.

Recommendation 9-5: Additional review and research is recommended on the key factors affecting induced land use change.

Recommendation 9-6: Beyond research on induced land use change and rebound effects, research should be done to identify and quantify the impacts of other indirect effects of biofuel production, including but not limited to market-mediated effects on livestock markets, land management practices, and dietary change of food type, quantity, and nutritional content.

Recommendation 9-7: Though the study of induced land use changes from biofuels has been the topic of intense study over the last decade, substantial uncertainties remain on many key components of economic models used to assess these impacts. Further work is warranted to update these estimates of market-mediated land use change and the models so as to inform the development and implementation of an LCFS.

Recommendation 9-8: Assessment of the consequential effects from a future proposed policy, such as induced land use change, should be further developed in order to assess the risk of market-mediated effects and emissions attributable to the policy. Consequential assessment can inform the implementation of safeguards within policies such as limits on high-risk feedstocks, can inform the development of supplementary policies, identify hotspots, and reduce the likelihood of unintended consequences.

Recommendation 9-9: To improve understanding of market-mediated effects of biofuels, research should be supported on different modeling approaches, including their treatment of baselines and opportunity costs, and to investigate key parameters used in national and international modeling based on measured data, including various elasticity parameters, soil carbon sequestration, land cover, and emission factors and others.

Recommendation 9-10: Because other market-mediated effects of biofuel production, such as livestock market impacts, land management practices, and changes in diets and food availability may be linked to land use and biofuel demand assessed using induced land use change models, additional research should be done and model improvements undertaken to include these effects.

Recommendation 9-11: Current and future low-carbon fuel policies should strive for transparency in their modeling efforts.

ELECTRICITY AS A VEHICLE FUEL

Recommendation 10-1: Regulatory impact assessment or other analyses estimating the emissions implications of a change in PEV charging load should use a CLCA approach to estimate the implications of power grid emissions and clearly characterize uncertainty of estimates due to assumptions, especially for future scenarios.

Recommendation 10-2: Research should be conducted to estimate how upstream emissions in the power sector change in response to changes in generation.

Recommendation 10-3: Analyses that estimate the emissions implications of changing PEV adoption or PEV policy should provide a transparent assessment of how sensitive or robust the results of the analyses are to reasonable variations in modeling assumptions and future scenarios.

Recommendation 10-4: Analyses estimating the emissions implications of PEV adoption in future power grid scenarios should consider changes in power grid emissions caused by PEV charging in each power grid scenario.

Recommendation 10-5: LCA to estimate the change in GHG emissions induced by a policy or a change in technology adoption should consider how interaction with existing policies may affect outcomes. For cars and trucks, national fleet standards are key to understanding the net GHG outcomes of technology or policy actions.

Recommendation 10-6: Methods for LCAs of low-carbon transportation fuels can evaluate co-benefits and tradeoffs of transportation policies in terms of climate impact, human health, and other factors.

Recommendation 10-7: Continuing and improved data are needed to support evaluation of the GHG emissions of electricity used as a transportation fuel.

Appendix B
Committee Members' Biographical Sketches

Valerie M. Thomas (*Chair*) is the Anderson-Interface Chair of Natural Systems in the H. Milton Stewart School of Industrial and Systems Engineering at the Georgia Institute of Technology, College of Engineering. She holds a secondary appointment as professor in the School of Public Policy. She received a B.A. in physics from Swarthmore College in 1981, a Ph.D. in high-energy physics from Cornell University in 1987, and completed post-doctoral training at Carnegie Mellon University in the Department of Engineering and Public Policy. Before coming to Georgia Tech she worked at Princeton University in the Center for Energy and Environmental Studies and at the Princeton Environmental Institute. She has also served as the American Physical Society Congressional Science Fellow. Her research is in the area of environmental life-cycle analysis, energy systems analysis, industrial ecology, and sustainability. Recent work has included evaluation of the environmental impacts of biofuels, direct air capture of carbon dioxide, and production of platform chemicals from biomass; comparison of electric vehicle technologies; smart-grid approaches to reducing health impacts from power plants; infrastructure resilience; and international electricity development pathways. She has also worked on leaded gasoline, dioxin, and nuclear issues. In the last several years she has had funding from the U.S. Department of Energy, Algenol Biotech, the Natural Resources Defense Council, Exxon Mobil Corporate Strategic Research, the National Science Foundation, Green Seal, and the JPB Foundation. She is a fellow of the American Physical Society and of the American Association for the Advancement of Science. She has previously served on the U.S. Environmental Protection Agency (EPA) Science Advisory Board and on the U.S. Department of Agriculture and Department of Energy's Biomass Research and Development Technical Advisory Committee.

Amos A. Avidan has retired as Senior Vice President and manager of Corporate Engineering & Technology from Bechtel Corporation. His primary interest has been in energy systems, including transportation fuels, from conventional hydrocarbons fuels, synthetic fuels, biofuels, reformulated fuels, and various proposed alternative fuels. He started his career working on developing synthetic fuels, then moved on to the production of conventional and reformulated fuels, and then on to natural gas based fuels, especially liquefied natural gas (LNG). Amos holds a Ph.D. in chemical engineering from the City University of New York. He has authored and co-authored numerous technical publications and patents and was elected to the National Academy of Engineering in 2009.

Jennifer B. Dunn is an associate professor of chemical and biological engineering at Northwestern University. She studies emerging technologies, their energy and environmental impacts, and their potential to influence greenhouse gas and air pollutant emissions, water consumption, and energy consumption at the economy-wide level. Particular technologies of interest include biofuels and bioproducts, automotive lithium-ion batteries, waste plastics recycling and utilization, advanced manufacturing, and fuels and chemicals made from natural gas liquids. Techno-economic and life cycle analyses are primary tools in her research. She has published extensively on the life-cycle assessment of low-carbon transportation fuels and has served on a National Academies Committee on Developing a Research Agenda for Utilization of Gaseous Carbon Waste Streams (2017–2018). Jennifer enjoys leadership roles in sustainability research and practice at Northwestern as the director of the Center for Engineering Sustainability and Resilience and the co-chair of the Sustainability Council. Jennifer holds a Ph.D. and M.S. in Chemical Engineering from the University of Michigan and a B.S. in Chemical Engineering from Purdue University.

Patrick L. Gurian is a professor in the Department of Civil, Architectural, and Environmental Engineering at Drexel University. He has an A.B. from Harvard in chemistry, an M.S. in environmental engineering from Stanford University, and a Ph.D. in engineering and public policy and civil and environmental engineering from Carnegie Mellon University. He has over 20 years of experience in policy analysis of technological systems with particular attention to supporting regulatory decision making under uncertainty. He has studied how to address the uncertainty in the life-cycle impacts of biofuels, given the variety of conditions under which the feedstocks for these fuels are cultivated.

Jason D. Hill is professor in the Department of Bioproducts and Biosystems Engineering at the University of Minnesota. He also serves as a resident fellow of the University's Institute on the Environment. His research focuses on the environmental consequences of food, energy, agriculture, and natural resource use from a life-cycle perspective. He served on the National Research Council's Committee on the Economic and Environmental Impacts of Increasing Biofuels Production and on the U.S. EPA Science Advisory Board Biogenic Carbon Advisory Panel. Dr. Hill received his A.B. in biology from Harvard College and his Ph.D. in plant biological sciences from the University of Minnesota.

Madhu Khanna is the ACES Distinguished Professor of Environmental Economics and the Alvin H. Baum Family Chair and Director of the Institute for Sustainability, Energy, and Environment at the University of Illinois at Urbana-Champaign. She is internationally renowned for developing economic models to analyze the direct lifecycle greenhouse gas savings, indirect land use change effects, and biogenic carbon emission effects of first- and second-generation biofuels and of bioelectricity. She has deep expertise in data needs, modeling assumptions, integrating economic and ecosystem models, and uncertainty analysis to study the implications of various methods for life-cycle analysis on the estimates of greenhouse gas savings and land use change due to low carbon fuels under a range of climate and renewable energy policies, including the Low Carbon Fuel Standard, the Renewable Fuel Standard, carbon taxes, and others. Dr. Khanna has served on the Chartered Science Advisory Board of the U.S. EPA and chair of its Environmental Economics Advisory Committee. She is a fellow and past President of the Agricultural and Applied Economics Association. She obtained her Ph.D. in Agricultural and Resource Economics from the University of California at Berkeley in 1995.

Annie Levasseur is a professor at the Department of Construction Engineering of École de technologie supérieure, an engineering faculty based in Montreal, Canada. She is also the chairholder of the Canada Research Chair in Measuring the Impact of Human Activities on Climate Change and the Scientific Director of the Centre of Intersectoral Studies and Research on Circular Economy. She is a chemical engineer by training (1999, Polytechnique Montreal) and worked in the oil refining industry for 8 years. She then got a Ph.D. degree (2011, Polytechnique Montreal) with a thesis about the development of a dynamic life-cycle assessment approach for global warming impact assessment. She teaches the integration of environmental and sustainability aspects in engineering projects. Her research expertise is about environmental impact assessment using different methods or combinations of approaches such as life-cycle assessment, greenhouse gas emission inventories, and so on. She works on projects in different sectors, especially in forestry, energy, and construction. She is a recognized expert on topics related to biofuels LCA such as the consideration of biogenic carbon fluxes and the timing of greenhouse gas emissions. She often acts as an expert in the field of life-cycle assessment of bioenergy. For instance, she was the chair of the critical review panel for the Fuel LCA Model Methodology developed by Environment and Climate Change Canada for the coming Canadian Clean Fuel Standard. A few years ago, she was part of the National Academy of Science project "Negative Emissions Technologies and Reliable Sequestration: A Research Agenda."

Jeremy I. Martin is director of fuels policy and a senior scientist in the Clean Vehicles Program at the Union of Concerned Scientists (UCS). Dr. Martin works on state and federal fuels policies and has testified before Congress and state legislatures and briefed legislators and regulators on key fuel policies. Dr. Martin is the author of more than 15 technical publications and 13 patents on topics including biofuels, autonomous vehicles, semiconductors, and polymer physics. His recent reports for UCS cover transportation fuels and fuel policy, ride-haling, and automated vehicles. Before coming to UCS, Dr. Martin worked in research and development and manufacturing of computer chips at Advanced Micro Devices. Dr. Martin has a Ph.D. in chemistry and a minor in chemical engineering from the California Institute of Technology and a bachelor's degree in chemistry and English literature from Haverford College.

Jeremy J. Michalek is a professor of engineering and public policy, professor of mechanical engineering, and director of the Vehicle Electrification Group at Carnegie Mellon University. His research focuses on technical, economic, environmental and policy aspects of vehicle electrification, automation, sharing, and other trends in advanced vehicle technologies. Michalek's awards include the ASME Thar Energy Design Award, the Fenves Award for Systems Research, and the National Science Foundation CAREER Award. He serves on the Alternative Transportation Fuels and Technologies Committee of the Transportation Research Board of the National Academies.

Steffen Mueller leads the Bioenergy and Sustainable Landscapes Research Group at the University of Illinois at Chicago. Research activities focus on life-cycle emissions analyses of fuel pathways including greenhouse gas emissions assessments from land use change related to biofuels and bioenergy production, quantification of carbon emissions and sequestration effects from production agriculture, and comparative emissions assessments between biofuel and electric vehicle use. Recently, Steffen has been serving as a technical expert to the U.S. Grains Council on the sustainability of U.S.-produced biofuels for export markets and has presented his research in Japan, Korea, China, India, Vietnam, Columbia, and Mexico. He served on the Expert Working Group on Land Use Change for the California Low Carbon Fuel Standard development in 2010 and has published widely on the use of remote sensing technologies for land cover and land use change assessments. Steffen is also a board member of International Sustainability and Carbon Certification (ISCC). Dr. Mueller has published over 20 peer-reviewed papers on life-cycle analysis with a wide network of collaborators including Argonne National Laboratory and the U.S. Department of Agriculture. Prior to joining the University of Illinois in 2001, he worked in the private sector as director of business development for Skygen Energy and Calpine Corporation, a developer of electric generation power plants. Steffen holds a Ph.D. in energy policy from the University of Illinois at Chicago, and an M.B.A. and a B.S. in environmental engineering from Karlsruhe, Germany.

Nikita Pavlenko is a senior researcher with the ICCT's Fuels team. He evaluates the climate implications and techno-economics of advanced alternative fuels, with a focus on the aviation sector. He has been nominated as a participant at the International Civil Aviation Organization's Fuels Task Workgroup to develop the implementation of alternative fuels crediting under CORSIA, and has presented his research on life-cycle assessment at Coordinating Research Council workshop at Argonne National Laboratory. Mr. Pavlenko has a background in carbon accounting and comparative life-cycle analysis, particularly as applied to fuel production. Prior to joining the ICCT, Mr. Pavlenko supported the EPA's life-cycle material management model, the Waste-Reduction Model (WARM), as well as contributed life-cycle assessment expertise to a variety of environmental impact statements on fossil fuel infrastructure and vehicle efficiency.

Donald W. Scott has 14 years of experience in life-cycle analysis of transportation fuels. He has contributed to the U.S. Life Cycle Inventory Database and numerous publications quantifying the direct and consequential life-cycle impacts of biofuels including land management and land use change. As the director of sustainability for the National Biodiesel Board (Aug. 2008–Aug 2020), Scott led consensus on the development of sustainability principles adopted by the U.S. industry. Scott also chairs the subcommittee on biomass sustainability under ASTM International. Scott currently audits producers and purveyors of biofuels and circular materials under the International Sustainability and Carbon Certification framework. Scott is a licensed professional engineer with a degree in civil engineering from the University of Missouri–Columbia, where he graduated in 1995. Scott also served as chief of surface water for the Missouri Water Resources Center and served 12 years enhancing watershed protection for the Missouri Department of Natural Resources.

Corinne D. Scown is a staff scientist and deputy division director in the Energy Analysis & Environmental Impacts Division at Lawrence Berkeley National Laboratory, jointly appointed in the Energy & Biosciences Institute at UC Berkeley. She also leads the Life-cycle, Economics, and Agronomy Division at the Joint BioEnergy Institute (JBEI). Dr. Scown is an expert in life-cycle assessment and technoeconomic analysis, with a focus on bio-based transportation fuel, waste-to-energy systems, and bioproducts. Her prior awards include the NSF Graduate Research Fellowship, ACS Sustainable Chemistry & Engineering Lectureship, and the Secretary of Energy's Achievement Award. She has published more than 60 articles in journals including PNAS, Energy & Environmental Science, Science Advances, and Nature. She also serves on the editorial boards at ACS Sustainable Chemistry & Engineering, Agronomy Journal, and Environmental Research: Infrastructure and Sustainability. Dr. Scown received a B.S. in civil and environmental engineering with a double major in engineering and public policy in 2006 from Carnegie Mellon University. She received her M.S. in 2008 and Ph.D. in 2010 in civil and environmental engineering at UC Berkeley.

Dev S. Shrestha is a professor at the University of Idaho Department of Chemical and Biological Engineering. The University of Idaho has a long history of biofuel research since 1979. Dr. Shrestha joined University of Idaho in 2004 after receiving his Ph.D. in agricultural engineering from Iowa State University. One of his research areas includes life-cycle analysis of agricultural systems, primarily biofuel. Dr. Shrestha has published several research papers related to the topic area including indirect land use change and food versus fuel issues. Currently Dr. Shrestha maintains a national website for biodiesel education called BiodieselEducation.org, which he and his colleagues developed during last 15 years as a part of the U.S. Department of Agriculture's National Biodiesel Education program. He teaches a course at the university related to energy and environmental policies.

Farzad Taheripour is a research professor in energy economics in the Department of Agricultural Economics of Purdue University. He received his Ph.D. in agricultural economics from the University of Illinois at Urbana-Champaign in 2006. Professor Taheripour's research interests are in the areas of energy, agriculture, policy analysis, economic modeling, and life-cycle assessment. He is the leading scholar in assessing biofuels induced land use change emissions. He collaborates with several national and international organizations and institutions. Currently, Dr. Taheripour is a U.S. delegate in the Alternative Fuels Task Force (AFTF) group of the Committee on Aviation Environmental Protection of the International Civil Aviation Organization (ICAO). As a delegate in the AFTF, Dr. Taheripour leads the efforts to assess induced land use change values for aviation biofuels, a major component of the life-cycle assessment for aviation biofuels. He is the GTAP Research Fellow for the term of 2017–2020. He has over 110 professional publications including journal papers, book chapters, conference papers, and reports with more than 2,800 citations. His i10 publication index is 53, meaning that he has 53 books, book chapters, or papers with at least 10 citations each. His h index is 29, meaning he has 29 publications with 29 or more citations.

Yuan Yao is an assistant professor of industrial ecology and sustainable systems at the Yale School of the Environment. Before joining Yale, she was an assistant professor of sustainability science and engineering at North Carolina State University. She has expertise in life-cycle assessment, technoeconomic analysis, carbon footprint accounting, data analytics, and systems modeling. Her current research focuses on carbon emissions and life-cycle assessment modeling of biofuels and biomass-based products. She received the National Science Foundation Faculty Early Career Development Award (CAREER), and the Laudise Medal from the International Society of Industrial Ecology. She has been named to the American Institute of Chemical Engineers "35 Under 35" list for emerging chemical engineering leaders. Yao serves as the associate editor for the journal Resources, Conservation & Recycling. Yao published papers in leading journals such as Science, Nature Sustainability, and Environmental Science and Technology. She also served on the Technical Advisory Group for LEAP Partnership in the Food and Agriculture Organization of the United Nations. She received her Ph.D. in chemical engineering from Northwestern University.

Appendix C
Open Session Agendas

MEETING #1, PART II
June 1, 2021

OPEN SESSION—Public welcome

11:00 Welcome and introductions; Purpose of the open session; disclaimer—Valerie Thomas, Committee Chair

11:05 Context and expectations from the study—Maria Martinez, Breakthrough Energy
The sponsor will address the following questions:
- Why did your organization commission this study?
- Can you explain how you plan to use the results from this study?
- What types of recommendations would be most/least helpful?
- What do you consider to be outside the statement of task?
- What parts of the statement of task are most important, and why?
- What kinds of information do you have/could provide that would help the committee do its work?

11:20 Speakers from federal/state agencies (max. 10 minutes per agency rep.)
Environmental Protection Agency—Aaron Levy & Sharyn Lie
Department of Energy—Valerie Reed
California Air Resources Board—Anil Prabhu
Oregon Department of Environmental Quality Clean Fuels Program—Kiara Winans
- Brief presentation on agency's work on assessment of greenhouse gas emissions from transportation fuels
- Comment on Statement of Task/how the study/report could be useful to them

12:00 Q&A speakers and committee

12:15 Stakeholder and public comments— Stakeholders and members of the public can send written comments; staff will read the comments aloud until the adjournment of the session.

12:30 Adjourn open session

Committee Meeting #3 (Virtual with open and closed sessions)
July 20, 2021 at 12:00 to 3:00 PM Eastern

OPEN SESSION—Public welcome

12:00 Welcome/open session goals and agenda—Valerie Thomas, Committee Chair

12:05 Q&A with Dr. Bo Weidema

12:30 Comments from the public

12:40   Adjourn open session

Committee Meeting # 4, Part II (Virtual open and closed sessions)
August 31, 2021 at 1:00 PM to 4:00 PM Eastern Time

OPEN SESSION—Public welcome

1:00   Welcome/open session goals and agenda—Valerie Thomas, Committee Chair

1:05   Q&A with Richard Plevin, Tyler Lark and Seth Spawn, and John Field
Comments from the public

2:30   Adjourn open session

Committee Meeting #5, Part II (Virtual Open and Closed Sessions)
September 28, 2021 at 11:00 AM to 2:00 PM Eastern Time

OPEN SESSION—Public welcome

11:00   Welcome/open session goals and agenda—Valerie Thomas, Committee Chair

11:05   Q&A with Amgad Elgowainy and Michael Wang

11:50   Comments from the public

12:00   Adjourn open session

Committee Meeting #6, Part I (Virtual)
October 20, 2021 at 2:00 PM to 2:30 PM Eastern Time

OPEN SESSION—Public welcome

2:00   Welcome/open session goals and agenda—Valerie Thomas, Committee Chair

2:05   Q&A with Joule Bergerson

2:30   Adjourn open session

Committee Meeting #6, Part II (Virtual)
October 26, 2021 at 11:00 AM to 12:00 PM Eastern Time

OPEN SESSION—Public welcome

11:00   Welcome/open session goals and agenda—Valerie Thomas, Committee Chair

11:05   James Hileman presentation
Q&A (committee and speaker)

11:50   Public comment period

12:00   Adjourn open session